# The Millennial City

Millennials have captured our imaginaries in recent years. The conventional wisdom is that this generation of young adults lives in downtown neighbourhoods near cafes, public transit and other amenities. Yet, this depiction is rarely unpacked nor problematized. Despite some commonalities, the Millennial generation is highly diverse and many face housing affordability and labour market constraints. Regardless, as the largest generation following the post-World War II baby boom, Millennials will surely leave their mark on cities.

This book assesses the impact of Millennials on cities. It asks how the Millennial generation differs from previous generations in terms of their labour market experiences, housing outcomes, transportation decisions, the opportunities available to them, and the constraints they face. It also explores the urban planning and public policy implications that arise from these generational shifts.

This book offers a generational lens that faculty, students and other readers with interest in the fields of urban studies, planning, geography, economic development, demography, or sociology will find useful in interpreting contemporary US and Canadian cities. It also provides guidance to planners and policymakers on how to think about Millennials in their work and make decisions that will allow all generations to thrive.

**Markus Moos**, PhD MCIP RPP, is Associate Professor in the School of Planning at the University of Waterloo in Canada. His research is on the economies, housing markets and social structures of cities, including youthification and the generational dimensions of urban change.

**Deirdre Pfeiffer**, PhD AICP, is Associate Professor in the School of Geographical Sciences & Urban Planning at Arizona State University in the USA. Her research focuses on housing strategies in the USA relevant to an aging and diversifying society, the outcomes of the foreclosure crisis, and the relationship between suburban growth and racial equity.

**Tara Vinodrai**, PhD, is Associate Professor in the Department of Geography & Environmental Management and the School of Environment, Enterprise, and Development at the University of Waterloo in Canada. Her research program focuses on the emerging and evolving geographies of the knowledge-based economy, including the dynamics of innovation, economic development and labour markets in cities.

**Global Urban Studies**
Series Editor: Laura Reese

Providing cutting-edge interdisciplinary research on spatial, political, cultural and economic processes and issues in urban areas across the USA and the world, volumes in this series examine the global processes that impact and unite urban areas. The organizing theme of the book series is the reality that behaviour within and between cities and urban regions must be understood in a larger domestic and international context. An explicitly comparative approach to understanding urban issues and problems allows scholars and students to consider and analyse new ways in which urban areas across different societies and within the same society interact with each other and address a common set of challenges or issues. Books in the series cover topics that are common to urban areas globally, yet illustrate the similarities and differences in conditions, approaches, and solutions across the world, such as environment/brownfields, sustainability, health, economic development, culture, governance and national security. In short, the Global Urban Studies book series takes an interdisciplinary approach to emergent urban issues using a global or comparative perspective.

Published:

**Governing Urban Regions Through Collaboration**
A View from North America
*Joël Thibert*

**From Local Action to Global Networks: Housing the Urban Poor**
*Edited by Peter Herrle, Astrid Ley and Josefine Fokdal*

**Cities at Risk**
Planning for and Recovering from Natural Disasters
*Edited by Pierre Filion, Gary Sands and Mark Skidmore*

**Negative Neighbourhood Reputation and Place Attachment**
The Production and Contestation of Territorial Stigma
*Edited by Paul Kirkness and Andreas Tijé-Dra*

**The Millennial City**
Trends, Implications, and Prospects for Urban Planning and Policy
*Edited by Markus Moos, Deirdre Pfeiffer and Tara Vinodrai*

# The Millennial City

Trends, Implications, and Prospects for
Urban Planning and Policy

**Edited by
Markus Moos, Deirdre Pfeiffer and
Tara Vinodrai**

LONDON AND NEW YORK

First published 2018
by Routledge

2 Park Square, Milton Park, Abingdon, Oxfordshire OX14 4RN
52 Vanderbilt Avenue, New York, NY 10017

*Routledge is an imprint of the Taylor & Francis Group, an informa business*

First issued in paperback 2019

*British Library Cataloguing-in-Publication Data*
A catalogue record for this book is available from the British Library

*Library of Congress Cataloging-in-Publication Data*
A catalog record for this book has been requested

ISBN: 978-1-138-63123-6 (hbk)
ISBN: 978-0-367-36204-1 (pbk)

Typeset in Times New Roman
by Apex CoVantage, LLC

# Contents

# Figures

# Tables

# Contributors

**Agarwal, Ajay**: Associate Professor, Department of Geography and Planning, Queen's University

**Filion, Pierre**: Professor, School of Planning, University of Waterloo

**Fulton, William**: Director, Kinder institute for Urban Research, Rice University

**Fry, Richard**: Senior Researcher, Pew Research Center

**Geobey, Sean**: Assistant Professor, School of Environment, Enterprise and Development, University of Waterloo

**Grant, Jill L.**: Professor, School of Planning, Dalhousie University

**Hawes, Emily**: PhD Student, Department of Geography & Planning, University of Toronto

**Henry, Jeff**: Master's Student, School of Planning, University of Waterloo

**Jiao, Jungfeng**: Assistant Professor, School of Architecture, The University of Texas At Austin

**Mallach, Alan**: Senior Fellow, Center for Community Progress, Washington DC.

**Mawhorter, Sarah L.**: Postdoctoral Scholar, Terner Center for Housing innovation, University of California, Berkeley.

**Moos, Markus**: Associate Professor, School of Planning, University of Waterloo

**Pearthree, Genevieve**: Master's Student, School of Geographical Sciences and Planning, Arizona State University

**Prayitno, Khairunnabila**: Master's Student, School of Planning, University of Waterloo

**Pfeiffer, Deirdre**: Associate Professor, School of Geographical Sciences and Planning, Arizona State University

**Ralph, Kelcie M.**: Assistant Professor, Edward J. Bloustein School of Planning and Public Policy, Rutgers University

**Revington, Nick**: PhD Candidate, School of Planning, University of Waterloo

**Shearmur, Richard**: Professor, School of Planning, McGill University

**Shelton, Kyle**, Director of Strategic Partnerships, Kinder institute for Urban Research, Rice University

**Simone, Dylan**: PhD Candidate, Department of Geography & Planning, University of Toronto

**Vinodrai, Tara**: Associate Professor, Department of Geography and Environmental Management & School of Environment, Enterprise and Development, University of Waterloo

**Walks, Alan**: Associate Professor, Department of Geography & Planning, University of Toronto

**Wegmann, Jake**: Assistant Professor, School of Architecture, The University of Texas At Austin

**Worth, Nancy**: Assistant Professor, Department of Geography and Environmental Management, University of Waterloo

# Acknowledgements

The editors would like to thank: Wendy Chen, Ritee Haider, Rachel Poon, Eric Rempel, Filiz Tamer, Robert Walter-Joseph, Mark Williamson, and Jon Woodside for research assistance at various stages of this research; participants at the Associate Collegiate School of Planning conferences in Philadelphia, Houston and Portland and the American Association of Geographers conference in San Francisco for input on earlier versions of some of the chapters; and the Social Sciences and Humanities Research Council of Canada (SSHRC) and the Province of Ontario's Early Researcher Awards program for funding that helped make this work possible.

# Part I
# Introduction

# 1 The Millennial city, shaped by contradictions

*Markus Moos, Deirdre Pfeiffer and Tara Vinodrai*

Interest in the Millennials has grown substantially over the past decade as this group of young adults enters housing and labour markets. Yet, we lack systematic insight into how Millennials are shaping how we live and plan our cities, and how planning from previous eras is shaping these patterns as well. This book fills this gap by bringing together novel research on 1) planning for generational change; 2) young adult labour and housing market transformations; 3) changing transportation patterns arising from generational change; and 4) the implications of generational change for the future of the North American city.

The Millennial generation exhibits contradictions. Millennials in the USA and Canada are the most highly educated generation in human history. Yet, Millennials are entering housing and labour markets when secure, long-term employment and low-cost housing are scarce, leading Millennials to live at home longer than previous generations. The Millennial generation is socially differentiated, shaped by increasing urban diversity, socio-economic divisions, and segregation (Pitter & Lorinc, 2016). For instance, a young urban professional and waiter may be similarly educated, but the young waiter may live in the young professional's basement apartment and struggle to pay the rent that builds the young professional's wealth.

Growing inequality within the Millennial generation is an outcome of neoliberal restructuring, which since the late 1970s has replaced large parts of the welfare state with institutions aimed at making individuals more robust to economic adversity in a more competitive economic landscape (Hackworth, 2007). Ironically, economic adversity and competition have heightened under neoliberal governance (Peck & Tickell, 2002; Peck, 2005; Peck, Theodore, & Ward, 2005). Education is a more important predictor of earnings for the Millennials than it was for the Baby Boomers, which reflects the ways in which neoliberal restructuring has contributed to greater intra-generational inequality (Moos, 2014).

Inequality is increasing across all generations, not solely the Millennials. Yet Millennials face special challenges by coming of age during a time of increasing societal change. Many in the Baby Boomer generation were able to comfortably afford homeownership and find secure work with only a high-school diploma. Over the course of the Baby Boomers' careers, skills requirements changed and the economy restructured from one based on physical labour (e.g. manufacturing) to one based more on emotional and knowledge-based work (e.g. higher and lower-end services). Many entry level jobs today require a bachelor's degree; many jobs with low educational requirements are now automated. Technological developments have amplified the pace of these changes (Moos, 2012). Generational analysis is gaining importance in this context (Vanderbeck, 2007; Moos, 2012), as two cohorts of young only ten years apart may face a city, economy, and labour and housing markets that not only offer different kinds of opportunities and risks but also have restructured in ways that change the qualifications and resources required for success (Furlong & Cartmel, 2007).

Urban and regional geographies reflect the contradictions of the Millennial generation. Growing social inequalities arising from changes in the economy and labour markets place pressures on opposite ends of the housing market. The influx of young adults into central cities through "youthification" (Moos, 2016) means cities have simultaneously become playgrounds for young professionals and refuges for low-income earners looking for cheap apartments. Transportation patterns have shifted in the wake of these migrations. Planners and urban policymakers have responded to these changes in an entrepreneurial manner by advocating for urban amenities, density, walkability and transit-oriented development in cities, and select accessible suburbs. These actions largely benefit well-to-do Millennials and reflect the broader neoliberal project, by restructuring central cities around entertainment, bars and restaurants, and other recreational amenities (Peck & Tickell, 2002; Brenner & Theodore, 2002; Moos, 2012). This competitive urban landscape, particularly in redeveloping central cities, rewards Millennials who had the privilege to prepare for the increasingly competitive housing and labour market through higher educational attainment, the luck of the birth lottery through access to parental support, or ever-increasing debt (Moos, 2012). This landscape shames Millennials who fall through the cracks as being individually responsible.

Millennials are not evenly distributed across cities in the USA and Canada. Figure 1.1 shows the metropolitan areas with the highest and lowest share of young adults. Cities with high concentrations are those with strong or emerging knowledge-based economies, but also those with strong natural resource-based economies. The young adult years are typically a time when people are most mobile, and are attracted to cities with job opportunities. However, a large share of these cities have an existing or growing concentration of urban amenities in their cores that attract Millennials. On the opposite end of the spectrum, the cities with the lowest share of young adults are primarily in the Rustbelt. Miami is likely on this list due to its sizeable older population rather than an absence of a younger population.

A city's regional context matters in planning for Millennials. Planning the Millennial city in Detroit, Buffalo or Cleveland is not the same thing as in Portland, San Francisco or New York. Nor would cities with similar shares of Millennials necessarily face similar kinds of issues. For example, Seattle and Houston have similar shares of Millennials; however, this has more to do with the young seeking an urban lifestyle in the former than it does in the latter. The Millennial city, and its local expression and contingencies, is more than just the revealed preferences of one particular generation: It is the outcome of years of restructuring in

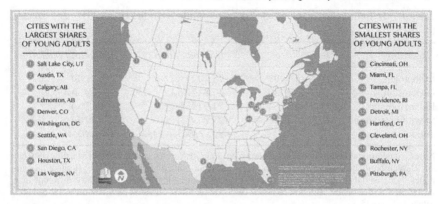

*Figure 1.1* Fifty-seven largest US and Canada metropolitan areas by share of young adults (25 to 34 years of age)

Source: Used with permission from generationedcity.uwaterloo.ca

governance, the economy and urban policy that will shape cities and planning for generations to come.

This book is the first to substantiate, debate and interrogate the past, present, and future of the Millennial city for planning and policy. Our 24 North American contributors include some of planning's most respected academics, emerging junior scholars and household names in public policy research. The book is organized by three overarching themes (Economy, Housing and Mobility), and concludes by looking forward to "what's next". Before we turn to the description of each of these themes, we first provide some definitions and justification for a generational (and demographic) lens on cities.

## Defining generations

"Generations" is a contentious concept. This is because the concept of generations relies on the idea that specific events and circumstances shape life experiences, outcomes, and even personality traits amongst those born during a similar time period (Ryder, 1965; Moos, 2012). What constitutes a generation, and whether there is any sort of internal uniformity that overrides other factors of social differentiation, are common criticisms of generational analysis. The research presented in this book adopts the view that each cohort of young people (defined as those who were born during the same time period in similar geo-political contexts, grow up, go to school, and enter housing and labour markets under similar conditions) face more similar opportunities and challenges compared to another cohort of young adults that precedes or follows them. Of course, these opportunities and challenges are still experienced differently by young adults, due to their varying socio-economic characteristics, gender identities, sexual orientations, ethnicities and other dimensions of social differentiation, which need consideration.

Thus, more narrowly defined, generational analysis is an entry point to examining how a cohort of people responds to particular circumstances and may share

*Table 1.1* Definition and size of generations (in millions) in the USA (2015) and Canada (2016)

| Generation | Born | US Population | Canadian Population* |
| --- | --- | --- | --- |
| Millennials | 1981 to 1997 | 75.4 (23%) | 7.5 (21%) |
| Generation X | 1965 to 1980 | 65.8 (20%) | 7.2 (20%) |
| Baby Boomers | 1946 to 1964 | 74.9 (23%) | 9.6 (27%) |
| Silent Generation | 1928 to 1945 | 27.9 (9%) | 3.7 (10%) |
| Total | | *321 (100%)* | *36 (100%)* |

Notes: Generational definitions and US data from Pew Research Center (Fry, 2015) using US Census Bureau population projections. Canadian data from Statistics Canada (2016) population estimates. *Due to data constraints, definitions of generations for Canada are slightly different than those cited in the table: Millennials 1981–1995, Generation X 1966–1980, Boomers 1946–1965, Silent 1927–1945.

similar constraints, opportunities and/or preferences. It is not a de facto assertion that "all Millennials are the same". This approach requires careful elaboration of contextual factors alongside the study of generational differences. This approach does not support a global view of generational analysis that, for example, the Millennial generation in Asia and Africa can be lumped with those in Europe and North America. Our book is thus restricted to analysis of generational change in the USA and Canada. If carefully contextualized, the findings likely do have some applicability to other Western contexts, particularly the United Kingdom or Australia that have some similarities in their housing and labour markets.

Since the concept of generations is socially constructed, there is therefore no "exact" boundary. Definitions of Millennials vary throughout the book due to data constraints and research objectives but generally follow those outlined in Table 1.1 where Millennials are those born between 1981 and 1997. This age range follows the definition derived by the Pew Research Center (Fry, 2015), which analyses similarities in "demographic, attitudinal and other evidence on habits and culture" to determine when a generation begins and ends.

Some researchers consider all Millennials, while others chose to focus on young adults 25 to 34 years of age, the latter being a useful way to study Millennials who have in general completed post-secondary education, and are likely entering housing and labour markets for the first time. This approach suggests that young adults are "canaries in the mine", in that comparison of young adult cohorts under different economic circumstances can provide insight on the negative socio-economic developments that require policy intervention (Myles, Picot, & Wannell, 1993). Younger households are more exposed to economic shocks than established households.

## Demography is not destiny, but it matters

It is often said that "demography is destiny". Overused and ill-defined, the phrase nonetheless provides a convenient point of departure for considering the dialectic relationship between generational change and the city. The phrase has at least two meanings. The first meaning of the phrase is that being born into a particular generation shapes economic prospects. Easterlin (1987), for example, has suggested that smaller birth cohorts have higher economic prospects than larger ones due to the resulting labour shortages. However, decades of empirical work in geography, planning, and urban and social studies suggest that social class, gender and race/ethnicity are more critical causal drivers of individual prospects than cohort size. The second meaning of the phrase is that demographic trends shape urban spatial structure, such as housing and labour markets and transportation behaviours (e.g. Foot & Stoffman, 1996). Existing research suggests that cohort size helps to shape urban spatial structure (Chapter 2). There is growing evidence that our cities are becoming more "generationed", socially and spatially, as each generation of young adults face different opportunities and constraints and as lifestyle and age have become more important determinants of cities' social geographies (Moos, 2014).

This book attempts to advance understanding of the generational drivers of urban social and economic differentiation. Our position is that demography as generational change matters but only as one of several broader socio-economic structures, all historically produced and reproduced, which determine future outcomes. Accounts of class, gender and racial dimensions of cities often end without consideration of their intersection with generational dimensions. Adding a generational lens ends great value to social science research. We must thus consider the ways class, gender, race/ethnicity and generational factors intersect (Furlong & Cartmel, 2007). Demography is not destiny but it matters, and it likely will continue to matter more and more in shaping cities.

**Book themes and organization**

Our book explores various facets of the generationed city, with a specific emphasis on Millennials, their experiences and preferences, and how those shape and are shaped by the city. The two remaining introductory chapters continue to build a conceptual framework for the book that we have started to describe here. The key tenets of this framework are recognition of: 1) the importance of generational change as *one* among several processes of urban change; 2) the dialectical relationship between urban and generational change; 3) the ways planning responds to and contributes to generational change; 4) the local *and* the more global impacts on cities and Millennials; and 5) the internal diversity of Millennial experiences despite sharing similar structural conditions.

Chapter 2, by Pierre Filion and Jill Grant, starts us off with a big picture discussion of the age-related factors that impact the evolution of cities. In particular, their work is context setting for the chapters that follow later on Millennial housing and mobilities (Parts 3 and 4 of the book). Filion and Grant define generation further as a concept distinct from cohort and other more generic demographic variables. The chapter considers in particular the predominant planning paradigms that have shaped the evolution of the urban structures across the North American landscape, and the ways these structures have both shaped and have been shaped by generational change. Importantly, Filion and Grant demonstrate the ways in which several factors must align for generational change to actually materialize in urban (and societal) change.

In Chapter 3, Tara Vinodrai considers the ways cities have been built for particular kinds of Millennials under the "creative cities" paradigm. She outlines the urban economic factors that are contributing to generationed spaces that are also simultaneously highly economically uneven. Vinodrai lays the groundwork for subsequent chapters on Millennial economies in Part 2 of the book.

## Millennial economies

The next set of chapters offer insights into how changes in the economy and labour markets have affected Millennials' economic well-being and future prospects. Chapters 4 and 5 consider several indicators of economic well-being, such as debt and wealth, to measure differences among generations. In Chapter 4, Richard Fry offers a generational comparison of income, education, poverty and wealth for the USA. In Chapter 5, Walks et al. focus on indebtedness more specifically, comparing the assets and debts held by different generations and age cohorts in Canada, and specific Canadian cities. Together, the two chapters illustrate various ways to measure differences in economic well-being.

In Chapter 6, Richard Shearmur offers a novel re-conceptualization of urban location theory in the context of dissolving workplaces. He discusses the declining importance of having a designated place of work among Millennials due to workplace (and technological) restructuring, and the impact this has for the urban spatial structure and planning. In Chapter 7, Nancy Worth offers a more specific analysis of one outcome of workplace restructuring, primarily the ways insecure work as a facet of neoliberal ideologies impacts the intergenerational transfer of wealth and resources. Finally, in Chapter 8, Sean Geobey highlights the growth of the sharing economy, arguing that planners and other policymakers should consider the potential implications of sharing platforms for governance. He also considers how the attitudes toward and uptake of sharing economy models are become structuring entities of the Millennial city.

## Housing Millennials

How Millennials are reshaping urban housing markets is the focus of the next five chapters of the book. Chapter 9 by Markus Moos offers a lens for considering the generational and intergenerational aspects of housing markets. He provides an in-depth empirical analysis of intergenerational wealth transfer in the housing market via downpayments, and its impact on housing demand, including the potential spatial implications. While Moos provides largely a demand-side perspective on housing, in Chapter 10, Deirdre Pfeiffer and Genevieve Pearthree provide an analysis of supply-side dimensions in two understudied metropolitan markets. They conduct a rigorous empirical analysis to demonstrate how the real estate sector, by virtue of development decisions about where and what to build, are "cementing" young adult preferences in the inner city.

The analysis offered in Chapters 9 and 10 is usefully put into perspective by Sarah Mawhorter's detailed quantitative comparison, in Chapter 11, of the housing opportunities of Baby Boomers and Millennials at similar life stages. She highlights the path-dependency of housing market decisions, and the diverse constraints faced by different generations of households. This is followed by an analysis of "peak millennials" by Jeff Henry and Markus Moos in Chapter 12. They provide a more nuanced, local analysis of Myers (2016) "peak millennial" thesis, which claims pressures from young adults on housing markets will start to ease as the size of the cohort entering housing markets begins to decline. Henry and Moos' work suggests that immigration and net migration at the metropolitan level are such large determinants of cohort size, and thus housing market demand, thereby raising questions about the narrative of peak millennial based on national level analysis alone. Their research suggests that the pressures on price that constrain Millennials' housing choices are not going to ease as quickly as prior research has suggested (Myers, 2016).

Continuing with the theme of price pressures, Chapter 13 by Junfeng Jiao and Jake Wegmann provides an empirical analysis of one strategy used by some Millennials to deal with increasingly high housing costs in large cities where many of them reside. They use a unique dataset to examine the rise and use of urban vacation rentals, such as Airbnb, in five major US cities by people in different generations, and the ways in which this aspect of the sharing economy influences housing market decisions.

### Millennial mobilities

This section of the book deals with mobilities in two ways: Residential location and moves, and commuting patterns. The analysis in Chapter 14 by Markus Moos and Nick Revington examines whether we can expect young adults who are currently residing in urban neighbourhoods to remain in these neighbourhoods as they grow older. In other words, can we expect a surge in demand for suburban single-family homes once Millennials age and more of them start to have children, or will demand for higher density living continue? They conduct this analysis using primary survey data on residential preferences for young adults in the USA and Canada. They find strong preferences for urban living amongst survey respondents, in line with previous research in this area. Specifically, they find linkages between young adults' current residential location in cities and their future preferences to remain there. However, they caution that a host of factors, including housing affordability and suitability, as well as planners' ability to respond to demand for family infrastructure in central cities, will determine whether Millennials will continue to reside in inner cities in high proportions as they enter later stages of life.

Chapters 15, 16 and 17 examine commuting and transportation patterns. Chapter 15 and 16 add useful nuance and critique to the discourse in the mainstream media that posits all Millennials as heavy transit-users, cyclists or pedestrians. In Chapter 15, Kelcie Ralph crafts a typology of different traveler types, and investigates their prevalence in the USA, and in specific US cities. She finds that while some Millennials fit the image of the car-free urbanite, nationally most Millennials still drive. Chapter 16 by Ajay Agarwal follows with a Canadian analysis, focusing primarily on Toronto. He carefully lays out the various hypotheses as to how we might expect Millennial travel behaviour to differ from that of prior generations when they were at the same age. He finds evidence that Millennials are less likely to drive but offers caution that it may be too early to know whether this will be a lasting effect.

Finally, in Chapter 17, Markus Moos, Nabila Prayitno and Nick Revington study a specific aspect of Millennial commuting behaviour, that of multi-modality (defined as using more than one commute mode in a given week). They ask about the connection between multi-modality and central city gentrification. Perhaps not too surprisingly, they find that it is the young, well-educated higher income Millennials in their sample that are able to afford locations in walkable, transit-oriented neighbourhoods, all the while still being able to afford a car.

## What's next?

The final section of the book looks beyond Millennials in proposing paths forward for US and Canadian cities. Kyle Shelton and William Fulton advocate for an intergenerational approach to urban planning and policymaking. Their approach suggests that analysis of individual generations is warranted to better understand their needs but developing strategies to help foster intergenerational connections and spaces and further well-being across generations is the ultimate goal. Shelton and Fulton identify green space, transportation and housing, as well as their intersection, as key structuring factors of the successful Millennial city. They demonstrate how progressive planning in each of these areas can be attained by harnessing the demands of particular generations.

In Chapter 19, Alan Mallach focuses on the geography of Millennials and their migration patterns as a means to study current trends and plausibly predict future trends. He examines the ways polarization, poverty and prosperity intersect spatially, and the resulting conflicts and contradictions that become embedded within the Millennial city, like racial tensions and gentrification. Mallach concludes on an optimistic note by showing how to direct Millennials' social capital to creating stronger and more inclusive cities.

In the concluding chapter, we synthesize the findings of our colleagues and pull together the various threads and themes emerging from their rich and insightful analysis. We bring together the diverse and wide-ranging policy and planning conclusions offered throughout the volume to provide an agenda for city building for Millennials and subsequent generations. Finally, we outline a research agenda for future work on the Millennial city.

## References

Brenner, N., & Theodore, N. (2002). Cities and geographies of "actually existing neoliberalism". *Antipode*, *34*(3), 350–379.

Easterlin, R. (1987). *Birth and fortune: The impact of numbers on personal welfare*. Chicago, IL: University of Chicago Press.

Foot, D., & Stoffman, D. (1996). *Boom, bust and echo: How to profit from the coming demographic shift.* Toronto, ON: Macfarlane, Walter & Ross.

Fry, R. (2015). *Millennials overtake baby boomers as America's largest generation.* Pew Research Center. Retrieved from www.pewresearch.org/fact-tank/2016/04/25/millennials-overtake-baby-boomers/.

Furlong, A., & Cartmel, F. (2007). *Young people and social change: New perspectives* (2nd ed.). Berkshire: McGraw-Hill.

Hackworth, J. (2007). *The neoliberal city: Governance, ideology, and development in American urbanism.* Ithaca, NY: Cornell University Press.

Moos, M. (2012). *Housing and location of young adults, then and now: Consequences of urban restructuring in Montreal and Vancouver.* PhD Dissertation, Department of Geography, University of British Columbia, Vancouver, BC.

Moos, M. (2014). "Generationed" space: Societal restructuring and young adults' changing location patterns. *The Canadian Geographer, 58*(1), 11–33.

Moos, M. (2016). From gentrification to youthification? The increasing importance of young age in delineating high-density living. *Urban Studies, 53*(14), 2903–2920.

Myers, D. (2016). Peak Millennials: Three reinforcing cycles that amplify the rise and fall of urban concentration by millennials. *Housing Policy Debate, 26*(6), 928–947.

Myles, J., Picot, G., & Wannell, T. (1993). Does post-industrialism matter? The Canadian experience. In G. Esping-Andersen (Ed.), *Changing classes: Stratification and mobility in post-industrial societies* (pp. 171–194). London: Sage.

Peck, J. (2005). Struggling with the creative class. *International Journal of Urban and Regional Research, 29*, 740–770.

Peck, J., Theodore, N., & Ward, K. (2005). Constructing markets for temporary labour: Employment liberalization and the internationalization of the staffing industry. *Global Networks, 5*(1), 3–26.

Peck, J., & Tickell, A. (2002). Neoliberalizing space. *Antipode, 34*(3), 380–404.

Pitter, J., & Lorinc, J. (2016). *Sub-divided: City-building in an age of hyper-diversity.* Toronto: Coach House Books.

Ryder, N. (1965). The cohort as a concept in the study of social change. *American Sociological Review, 30*(6), 843–861.

Statistics Canada. (2016). Table 051–0001: Estimates of population, by age group and sex for July 1, Canada, provinces and territories, annual (persons unless otherwise noted). CANSIM (database). Retrieved from http://www5.statcan.gc.ca/cansim/a26?lang=eng&retrLang=eng&id=0510001&&pattern=&stByVal=1&p1=1&p2=31&tabMode=dataTable&csid=.

Vanderbeck, R. M. (2007). Intergenerational geographies: Age relations, segregation and re-engagements. *Geography Compass, 1*(2), 200–221.

# 2 The impact of generational change on cities

*Pierre Filion and Jill L. Grant*

In 1964, director Michael Apted's documentary on 14 seven-year-old British children from different social classes, titled *7 Up*, described their respective living circumstances. It consisted mostly of interviews where the young participants aired their feelings about school, family, and friends, and described their aspirations for the future. Every seventh year a subsequent documentary mixed materials from previous episodes with new content to follow the life trajectories of these children. In the latest release of the series in 2012, the eighth episode, the original participants were 56 years old. Originally the study focussed on class differences to understand how these reverberated throughout their lives. The children had different accents, lived in diverse settings and went to public or private schools. As expected, their experiences and ambitions varied widely.

But something unexpected happened as the *Up* series progressed. As the 14 children moved through life, similar experiences related to common life stages, and in some cases overshadowed class differences (Thorne, 2009). The participants dated, had children, dealt with middle age, cared for aging parents, and so forth. Not only did the *Up* series participants pass through life stages together, but, by virtue of the favourable economic circumstances of the decades following World War II, the emphasis this particular generation placed on consumption meant their life circumstances had many similarities. For instance, participants commonly compared their cars, holidays, and the size and quality of their homes to those of their less prosperous parents.

By focussing on how age-related factors affect the evolution of cities, this chapter adopts a perspective similar to that of the *Up* series. The series and the chapter both stress the importance of age-related factors as determinants of life trajectories and societal (including urban) change. Moos (2014, 2016) has previously raised awareness of the generationed nature of urban social space, coining the terms "generationed city" and "youthification" to highlight the ways in which generational dimensions and age shape urban socio-spatial structures.

The focus on generational status is warranted as people within a specific age group share much in common in experiences, values, behaviour and societal impacts. We highlight values and common identifiers inherent in the concept of generation, and differentiate generation as a category from the exclusively demographic nature of other age-related categories such as cohort and life stage. Having defined generation in these terms, the chapter queries the persistence of generation-based features through successive life stages. Do characteristics that define a generation predominate over teenage and young adult years, when identification with the concept of generation is intense, only to dissipate in later life stages? Or do they persist over lifetimes?

Several findings emerge from our discussion. First, we note the transitory nature of the influence of generations on cities. Second, we observe differences in the direction of transitions initiated by different generations, and thus the absence of linear progression in generation-induced societal change. Third, we identify the need for generations to intersect with other social institutions and processes to provoke urban change. Finally, the chapter depicts the relation between generations and cities as bi-directional. As generations play a role in shaping cities, so too do cities in the self-awareness of generations.

Before exploring the relation between generation and urban phenomena, we recognize the geographic and cultural specificity in our use of the concept of generation. In the global North and more specifically North America, generation has been a recognized factor of societal change for nearly 100 years (Mannheim, 1952 [1925]). Our reflections here focus on the global North.

## Generation as a factor of societal change

Several concepts are associated with age groups: cohort, life cycle, life stage and generation (Alwin & McCannon, 2003; Elder, 1994; Marshall, 1983; Ryder, 1965). "Cohort" refers to people born during a specified period. "Life cycle", on the other hand, connotes the succession of "life stages" to which most people conform over the course of their existence (Kertzer, 1983; Riley, Foner, & Waring, 1988). Sociologists note that each life stage is associated with specific experiences, values and lifestyles (e.g. Morgan & Kunkel, 2007). A life cycle approach thus predicts value and lifestyle changes as one advances through life and transitions from one stage to another.

The generation-based perspective pertains to differences between cohorts at the same life stage. According to Dunham and Bengtson (1986, p. 5): "generation is a generic term that applies to a general class of groups or individuals who share a common experience, that is, their emergence in a unique period of biological, social, developmental, or historical time." The characteristics specific to a generation result from an age group being exposed to the same life circumstances, which favour adoption of similar values. These shared circumstances can take the form of parenting and educational trends, the social norms of the time, the cultural scene and the impact of extreme events such as wars and economic crises (Corsten, 1999; Eyerman & Turner, 1998). Such circumstances generate socialization processes leading to the constitution of a "habitus", whereby members of a same generation may share world views, values, as well as common communication and behavioural patterns (Bourdieu, 1977). A generational habitus can be enduring, allowing for the rise and persistence of a distinct social group or entity. These conditions can give rise to generational self-awareness and agency, allowing generations to function as collective social actors with potential for societal transformation.

Our interest in the concept of generation stems from the value, self-awareness and agency dimensions of generation, which contrast with the purely demographic attributes of cohorts, life cycles and life stages. Generational self-awareness and agency can operate at political, social and cultural levels. They can manifest through political movements defending the interests of a generation, or segment thereof, art and generation-specific social networks. Generations take shape around the framing and broadcasting of discourses defining their identity (Foster, 2013, p. 7).

Two characteristics of generations set them apart from other social categories. First, a generation tends to define itself in opposition to the previous generation. A new generation commonly challenges the worldviews and contributions of previous generations so it can substitute its own values and shape society accordingly. Hence the reactive nature of generations and the lack of linear progression in the changes they bring about (Wyn & Woodman, 2006). Generation-induced societal transformations tend to adopt pendulum-like patterns, as one generation reacts to the previous one, a process repeated by subsequent generations.

Second, generations are distinguished because their characteristics are generally shaped by major events their members experience in their teenage and young adult years (Rudolph & Zacher, 2016). Post-war prosperity and the Vietnam War influenced the Baby Boomers, while the Great Recession and its enduring economic consequences appear to be shaping the Millennials, as will the recent presidential election in the USA (Elder, 1974; Elder & Liker, 1982; Nesselroade & Baltes, 1974). As individuals construct their identity and try to secure a place in society during their youth, they prove unusually sensitive to prevailing societal circumstances. Thus, a generation can develop its defining characteristics over a narrow window of time.

It is impossible to attribute a specific societal impact to a generation, for generations as collective societal actors always intersect with other social categories. Such impacts are never caused by generations on their own, but rather stem from interactions with other social entities. The intersection of generations with other social categories also explains why generations do not behave as a bloc, but form segments according to the social categories within which generations interconnect in different circumstances (Furlong & Cartmel, 2007, p. 8). For example, student movements bear the imprint of social class divisions (and of other social categories), which explains why responses to public sector underfunding of post-secondary education have failed to elicit a unified response from the student body (Spiegel, 2014).

The effect of generations on society is mostly felt through market processes (consumption and employment), political movements and innovations: These include new worldviews, inventions, and fashion. Generations have differential capacity to shape society. All else being equal, the larger a generation, the greater political and consumer market influence it wields. Baby Boomers and Millennials are advantaged in this regard, while lower numbers reduce the impact of Generation X (Foot, 1996). Variation in the societal impact of generations is also a function of their uneven ability to achieve self-awareness, identify and mobilize around issues of common concern, and deploy strategies addressing these issues.

Society-wide circumstances, external to generations, also affect their influence. Such is the case with the performance of the economy, which determines the ease with which young adults can access jobs and the influence a generation exerts on the consumer market. Another societal factor affecting the influence of a genera-tion is the legitimacy of societal systems: institutional, economic, ideological and cultural. When legitimacy is under threat, the situation is ripe for critiques from, and alternatives advanced by, upcoming generations.

## The Millennials

Much has been written about the Millennial generation and for good reasons. It is the largest living generation (in the USA if not yet in Canada), yet its mem-bers struggle to gain an economic footing in a stagnant job market (Fry, 2016; also Haugen & Musser, 2013; Howe & Strauss, 2000; Winograd & Hais, 2011). Although debate persists about the precise age range that defines Millennials, most studies suggest that the generation comprises people born from the early 1980s to the late 1990s (see Chapter 1).

Millennials are differentiated from previous generations, although less from Gen Xers than Baby Boomers, by their high educational attainment (highest ever), information technology abilities, celebration of diversity and strong adherence to an environmental ethos. Millennials are also distinguished from prior generations by a greater predilection for urban rather than suburban locations (Chapter 14), rejecting accumulation of durable goods, reduced reliance on the automobile among some (Chapters 15 and 16), loyalty to friends and political activism (out-side formal institutional channels) (Delbosc & Currie, 2013; Kalita & Whelan, 2011; Kuhnimhof et al., 2012; McDonald, 2015; Moos, 2016; Schoettle & Sivak, 2014). Distrust of formal institutions on the part of Millennials may contribute to disengagement from political parties, aversion towards power hierarchies and lower (or at least delayed) marriage rates (Deal, Altman, & Rogelberg, 2010; Jang & Maghelal, 2016).

Life was good for Millennials until they entered the job market (Parment, 2012). From their births to the end of their education, Millennials were cherished, cared for by dedicated, often over-involved "helicopter" parents. Given their educational credentials, Millennials had every reason to be optimistic about their career outlooks. Then circumstances turned. First, employers increasingly came to rely on flexible work arrangements, which made access to permanent employ-ment with benefits less likely. Meanwhile, large swaths of well-paid employment vanished due largely to outsourcing. Worst of all, the entry of the Millennials on the job market coincided with the Great Recession and the subsequent anaemic recovery, which failed to create the quantity and quality of jobs needed to match the size, expectations and educational credentials of the generation (Fry, 2013a, 2013b; Lee & Painter, 2013; Mykyta, 2012; Rogers & Winkler, 2014).

Consequently, Millennials register disproportional unemployment rates and, when in the labour force, often find themselves under-employed (*The Economist*, 2016). At the same time, they confront expensive housing markets. The hous-ing cost problem is a consequence of the tendency for jobs, especially the type of employment appealing to well-educated Millennials, to concentrate in large urban centres with strong global connections. In these centres housing prices have risen dramatically. The costly housing of urban centres combined with polarized income distributions result in highly fragmented urban social geographies, now the reality confronting many Millennials (Hulchanski, 2010; Sassen, 1991).

This precarious economic situation curtails the Millennial generation's impact on society, causing it to punch below its demographic weight. For example, given their large numbers Millennials could modify the social geography of urban areas, except that economic circumstances force some of them to reside with their parents, or with roommates (Biggart & Walther, 2006; Mitchell, 2006; Qian, 2012). Confronted with difficult employment and housing conditions, a high proportion of Millennials extend the young adult phase of their lives and thus delay, or eschew in some cases, the family/child-rearing life stage (Astone, Martin, & Peters, 2015; Giroux, 2014; Martin, Astone, & Peters, 2014; Mettle, 2014). Watters (2003) sees the importance of networks of friends for this generation (already noted among Gen Xers), and urban living, as a manifestation of its stretched young adult phase.

Yet, many Millennials remain in suburban areas, of course. So, we need to close the section with a caveat. Clearly, not all Millennials conform to typical traits characterizing their generation. As is the case for all generations, features of the generation pertain only to some segments of its members. Indeed, Millennials may be more heterogeneous than preceding generations, due to profound income and employment polarization, which reverberates on their values and lifestyles. The fragmented nature of the Millennials generation is likely to intensify. Shaped by the advancing inequality defining neoliberal/globalized societies, as they progress through the employment stage of their lives, Millennials will likely split into fractions experiencing contrasting work and economic circumstances, and developing self-awareness and values consistent with these polarizing conditions.

## The dialectic relationship between generational and urban change

In order to understand the impact generations can have on urban change we offer a brief historical portrait of the connection between generational transition and urban transformation since the 1950s. Having experienced the restrictions of the Great Depression and World War II, young adults in the 1950s enthusiastically joined a consumption wave targeting durable goods. This generation adhered to modernist values, rejecting traditional, frugal lifestyles. Transposed to the urban sphere, these values propelled the massive deployment across North America of car-oriented suburbs and houses with large yards. The conformist climate of the period engendered broad support, at least at the surface, for this urban form and attendant lifestyle. But novels of the time exposed turmoil below the complacent veneer of post-war suburban life (e.g. Yates, 1961; Keats, 1956). A large proportion of the Baby Boom generation grew up in this environment of newly built bungalows and split-level houses.

As they reached teenage years and young adulthood, Baby Boomers enjoyed the benefits of a prosperous economy providing easy access to jobs and self-expression through consumption. In their teenage and young adult years, Baby Boomers became the target of consumer markets, especially clothing fashion and the music industry. As a generation, they became defined by a crisis of values stoked by reactions against conformity, acquisitiveness, environmental degradation, social inequality (especially in regards to minorities) and the Vietnam War.

The Baby Boom generation demonstrated a renewed interest in inner-city living, a reaction to the suburban preferences of their parents. Indeed, iconic images of this generation from the late 1960s include hippies hanging out in the San Francisco Haight-Ashbury neighbourhood and crowds of young people in Greenwich Village (New York City), Kitsilano (Vancouver), or Yorkville (Toronto). By creating an active inner-city cultural scene, the Baby Boomers helped ignite the gentrification of older neighbourhoods.

But the resettlement of the inner city is only a small part of the impact of Baby Boomers on the city. By virtue of its exceptionally large size, this generation affected all aspects of the urban phenomenon. In practice, only a small minority of Baby Boomers opted for the inner city and, even among these urban enthusiasts, many chose suburban living for the family-rearing stage of their lives. And in fact, young Baby Boomers were most concentrated in suburban settings (Moos, 2014, 2016). Most Baby Boomers lived their entire lives in suburban environments, thus following location patterns adopted by most North Americans. For all the transformative discourse and behaviour characterizing their teen and young adult years, Baby Boomers generally extended rather than transformed the urban form they inherited. As its members traversed the life cycle, the Baby Boom generation forced major extensions of facilities within urban areas, and of these areas themselves: schools, universities, apartments and then suburban homes, and finally condos, apartments and retirement homes (Foot, 1996). Paradoxically, the more conformist generation of their parents is the one that most profoundly modified urban patterns by being first to buy massively into the car-oriented suburban model, which set the subsequent trajectory for North American urban development.

Generation X, which was born from the mid-1960s through the early 1980s, is more difficult to define, hence the "X" moniker (Howe & Strauss, 1993). It is a much smaller generation than the Baby Boomers due to plummeting birth rates from the mid-1960s onwards. It was shaped by the neoliberal messages broadcast by the Reagan, Thatcher and Mulroney administrations, and the enduring effect of this ideology on society; hence the lower expectations of Gen Xers than those of previous generations, their market orientation and self-reliance (Economic and Social Research Council, 2016; Holtz, 1995). Generation X experienced unstable economic times, marked by frequent alternations between growth and recession, as well as severe public sector cutbacks including those affecting the educational system. Gen X is the first generation since World War II to underperform the previous generation in terms of personal income and wealth accumulation at comparable points in their lives. Finally, Gen Xers grew up with personal computers and other information technology devices, and have thus been computer literate from the early stages of the technology.

In terms of urban impact, Generation X is comparable to the Baby Boomers in that its residential patterns took place within the urban structure it inherited. Some of its members opted for inner-city neighbourhoods while the majority settled in suburban areas, thereby reinforcing the existing urban structure. Due to its smaller size and the economic difficulties it encountered, Gen X did not have the explosive effect on urban growth that the Baby Boomers did. Generation X can be seen as a transitional generation between the large and generally prosperous Baby Boom generation and the large but less prosperous Millennials.

The urban image Millennials evoke is that of young adults living in edgy inner-city neighbourhoods filled with cafés, micro-breweries, pastry shops, and gyms and providing an active music scene (Grant & Gregory, 2015; Smith, 2005; Figure 2.1). The ideal representation of Millennials is that of young adults living in apartments, often in high-rise condos, sharing their living space with members of their generation in trendy neighbourhoods. Of course, as in the case of the other generations, this urban image is an oversimplified representation of the living conditions of Millennials. Although young adults represent a higher percentage of residents in urban neighbourhoods than suburban settings (Moos, 2016), members of this generation who have stable and relatively well-remunerated employment may be married, have children, and/or live in suburban subdivisions; these groups are obviously not captured by this edgy neighbourhood stereotype.

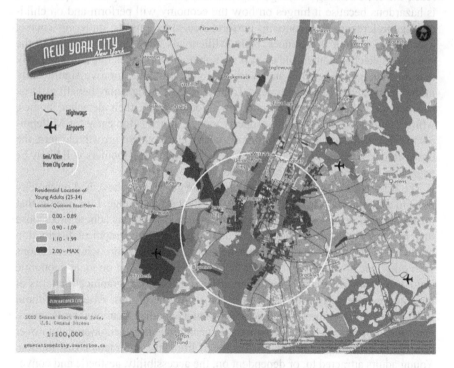

*Figure 2.1* The concentration of young adults in inner city New York

Source: Used with permission from generationedcity.uwaterloo.ca

In some circumstances, the urban context in which Millennials reside and carry out their activities reflects their values and economic circumstances. This is notably the case of edgy neighbourhood Millennials. They often delay child rearing by choice (in a fashion that reflects their values), because of economic circumstances and/or due to higher educational attainment. This mix of values, education and economic reality also explains their tendency to opt for shared living arrangements and rely on friends as sources of emotional and material support. The location and makeup of inner urban areas are consistent with the preference of members of this generation for walking- and public transit-conducive urban environments (Mallach, 2014). Single Millennials living with roommates in edgy neighbourhoods may compromise their preferences because of their precarious economic situations, while Millennials with children may not be able to access their inner city preferences due to high prices and thus opt for suburban housing. Observable residential distributions are not, on their own, a good indicator of actual desire when prices inherently restrict access to a desired set of urban amenities.

Will the segment of this generation identified with edgy neighbourhoods maintain its urban identity, or lose it when moving through life stages, as observed in previous generations (Chapter 14)? Some predict that, like Baby Boomers and Gen Xers, the Millennials will opt in mass for the suburbs when they (belatedly) reach the child-rearing life stage (e.g. Kotkin, 2016). Making such a prediction is hazardous because it hinges on how the economy will perform and on child-rearing preferences. If economic performance maintains its stuttering course, irrespective of their preferences, many Millennials will be denied access to the suburban detached home. And even if economic growth accelerates, it is unlikely that income, social benefits and working conditions will improve significantly in the globalized neoliberal context. To be sure, Millennials may benefit as large numbers of aging and downsizing Baby Boomers sell their detached homes, which could depress residential property values. But it is too early for Millennials to rejoice, for in large urban regions where employment growth clusters, the downward residential price effect created by retiring Baby Boomers will likely be negated by continuing in-migration (also see Chapter 12).

The brief depictions of the impact successive generations have had on the city point to different ways in which generations affect urban change. First, we see that locational preferences may shift. Values to which generations adhere predispose their members to live in certain parts of urban areas. The geographical distribution of members of a generation illustrates the processes at work through intersections with other social categories (such as household composition or socio-economic status) to produce societal – including urban – effects. Geographical patterns of generations necessarily overlap with those of income groups: After all, without financial means, members of a generation cannot actualize locational preferences. Gentrification reflects such phenomena. Especially as it gathers momentum over time, gentrification reveals the confluence of life cycle and income distribution. Young adults attracted to, or dependent on, the accessibility, aesthetic and convenience attributes of inner-city neighbourhoods generally trigger gentrification in its early stages. These individuals, whose income is often limited and who have accordingly been labelled "marginal gentrifiers", have to make living arrangements and space trade-offs to settle in attractive neighbourhoods (Lees, Slater, & Wyly, 2008, pp. 3–38; Rose, 1984). Over time, the interest of higher income groups for inner-city living and the walkability and animation of inner-city neighbourhoods compromises the possibility for less affluent young adults to settle there, except perhaps as renters and in shared accommodation (Chapter 13).

A second way in which a generation contributes to urban change has to do with the reshaping of the urban areas it settles, in agreement with its values and preferences. Such a process concerns retail and services, both mirroring consumption preferences. A generation also has other means of transforming its urban environment. One such instrument is reliance on political pressure to ensure that governments provide services and infrastructures conforming to the values of a generation. For example, in neighbourhoods occupied by Millennials, we could expect to encounter requests for more cycling lanes, as well as walking and running tracks. A generation can also rely on community involvement to leave its mark on urban settings, as when volunteers participate in naturalizing green spaces or painting intersections in their neighbourhood.

Generations can shape the urban environment at different scales. While neighbourhood-scale transformations are most easily associated with generational changes, we cannot exclude the possibility of generations impacting larger geographical areas. Indeed, the generations that fuelled suburban development were instrumental in transforming metropolitan-scale morphologies and dynamics by ensuring their adaptation to near universal automobile use. Generations can also affect cities when their members (typically young adults who most identify with the values defining a generation) formulate, and rally around, new forms of planning and development. The influence of these ideas can be felt on the supply side when adopted by planning agencies and developers (Chapter 10), and on the demand side when they shape consumer preference (Chapter 14).

## Transitory, yet potentially long lasting: the urban dimension of generational change

We conclude here with a brief reflection on our main arguments and their implications. We note that the influence of generations on their urban context tends to be transitory, because generations succeed each other and tend to advance ideas and adopt behaviours that challenge those of previous generations. Another factor accounting for transitory impacts is the relatively short part of the life cycle of the members of a generation devoted to the formulation and promotion of the views that differentiate generations. As members age, generations tend to lose part of their distinctiveness. The influence of a generation on the city is therefore likely fleeting, as illustrated by transitions in the retail makeup of commercial streets or spaces. In some circumstances, however, as when the car-oriented suburban model rallied young adults in the post-World War II decades and became the prominent North American urban form, generational influence can result in structural – and therefore persistent – urban change. The effect of some technologies, such as the car, on cultural practices ultimately exceeds the influence of generational change.

Thus, we see that understanding the impact of generations on cities is complicated by the intermingling of generation with other factors driving urban change. It is impossible to single out the specific contribution of generation to urban change.

Finally, the chapter exposed the dialectic relation between the city and generation. If, at least at the early stages of their members' life cycles, generations leave their mark on the city, so does the city contribute to a generation's self-definition. The city provides settings that position the cultural practices and values of a generation, linking generational identity to particular places and interactions.

Returning to the *7 Up* series discussed at the outset of this chapter, we acknowledge the role of life stages; however, we might propose a modified series better able to examine the relation between generation and the city. Our version of the series would use narratives from the 14 participants to explore their living circumstances and compare them with those of their parents and children. It could also examine how, at different stages of their life cycle, participants negotiated the job market and their living environments such as home and neighbourhood. And the new series could deal with how these circumstances relate to their aspirations for how they might live their lives in the city. Such a series would offer useful insights into how factors such as generation, class, and life experiences leave their mark on the cities we inhabit.

## References

Alwin, D. F., & McCannon, R. J. (2003). Generations, cohorts, and social change. In J. T. Mortimer & M. J. Shanahan (Eds.), *Handbook of the life course* (pp. 23–49). New York: Springer.

Astone, N. M., Martin, S., & Peters, H. E. (2015). *Millennials childbearing and the recession*. Washington, DC: Urban Institute.

Biggart, A., & Walther, A. (2006). Coping with yo-yo transitions: Young adults' struggle for support, between family and state in comparative perspective. In C. Leccardi & E. Ruspini (Eds.), *A new youth? Young people, generations and family life* (pp. 41–62). Aldershot Hampshire: Ashgate.

Bourdieu, P. (1977). *Outline of a theory of practice*. Cambridge: Cambridge University Press.

Corsten, M. (1999). The time of generation. *Time and Society, 8*, 249–272.

Deal, J. J., Altman, D. G., & Rogelberg, S. G. (2010). Millennials at work: What we know and what we need to do (if anything). *Journal of Business and Psychology, 25*, 191–199.

Delbosc, A., & Currie, G. (2013). Cause of youth licensing decline: A synthesis of evidence. *Transport Reviews, 33*, 271–290.

Dunham, C. C., & Bengtson, V. L. (1986). Conceptual and theoretical perspectives on generational relations. In N. Datan, A. L. Greene, & H. W. Reese (Eds.), *Life-span development psychology: Intergenerational relations* (pp. 1–28). New York: Psychology Press.

Economic and Social Research Council. (2016, March 11). *Thatcher's children: The lives of generation X*. Retrieved from www.esrc.ac.uk/news-events-and-publications/news/news-items/thatcher-s-children-the-lives-of-generation-x/.

*The Economist* (2016, January 23). The Millennial generation: Young, gifted and held back – the world's young are an oppressed minority, unleash them.

Elder, G. H., Jr. (1974). *Children of the great depression: Social change in life experiences*. Chicago, IL: University of Chicago Press.

Elder, G. H. Jr. (1994). Time, human agency, and social change: Perspectives on the life course. *Social Psychology Quarterly, 57*, 4–15.

Elder, G. H. Jr., & Liker, J. K. (1982). Hard times in women's lives: Historical influences across forty years. *American Journal of Sociology, 88*, 241–269.

Eyerman, R., & Turner, B. S. (1998). Outline of a theory of generations. *European Journal of Social Theory, 1*, 91–106.

Foot, D. (1996). *Boom, bust and echo: How to profit from the coming demographic shift.* Toronto: Macfarlane Walter and Ross.

Foster, K. R. (2013). *Generation, discourse, and social change.* New York: Routledge.

Fry, R. (2013a, February 21). *Young adults after the recession: Fewer homes, fewer cars, less debt.* Pew Research Centre: Social and Demographic Trends. Retrieved from www. pewsocialtrends.org/2013/02/21/young-adults-after-the-recession-fewer-homes-fewer-cars-less-debt/.

Fry, R. (2013b, August 1). *A rising share of young adult live in their parents' home.* Pew Research Centre: Social and Demographic Trends. Retrieved from www.pewsocial trendsorg/2013/08/01/a-rising-share-of-young-adults-live-in-their-parents-home/.

Fry, R. (2016, April 25). *Millennials overtake baby boomers as America's largest generation.* Washington, DC: Pew Research Center.

Furlong, A., & Cartmel, F. (2007). *Young people and social change: New perspectives* (2nd ed.). Maidenhead: Open University Press.

Giroux, H. A. (2014). *Neoliberalism's war on higher education.* Chicago, IL: Haymarket Books.

Grant, J. L., & Gregory, W. (2015). Who lives downtown? Neighbourhood change in central Halifax, 1951 to 2011. *International Planning Studies, 21*(2), 176–190. doi:10.108 0/13563475.2015.1115340.

Haugen, D. M., & Musser, S. (Eds.). (2013). *The Millennial generation.* Detroit, MI: Greenhaven Press.

Holtz, G. T. (1995). *Welcome to the jungle: The why behind "Generation X".* New York: St. Martin's Griffin.

Howe, N., & Strauss, B. (1993). *13th Gen: Abort, retry, ignore, fail?* New York: Vintage Books.

Howe, N., & Strauss, W. (2000). *Millennials rising.* New York: Vintage Books.

Hulchanski, J. D. (2010). *The three cities within Toronto: Income polarization among Toronto's neighbourhoods, 1970–2005.* Toronto: Cities Centre, University of Toronto.

Jang, H. S., & Maghelal, P. (2016). Exploring Millennial generation in task values and sector choice: A case of employment in planning. *International Journal of Public Administration, 39*, 173–183.

Kalita, S. M., & Whelan, R. (2011, January 14). No McMansions for Millennials. *Wall Street Journal.* Retrieved from www.cobaltdbs.com/app/webroot/pdf/Generation%20 y%20new%20home.pdf.

Keats, J. (1956). *The crack in the picture window.* New York: Houghton.

Kertzer, D. I. (1983). Generation as a sociological problem. *Annual Review of Sociology, 9*, 125–149.

Kotkin, J. (2016). *The human city: Urbanism for the rest of us.* Evanston, IL: Agate B2 Publishing.

Kuhnimhof, T., Armoogum, J., Buehler, R., Dargay, J., Denstadli, M., & Yamamoto, T. (2012). Men shape a downward trend in car use among young adults – evidence from six industrialized countries. *Transport Reviews, 32*, 761–779.

Lee, K. O., & Painter, G. (2013). What happens to household formation in a recession? *Journal of Urban Economics, 76*, 93–109.

Lees, L., Slater, T., & Wyly, E. (2008). *Gentrification.* New York: Routledge.

Mallach, A. (2014). *Who's moving to the cities, Who isn't: Comparing American cities.* Flint, MI: Center for Community Progress.

Mannheim, K. (1952 [1925]). The problem of generation. In K. Mannheim (Ed.), *Essays on the sociology of knowledge.* New York: Oxford University Press.

Marshall, V. W. (1983). Generations, age groups and cohorts: Conceptual distinctions. *Canadian Journal on Aging, 2,* 51–62.

Martin, S. P., Astone, N. M., & Peters, H. E. (2014). *Fewer marriages, more divergence: Marriage projections for Millennials to age 40.* Washington, DC: Urban Institute.

McDonald, N. C. (2015). Are millennials really the "go-nowhere" generation. *Journal of the American Planning Association, 81,* 90–103.

Mettle, S. (2014). *Degrees of inequality: How the politics of higher education sabotage the American dream.* New York: Basic Books.

Mitchell, B. A. (2006). *The Boomerang age: Transitions to adulthood in families.* New Brunswick, NJ: Transaction Publishers.

Moos, M. (2014). "Generationed" space: Societal restructuring and young adults' changing location patterns. *The Canadian Geographer, 58*(1), 11–33.

Moos, M. (2016). From gentrification to youthification? The increasing importance of young age in delineating high-density living. *Urban Studies, 53*(14), 2903–2920. http://usj.sagepub.com/content/early/2015/09/15/0042098015603292.abstract.

Morgan, L. A., & Kunkel, S. R. (2007). *Aging, society, and the life course* (3rd ed.). New York: Springer.

Mykyta, L. (2012). *Economic downturns and the failure to launch: The living arrangements of young adults in the US 1995–2011.* Paper given at the Transition to Adulthood after the Great Recession Conference, Milan, Italy, October 25–26.

Nesselroade, J. R., & Baltes, P. B. (1974). *Adolescent personality development and historical change: 1970–1972.* Monographs of the society for research in child development, pp. 1–80.

Parment, A. (2012). *Generation Y in consumer and labour markets.* New York: Routledge.

Qian, Z. (2012). *During the great recession, more young adults lived with parents.* Providence, RI: Russel Sage Foundation and American Communities Project of Brown University.

Riley, M. W., Foner, A., & Waring, J. (1988). Sociology. In N. J. Smelser (Ed.), *Handbook of sociology* (pp. 243–290). Newbury Park, CA: Sage.

Rogers, W. H., & Winkler, A. E. (2014). *How did the housing and labor market crises affect young adults' living arrangements?* IZA Discussion Paper No. 8568, The Institute for the Study of Labor, Bonn.

Rose, D. (1984). Rethinking gentrification: Beyond the uneven development of Marxist urban theory. *Environment and Planning D: Society and Space, 2,* 47–74.

Rudolph, C. W., & Zacher, H. (2016). Considering generations from a lifespan developmental perspectives. *Work, Aging and Retirement.* Retrieved November 2016, from http://workar.oxfordjournals.org/content/workar/early/2016/05/20/workar.waw019.full.pdf.

Ryder, N. B. (1965). The cohort as a concept in the study of social change. *American Sociological Review, 30,* 843–861.

Sassen, S. (1991). *The global city.* Princeton, NJ: Princeton University Press.

Schoettle, B., & Sivak, M. (2014). The reasons for the recent decline in young driver licensing in the United States. *Traffic Injury Prevention, 15,* 6–9.

Smith, D. P. (2005). Studentification: The gentrification factory? In R. Atkinson & G. Bridge (Eds.), *Gentrification in a global context: The new urban colonialism* (pp. 73–90). Oxford: Routledge.

Spiegel, J. B. (2014). Performing "in the red": Transformations and tensions in repertoires of contention during the 2012 Quebec student strike. *Social Movement Studies, 15,* 531–538.

Thorne, B. (2009). The Seven Up! films: Connecting the personal and the sociological. *Ethnography, 10,* 327–340.

Watters, E. (2003). *Urban tribes: A generation redefines friendship, family and commitment.* London: Bloomsbury.

Winograd, M., & Hais, M. D. (2011). *Millennial momentum: How a generation is remaking America.* New Brunswick, NJ: Rutgers University Press.

Wyn, J., & Woodman, D. (2006). Generation, youth and social change in Australia. *Journal of Youth Studies, 9,* 495–514.

Yates, R. (1961). *Revolutionary road.* New York: Little Brown.

# 3 Planning for "cool"

## Millennials and the innovation economy of cities

*Tara Vinodrai*

As Millennials are entering housing and labour markets, they are doing so at a time when firms, governments, universities and other actors are placing a premium on entrepreneurship, innovation and creativity in their policies, programmes and practices. As important local actors involved in creating and shaping the urban environments where these activities take place, planners are by no means an exception. Plans for growth and development emphasize the importance of making cities and places the crucibles of the next economy: Attracting venture capital and investment, supporting start-ups, and creating "cool", amenity-rich environments that are conducive to innovation processes and the human capital involved in them. Planners must plan for "cool".

However, the "cool", creative city did not emerge in a vacuum. This chapter provides a contextual overview of recent literature that helps us to situate and understand the urban economies that Millennials are encountering. First, this chapter highlights the ongoing debate about the importance of creativity and innovation in cities. Second, the changing nature and organization of work and labour markets is discussed, highlighting the growing level of precariousness experienced by workers. Related to these shifts, the chapter identifies the growing emphasis on entrepreneurship across advanced economies and how this is being articulated in the urban landscape. Specifically, the chapter discusses how these ideas have been mobilized in planning, policy and practice in cities via innovation districts, co-working spaces and maker spaces. In doing so, this chapter takes stock of recent trends and puts into context the urban labour markets and economies in which Millennials are participating.

### Innovation, creativity and the city

Urban economies are necessary for generating new ideas. This observation has been made in the planning literature by Jane Jacobs and others, but is also the received wisdom in the wider literature in urban economics, urban and regional studies, economic geography, and innovation studies. The sheer diversity, density, and volume of interactions between economic actors that are possible in urban environments means that cities are viewed as important platforms for knowledge circulation and exchange between firms and within industries or clusters (Duranton & Puga, 2000; Audretsch, 2002; Storper & Venables, 2004). For this reason, cities are seen as playing a critical role in fostering innovation and are viewed as engines of economic growth (c.f. Glaeser, 2011; Storper, 2013). Within this literature, there is widespread acceptance that human capital contributes to the growth, productivity and economic development of urban economies. Human capital is vital in innovation and knowledge-creation processes, in part due to workers' mobility between projects, firms, and places, which facilitates the transfer of ideas and the exchange and circulation of knowledge (Saxenian, 1996, 2007). However, there has been an ongoing debate regarding what attracts these critical agents of the innovation economy to particular urban economies.

First, there is a body of scholarship that argues that highly skilled workers are drawn to particular locales because of the widespread availability of jobs. Highly skilled workers seek urban locations with an abundance of meaningful employment opportunities related to their knowledge and skill sets. In this view, the presence of thick labour markets is paramount to the locational calculus of individuals, overriding the importance of other place characteristics (Shearmur, 2007). Related to this perspective, some authors have emphasized the presence of universities and other higher education institutions as key attractors of talent, either directly through the attraction of star scientists, researchers, and students but also indirectly since firms will locate in proximity to post-secondary institutions to gain access to cutting-edge knowledge, laboratory research, intellectual property, as well as the knowledge embodied in talent itself. For example, studies have demonstrated that universities and other post-secondary institutions are important actors anchoring specialized industrial clusters (Wolfe & Gertler, 2004). And although universities are often critical for anchoring specialized industrial clusters, they can also play an important anchoring role for communities by contributing to social cohesion and inclusion (Gertler & Vinodrai, 2005).

Elsewhere, scholars argue that rather than thick labour markets and the institutions that support its (re)production, amenities are the key driver attracting talented individuals to particular places. For instance, Storper and Scott (2009) identify three variants in the research on human capital and urban environments that emphasize the role of amenities. First, Glaeser (2005) highlights the importance of physical climate, the presence of high quality school systems, infrastructure (including transportation systems) and services as critical to attracting highly skilled workers (see also Berry & Glaeser, 2005; Glaeser & Gottlieb, 2006). In a slightly different vein, authors such as Clark, Lloyd, Wong, & Jain (2002) emphasize the role of urban cultural amenities of the kind often appreciated by tourists: museums, restaurants, galleries, as well as attractive and authentic built environments and lifestyle amenities (see also Clark, 2011). Finally, Richard Florida (2002, 2014) and others have argued that highly skilled and educated labour, or what Florida (2002) refers to as the creative class, is attracted to places rich with the amenities described above, but that also offer diverse and tolerant environments.

In particular, Florida's (2002) work on the creative class has sparked debate and controversy amongst academics, policymakers and other observers of urban economies. For example, Storper and Scott (2009) raise fundamental questions about causality, while other scholars are concerned about the applicability of the creative class thesis beyond large US metropolitan regions (Sands & Reese, 2008; Lewis & Donald, 2010; see also Grant, 2014). Markusen (2006) notes that there is great differentiation amongst the so-called creative class in terms of their labour market and other social circumstances. Moreover, while Florida (2002) is hardly the first to make observations about the transformation of labour markets and the growth of knowledge-intensive work (c.f. Bell, 1973; Ley, 1996), the research has been accompanied by seemingly easy-to-implement planning and policy solutions that have been quickly (and often uncritically) adopted by local and regional governments. Following the policy prescriptions associated with the creative class thesis and other variants (c.f. Landry, 2000), cities from across North America and further afield developed creative city strategies as a means of attracting and retaining highly skilled labour. This form of "fast policy" is seen by some observers as contributing to inequality within the city and as an extension of neoliberal urban policy making (Peck, 2005).

While the overarching narrative of those places that adopted the creativity mantra was to attract any highly educated, creative workers, the emphasis on downtown urban living and urban amenities such as nightlife, bicycle paths, coffee shops and a "scene" has implicitly – if not explicitly – oriented the attention of these efforts towards young professionals. In this way, planners and policymakers working to attract knowledge workers have contributed to the process of what Moos (2016) describes as "youthification". The amenities emphasized in the innovation-oriented, talent-attraction strategies of many large US and Canadian cities catered to the preferences of well-educated, affluent Millennials rather than those of highly-educated individuals at other stages of life and belonging to other generations, such as childcare, high quality schools, and playgrounds or to the preferences of Millennials in other socio-economic strata.

## Rise of the gig economy?

While cities may offer attractive living environments for creative class workers, and especially cadres of newly minted Millennial new economy workers, dense urban living (and working) also provides the necessary spatial conditions to support the functioning of the labour markets underpinning innovation-related activity. For the most part, this work can be viewed as flexible, short term and contract-based. Over the last few years, observers have increasingly referred to these forms of work that deviate from full-time employment with a single employer as the "gig economy" (Friedman, 2014). While some analysts associate the gig economy explicitly with the employment practices of on-demand firms such as Uber, Lyft and Airbnb that use technology or sharing economy platforms to match customers with service providers (Hathaway & Muro, 2016; Kenney & Zysman, 2016; see also Chapter 8), the term is also used more expansively to include the wide variety of non-traditional work arrangements that are increasingly used by employers (Chapter 6). And, as Friedman (2014) notes, in the USA more than 80% of employment growth between 2005 and 2013 involved workers under alternative contract arrangements rather than traditional ones. In other words, as

Millennials enter the job market, they are more likely than their counterparts in previous cohorts to encounter gig work (Chapter 7).

It is important to recognize that these forms of non-standard work, which include contingent, freelance, part-time temporary, contract, and project-based work, are hardly new. Indeed, "gigs" have been an enduring feature and commonplace form work organization in many industries for quite some time. Industries such as advertising, film and television, design, fashion, and architecture, but also industries like construction are organized around forms of project-based and contract work. In particular, the drive for constant novelty and innovation in the cultural and creative industries – industries that have been central to the creative talent-attraction strategies discussed above – have demanded the constant reconfiguration of work teams and the ability to access just-in-time talent (Vinodrai & Keddy, 2015). This necessitates access to deep pools of highly skilled labour; such demands are best met when firms are able to cluster or agglomerate in particular locations. In this case, downtown and inner city neighbourhoods provide the necessary social and spatial infrastructure for finding work, learning and knowledge exchange, and the creation of localized networks (Hutton, 2010; Vinodrai, 2006, 2013; Vinodrai & Keddy, 2015).

While gig work may have clear business, innovation and economic performance-related benefits to firms, there are a number of downsides to workers. Gig work places a high degree of risk on individuals. In addition to a lack of job security, pensions, and healthcare benefits often afforded by full-time employment, workers participating in project-based (or gig) environments often experience higher levels stress, fatigue, anxiety and strains on personal relationships and work–life balance (Vinodrai & Keddy, 2015). Gig work also diminishes financial security, with potential implications for longer-term decision making regarding major household decisions related to life cycle, such as entering into home ownership or having children. Returning to the case of culture work, there is a constant need to build a personal brand or identity, develop and maintain a strong reputation, and constantly search out the next opportunities, blurring personal and professional identities. And so while Millennials entering the job market may see the benefit to building a portfolio of experiences and enjoy having some flexibility in their schedules through gig economy work, these types of jobs are highly precarious, coming with great personal, social and financial risks and costs.

## Entrepreneurial spaces in the city: innovation districts, maker-spaces and co-working spaces

As noted in the previous section, Millennials are entering a labour market characterized by higher levels of risk, uncertainty and precariousness. With this knowledge in hand, there is a growing sentiment amongst decision makers that Millennials should be encouraged to take matters into their own hands by forging their own career paths via entrepreneurship. Indeed, governments inspired by neoliberal ideologies have re-oriented many programmes and services towards institutions that emphasize the responsibility of individuals for their welfare, including through entrepreneurship (Hackworth, 2007; see also Chapter 1).

In terms of the growing emphasis on entrepreneurship, government-supported initiatives have included the expansion of industry work placements as part of educational programming at colleges and universities, as well as the growth of business development programmes that support entrepreneurial start-up firms, including business incubators and accelerator programmes. Universities have themselves become more prominent development agents in cities, taking on a more expansive economic and community development role (Bretznitz & Feldman, 2012).

Indeed, universities, governments and other actors have engaged in a series of urban experiments focused on equipping Millennials with entrepreneurial opportunities and creating spaces to encourage entrepreneurial activity. Across Canadian and US cities, observers have noted the proliferation of maker spaces, co-working spaces, and the development of innovation districts as popular policy instruments and tools for fostering innovation, creativity and entrepreneurship in urban environments, all while offering "authentic" urban experiences. In all of these cases, the physical design of the spaces and the surrounding built environment is predicated on promoting collaboration and unintended encounters to maximize the chances for innovation and knowledge exchange. And certainly, in planning for the spaces intended to generate the next wave of economic activity, there are implicit assumptions being made about the locational preferences of Millennials.

As noted, innovation districts have become one of the organizing tools for aligning economic development and urban planning objectives that simultaneously revitalize downtown areas for the purposes of entrepreneurship and innovation-related activities and reimagine these neighbourhoods as spaces for young professionals to live, work and play. Katz and Wagner (2014, p. 2) describe innovation districts as the

> ultimate mash up of entrepreneurs and educational institutions, start-ups and schools, mixed-use development and medical innovations, bike-sharing and bankable investments – all connected by transit, powered by clean energy, wired for digital technology, and fueled by caffeine . . . . [They] contain economic, physical, and networking assets. When these three assets combine with a supportive, risk-taking culture they create an innovation ecosystem – a synergistic relationship between people, firms, and place (the physical geography of the district) that facilitates idea generation and accelerates commercialization.

The most well-known innovation districts are in larger cities, where (public) research institutions (e.g. universities, research hospitals) or private companies act as anchor tenants and key agents in revitalization efforts. The Brookings Institute identifies several burgeoning innovation district developments including Kendall Square (Cambridge, MA), University City (Philadelphia), the Cortex Innovation Community (St. Louis), South Lake Union (Seattle) and the Boston Innovation District (Katz & Wagner, 2014). In these cases, a mix of public and private leadership, investment and programming has resulted in revitalized and reimagined urban spaces that are housing new firms and young talent. However, it should be noted that some projects and developments within these innovation districts have existed for a long time. For example, Philadelphia's University City Science Center (UCSC) was founded in the 1960s as the first urban research park in the USA; its business incubator programmes were initiated in the 1970s. In more recent history, the UCSC has expanded its focus and reach, including residential projects and greater interventions to improve transit infrastructure and the built environment that go well beyond its traditional role in supporting new ventures, tech transfer and commercialization.

While the adoption and implementation of innovation districts has led to adaptive reuse and redevelopment projects across large American and Canadian cities, mid-sized and smaller cities have also been able to engage in redevelopment efforts intended to bolster the urban innovation economy. For example, the city of Kitchener (a mid-sized city in Canada) has very deliberately employed urban planning and economic development strategies that focus on creating spaces for creativity, innovation and entrepreneurship (Vinodrai, 2016). For several decades, the city had experienced decline in its downtown core associated with manufacturing plant closures and the relocation of retail to suburban locations. As a result of this economic disinvestment and dislocation, the city had a high proportion of contaminated brownfield sites and abandoned industrial buildings in the downtown.

In the early 2000s, the city created an economic development fund to incent public and private investors to (re)locate knowledge-intensive activity to these areas. The University of Waterloo (UW) was one of the first organizations to access these funds, leveraged against provincial and federal funding, to create a School of Pharmacy atop the site of a former industrial site. Planners and local officials view this investment decision as the catalyst for the further redevelopment of Kitchener's "innovation district". In the years following this initial investment, private investors converted adjacent vacant factory spaces into industrial lofts (Kauffman Lofts) and a consortium of private and public partners created an innovation hub (Communitech Hub). This innovation hub was designed to bring together local and global tech players, including Google, as well as house a suite of programmes run by the local tech industry association and the university to promote entrepreneurship and start-up activity. These initiatives connect university programming that supports students creating their own start-up to the regional innovation system and the key actors within it, with the intent of embedding young entrepreneurs in the regional economy.

Elsewhere, the idea of the maker economy is taking hold and new spaces are emerging to support this movement. The maker economy is predicated on the rise of democratized access to 3D printing (3DP), additive manufacturing (AM) and related technologies. Optimistic accounts suggest that these disruptive organizational and technological phenomena will be transformative: re-localizing production, shortening supply chains, improving sustainability and spurring innovation. Moreover, these technologies have the potential to create new areas of urban competitiveness through the convergence of art, design, digital technologies and manufacturing. While some of the underlying digital technologies supporting this movement have existed for several decades, open source protocols and other advances have made these technologies widely accessible and affordable to inventors, entrepreneurs and users outside universities and large firms (Chapter 8). As the National League of Cities (2016) notes

> The maker movement is centered in cities. And this new, hyperlocal manufacturing environment holds potential not only for individual hobbyists but also for community-wide advances in local entrepreneurship and job creation. Cities have a great opportunity to catalyze this movement as a way to improve our local economies, diversify workforce opportunities, and support the creative economy.

Several recent studies highlight the programmes and developments emerging across North American cities (National League of Cities, 2016; Schrock, Doussard, & Wolf-Powers, 2016). Schrock et al. (2016) provide detailed, in-depth analysis of the emerging landscape of the urban maker economy in Portland, Chicago and New York. They find that there are three distinct types of maker-entrepreneurs operating in these places (micro-makers, emerging place-based manufacturers, and global innovators) that require (and benefit from) different types of urban policy and infrastructure supports. Schrock et al. also find that makers often rely on intermediary organizations to provide business and technical assistance, advocate with local and regional government, and provide access to affordable space, prototyping equipment and other maker technologies, and sales and marketing platforms.

Clark (2014) also identifies the emergence of local and regional intermediary organizations that are inserting themselves into the fabric of the urban built environment to facilitate and advocate for spaces that support the next generation of production and entrepreneurial activities. For example, SF Made (San Francisco) and Portland Made, both members of the Urban Manufacturing Alliance, act in this capacity in their respective cities.

While makerspaces have brought attention to entrepreneurial activity related to the intersections of micro-manufacturing, artisanal crafts and advanced technologies, elsewhere, planners and other local actors are creating and enabling collaborative working environments that are not tied to specific employers or technologies. As noted in this chapter, for a growing number of people, work is no longer tied to a specific location (Chapter 6). While some individuals find themselves working in spaces such as airport lounges, cafes or home offices, others are more likely to seek out spaces where they can work alongside others who share similar work practices related to freelancing, contract work and other such forms of "gig" work (Spinuzzi, 2015).

As a result, North American cities have witnessed a proliferation of co-working spaces. Lodato and Clark (2016) offer one of the first systematic examinations of the co-working industry in US cities, finding that an entire industry related to supporting this type of work has emerged. They find that co-working spaces generally provide space, community, access to a professional network, and attempt to address work–life balance issues. In their study of co-working spaces in the

Province of Ontario (Canada), Jamal and Grady (2016) find that the development of co-working spaces has become intertwined with downtown revitalization efforts, particularly in mid-sized and small cities. Beyond simply providing space to (young) new economy workers, in many instances they found that co-working spaces were also offering services to support collaboration and entrepreneurial business incubation activities.

In all of the cases described above, there are strong economic development, entrepreneurship and innovation imperatives, explicitly oriented towards young entrepreneurs and designed to encourage collaboration and interaction, viewed as central to the knowledge circulation and exchange processes that underpin innovation and creativity. These initiatives are intended to embed young people in the urban economy and – specifically – in downtown, central city locations. However, these efforts are not without controversy (Chapter 19). Deliberate planning interventions intended to support the urban innovation economy have attracted criticism due to the unintended (but arguably expected) outcomes related to displacement and pressures on land and housing markets.

Returning to the case of San Francisco, the city administration used tax and zoning regulations both to address threats made by the social media giant, Twitter, that it would relocate from San Francisco and to attract further tech-oriented investment. In these regards, the intervention can be viewed as a success: Twitter remains as a key anchor firm in the neighbourhood and a series of tech companies, including Yahoo, Uber, Square, Pinterest and Zynga have subsequently located their offices nearby. Yet, these inward investments, bringing with them an influx of highly skilled, young workers, have also brought extensive and highly racialized displacement and gentrification, even as planning documents reveal that city officials recognize and grapple with how to preserve the very social dynamics and spatial qualities that provide environments conducive to creativity and innovation in the first instance (Stehlin, 2016).

In other cases, observers are quick to note that inclusive growth objectives are central to these development plans, pointing to examples where innovation districts provide employment opportunities for low-income groups, as well as cadres of scientists, tech workers and the like (Katz, 2015). There is also a risk that in the rush to (re)create the conditions necessary for contemporary urban innovation, planners, real estate developers and policymakers have quickly and uncritically implemented innovation districts with little attention paid to the underlying physical, economic and networking assets necessary for an innovation economy (Katz, Vey, & Wagner, 2015). Even in the much-lauded case of Philadelphia's innovation district, there is evidence that real estate developers, planners and other local public and private sector actors have potentially created an oversupply of collaboration spaces that will not be easily filled (Saffron, 2016).

## Cool spaces, young places?

Cities have long been viewed as engines of growth and development. In the contemporary economy, this role has been further emphasized as scholars and policymakers alike recognize and grapple with the roles of innovation, creativity and entrepreneurship in the urban economy. Moreover, planning is deeply implicated in providing the urban foundations of the innovation economy. This chapter provides some signposts for understanding the role that cities play in the innovation economy and how this relates to current debates about the role of human capital in urban economies. It goes further to provide some tangible examples of how planners are planning for the next economy and shaping the city to support new and emerging forms of work and industry: The urban environment that Millennials are encountering as they enter job and housing markets.

As this chapter has shown, cities are actively constructing themselves in ways that they hope will attract and retain young, highly educated Millennials. They are creating physical spaces – such as co-working and maker spaces, business incubators, and innovation districts – that are explicitly designed to allow for the fluid exchange of ideas, the creation of entrepreneurial new ventures and startups, and ostensibly stimulate the creative process. In doing so, cities are creating the necessary infrastructure to allow for new forms of economic activity to emerge. Yet, absent new institutions and a wider social security net, the work ("gig" work) associated with these emerging economic activities is highly risky and precarious for individuals.

More broadly, the land and real estate development patterns associated with this activity also come with many risks related to social exclusion and socio-spatial inequalities (Chapter 19). Moreover, whether intentional or not, these developments may well contribute to the creation not just of youthified neighbourhoods, but of generationed cities as well (Moos, 2014, 2016): That is, young places that are not actually diverse in terms of their age or other social structures. It is against this backdrop of a growing emphasis on innovation, entrepreneurialism, the rise of the gig economy and precarious work, and the desire for cities to attract and retain the highly skilled that we must understand and contextualize the rise of the Millennial city.

## References

Anderson, C. (2012). *Makers: The new industrial revolution*. Toronto: McClelland and Stewart.

Audretsch, D. (2002). The innovative advantage of US cities. *European Planning Studies*, *10*(2), 165–176.

Bell, D. (1973). *The coming of post-industrial society*. New York: Basic Books.

Berry, C. R., & Glaeser, E. L. (2005). The divergence of human capital levels across cities. *Regional Science*, *84*(3), 407–444.

Bretznitz, S. M., & Feldman, M. P. (2012). The engaged university. *Journal of Technology Transfer*, *37*, 139–157.

Clark, J. (2014). Manufacturing by design: The rise of regional intermediaries and the re-emergence of collective action. *Cambridge Journal of Regions, Economy and Society*, *7*(3), 433–448.

Clark, T. N. (2011). *The city as an entertainment machine*. Lanham, MD: Lexington Books.

Clark, T. N., Lloyd, R., Wong, K. K., & Jain, P. (2002). Amenities drive urban growth. *Journal of Urban Affairs*, *24*(5), 493–515.

Duranton, G., & Puga, D. (2000). Diversity and specialization in cities: Why, where and when does it matter. *Urban Studies*, *37*(3), 533–555.

Florida, R. (2002). *The rise of the creative class*. New York: Basic Books.

Florida, R. (2014). The creative class and economic development. *Economic Development Quarterly, 28*(3), 195–205.

Friedman, G. (2014). Workers without employers: Shadow corporations and the rise of the gig economy. *Review of Keynesian Economics, 2*(2), 171–188.

Gertler, M. S., & Vinodrai, T. (2005). Anchors of creativity: How do public universities create competitive and cohesive communities? In F. Iacobucci & C. Tuohy (Eds.), *Taking public universities seriously* (pp. 293–315). Toronto: University of Toronto Press.

Glaeser, E. (2005). *Smart growth: Education, skilled workers and the future of cold-weather cities*. Cambridge, MA: Kennedy School, Harvard University.

Glaeser, E. (2011). *Triumph of the city*. New York: Penguin.

Glaeser, E., & Gottlieb, J. D. (2006). Urban resurgence and the consumer city. *Urban Studies, 43*(8), 1275–1299.

Grant, J. (Ed.). (2014). *Seeking talent for creative cities: The social dynamics of innovation*. Toronto: University of Toronto Press.

Hackworth, J. (2007). *The neoliberal city: Governance, ideology, and development in American urbanism*. Ithaca, NY: Cornell University Press.

Hathaway, I., & Muro, M. (2016). *Tracking the gig economy: New numbers*. Washington, DC: Metropolitan Policy Program, Brookings Institute.

Hutton, T. (2010). *The new economy of the inner city: Restructuring, regeneration and dislocation in the twenty-first-century metropolis*. London and New York: Routledge.

Jamal, A., & Grady, J. (2016). *The new economy: Co-working in mid-sized Ontario cities*. Guelph, ON: 10 Carden Shared Space.

Katz, B. (2015, January 26). *How innovation districts can be platforms for economic growth and opportunity*. [Web log] Retrieved from www.livingcities.org/blog/761-how-innovation-districts-can-be-platforms-for-economic-growth-and-opportunity.

Katz, B., Vey, J. S., & Wagner, J. (2015, June 24). *One year after: Observations on the rise of innovation districts*. Washington, DC: Metropolitan Policy Program, Brookings Institute. Retrieved from www.brookings.edu/research/one-year-after-observations-on-the-rise-of-innovation-districts/.

Katz, B., & Wagner, J. (2014). *The rise of innovation districts: A new geography of innovation in America*. Washington, DC: Metropolitan Policy Program, Brookings Institute.

Kenney, M., & Zysman, J. (2016). The rise of the platform economy. *Issues in Science and Technology, 32*(3).

Landry, C. (2000). *The Creative City: A Toolkit for Urban Innovators*. London: Earthscan.

Lewis, N. M., & Donald, B. (2010). A new rubric for creative city potential in Canada's smaller cities. *Urban Studies, 47*(1), 29–54.

Ley, D. (1996). *The new middle class and the remaking of the central city*. London: Oxford University Press.

Lodato, T., & Clark, J. (2016, October 31). *Flexible work, flexible work spaces: The emergence of the coworking industry in US cities*. Retrieved from https://gtcui.wordpress.com/2016/10/31/flexible-work-flexible-work-spaces-the-emergence-of-the-coworking-industry-in-us-cities/.

Markusen, A. (2006). Urban development and the politics of a creative class: Evidence from a study of artists. *Environment and Planning A, 38*, 1921–1940.

Moos, M. (2014). "Generationed" space: Societal restructuring and young adults' changing location patterns. *The Canadian Geographer, 58*(1), 11–33.

Moos, M. (2016). From gentrification to youthification? The increasing importance of young age in delineating high-density living. *Urban Studies, 53*(14), 2903–2920.

National League of Cities. (2016). *How cities can grow the maker movement.* Washington, DC: Center for City Solutions and Applied Research, National League of Cities.

Peck, J. (2005). Struggling with the creative class. *International Journal of Urban and Regional Research, 29*(4), 740–770.

Saffron, I. (2016, March 18). Changing skyline: Making sense of Philadelphia's many innovation districts. *The Philadelphia Inquirer and Daily News.* Retrieved from www.philly.com/philly/columnists/inga_saffron/20160318_Changing_Skyline__Making_sense_of_Philadelphia_s_many_innovation_districts.html.

Sands, G., & Reese, L. A. (2008). Cultivating the creative class: And what about Nanaimo? *Economic Development Quarterly, 22*(1), 8–23.

Saxenian, A. (1996). *Regional advantage: Culture and competition in Silicon Valley and Route 128.* Boston, MA: Harvard University Press.

Saxenian, A. (2007). *The new Argonauts: Regional advantage in a global economy.* Boston, MA: Harvard University Press.

Schrock, G., Doussard, M., & Wolf-Powers, L. (2016). *The maker economy in action: Entrepreneurship and supportive ecosystems in Chicago, New York and Portland.* Retrieved from http://www.urbanmakereconomy.org/

Shearmur, R. (2007). The new knowledge aristocracy: The creative class, mobility and urban growth. *Work, Organization, Labour and Globalization, 1*(1), 31–47.

Spinuzzi, C. (2015). *All edge: Inside the new workplace networks.* Chicago, IL: The University of Chicago Press.

Stehlin, J. (2015). The post-industrial "shop floor": Emerging forms of gentrification in San Francisco's innovation economy. *Antipode, 48*(2), 474–493. doi:10.1111/anti.12199.

Storper, M. (2013). *Keys to the city: How economics, institutions, social interaction and politics shape development.* Princeton, NJ: Princeton University Press.

Storper, M., & Scott, A. J. (2009). Rethinking human capital, creativity and urban growth. *Journal of Economic Geography, 9*(2), 147–167.

Storper, M., & Venables, A. J. (2004). Buzz: Face-to-face contact and the urban economy. *Journal of Economic Geography, 4*(4), 351–370.

Vinodrai, T. (2006). Reproducing Toronto's design ecology: Career paths, intermediaries, and local labor markets. *Economic Geography, 82*(3), 237–263.

Vinodrai, T. (2013). Design in a downturn? Creative work, labour market dynamics and institutions in comparative perspective. *Cambridge Journal of Regions, Economy and Society, 6*(1), 159–176.

Vinodrai, T. (2016). A city of two tales: Innovation, talent attraction and governance in Canada's technology triangle. In D. A. Wolfe & M. S. Gertler (Eds.), *Growing urban economies: Innovation, creativity, and governance in 21st century Canadian city-regions* (pp. 211–238). Toronto: University of Toronto Press.

Vinodrai, T., & Keddy, S. (2015). Projects and project ecologies. In C. Jones, M. Lorenzen, & J. Sapsed (Eds.), *The Oxford handbook of creative and cultural industries* (pp. 251–267). London: Oxford University Press.

Wolfe, D. A., & Gertler, M. S. (2004). Clusters from the inside and out: Local dynamics and global linkages. *Urban Studies, 41*(5), 1071–1093.

# Part II
# Millennial economies

# 4 Young adult household economic well-being

## Comparing Millennials to earlier generations in the United States

*Richard Fry*

Given the labour market struggles of young adults in the aftermath of the Great Recession, the Millennial generation is often portrayed as lacking the economic wherewithal to be the nation's next generation of homebuyers. Assessments of their economic situation sometimes do not go beyond references to student loan debt. This chapter sizes up the economic outcomes of households headed by young adults in the USA using basic measures of their well-being. It shows whether Millennials in the USA are in any more dire financial straits than members of Generation X, the Baby Boom, and the Silent generation were when they were young. The US Census Bureau has uniformly collected data on household income since the 1960s. This chapter analyses these and other long-running data series on household well-being to assess whether Millennials are faring better than prior generations when they were the same age as today's Millennials.

There are lots of ways to assess how a generation is faring. We could look in detail at outcomes in the labour market. A complete assessment of household well-being would include access to credit, credit distress, and measures of financial security or fragility (including rainy-day funds and ability to handle emergency expenses). The analysis below is limited in scope and simply uses education, income, poverty and wealth as measures of household economic well-being.

There is not a clear consensus among researchers on the definition of generations. I follow the Census Bureau and define the Baby-Boomers as those born 1946 to 1964 (Colby & Ortman, 2014). Where feasible I distinguish between early Boomers (born before 1955) and later Boomers (born after 1954). Following Pew Research Center definitions, Millennials are adults born after 1980. Members of Generation X were born 1965 to 1980. The members of the Silent Generation refer to those born 1928 to 1945.

Income and net worth are typically measured for households. Here the outcomes of households headed by Millennials are presented. Measuring outcomes at the household level means that some young adults are omitted. All Millennial households are captured, but not all Millennials are captured since some Millennials do not live in a household headed by a Millennial. Though not all young adults are included, for the purposes of thinking about housing needs and preparation to own a home, the well-being of households headed by young adults is a sensible unit of analysis. The analysis is limited, however, because Millennials not living independently would likely also be those with lower economic means. It is thus plausible that the analysis overestimates the economic well-being of all Millennials. Conclusions should therefore be limited to already independently living young adults, i.e. households headed by a Millennial.

In the analysis of US Census Bureau data sample sizes are larger, and thus the sample can be restricted to only include households headed by young adults age 25 to 34 without compromising sample size. It should be noted that college enrollment has grown in importance in the past decades, and thus some young adults have delayed their entry into the full-time work world. In recognition of this I used the age range 25 to 34 rather than 18 to 34 since the latter would include a large share of post-secondary students. Other data and methodological issues are discussed in the appendix included at the end of this chapter.

The most recent available data are analysed to assess how today's young adults are faring. But how young adults fare today is not necessarily indicative of how they will do in the future. As recent history attests, some assets (homes and stocks and bonds) experience sharp price swings and this does not necessarily impact generations equally since they are at different ages. Furthermore, the structure of the labour market can change. The rewards for education and job experience may change. Although the educational profile of the Millennial generation going forward is fairly set, their future economic success is yet to unfold.

## The economic well-being of Millennial households in comparative perspective

How young adults fare in the labour market and economy is increasingly influenced by the existing skills that they bring to the market. For example, Gale and Pence (2006) find that the key factors influencing changes in wealth across successive generations are changes in household demographic characteristics,

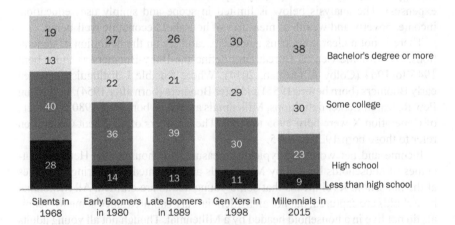

*Figure 4.1* Millennials are better educated than prior generations: educational attainment of 25- to 34-year-old head of household (%)

Source: Tabulations of the 2015, 1998, 1989, 1980, and 1968 Current Population Survey Annual Social and Economic Supplements

including education. Education is a proxy for skill and, at least in terms of formal degree attainment, Millennials are the best-educated generation of young adults in history. In 2015, 38% of Millennial household heads had completed at least a bachelor's degree. This is double the share of Silent generation households completing at least a bachelor's degree (19%) and even outpaces the share of Gen X households completing at least a bachelor's degree (30%) when they were age 25 to 34. About one-third of Millennial households have no education beyond high school compared to almost seven in ten of their Silent counterparts (Figure 4.1).

## Household income and poverty

So far, the educational accomplishments of US Millennials have not been matched in outsized differences in their measured economic well-being compared to earlier generations of young adults. In specific cities, for instance in the Canadian case, young adults in the early 2000s fared worse than those in the early 1980s (Moos, 2014). Now, turning first to household income, it is sensible to adjust household incomes for the size of the household. A given household income goes further in a household with fewer members.

In 2015, the typical household income (adjusted for household size) of Millennial households was about $61,000 for a three-person household. This is very similar to the adjusted household income received by households headed by Gen Xers (about $62,000) and Baby Boomers when they were young adults. The typical or median Millennial household income is substantially greater than the income obtained by households headed by Silents in 1968 (about $43,000) (Figure 4.2).

The increase in the monetary standard of living from the Silents in the late 1960s to more recent generations of young adults can be accounted for by several economic and demographic forces that have played out over the last half

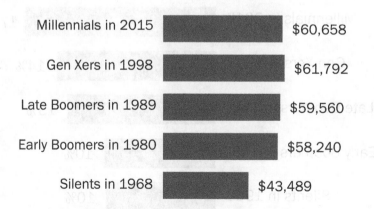

*Figure 4.2* Typical household income of Millennials on par with earlier generations: median adjusted household income in 2014 dollars

Notes: Household income for households headed by 25- to 34-year-olds. Income standardized to a household size of three. Source: Tabulations of the 2015, 1998, 1989, 1980, and 1968 Current Population Survey Annual Social and Economic Supplements

century. First, the household income of young adults has grown due to the rise in young women participating in the paid labour market (Aguiar & Hurst, 2007; Macunovich, 2010). In 1968 43% of female 25- to 34-year-olds were in the labour market. In contrast, 73% of Millennial females were in the labour force in 2015. The increase in women's employment was largely played out by the end of the 1970s, but it no doubt can partly account for the rise in adjusted household income from the Silent generation in 1968 to the Baby Boomers in the 1980s. Second, the inflation-adjusted earnings of young women have been rising over time. Young women have made outsized gains in the education they bring to the labour market. But even within an education category, young women's earnings have increased for all young women except those who did not finish high school (Autor & Wasserman, 2013). Third, the household income measure takes account of household size and the size of households has decreased over the decades. The average Millennial household had 2.8 members in 2015. By comparison, the average household headed by a Silent in 1968 had 3.9 members. So even if household incomes had not risen, the measured economic well-being of more recent generations of young adults would increase as a result of smaller families.

The median only reveals the income of the household in the middle of the distribution of young adult households. Contrary, the standard federal poverty measure is indicative of the well-being of households at the lower end of the distribution (Figure 4.3). A greater share of Millennial households have below poverty level incomes than earlier generations of young adults. In 2015, 17% of Millennial households lived in poverty. By comparison, 14% of Gen X households were in poverty and 13% of later Boomer households in 1989. About one in ten Silent households lived in poverty in 1968.

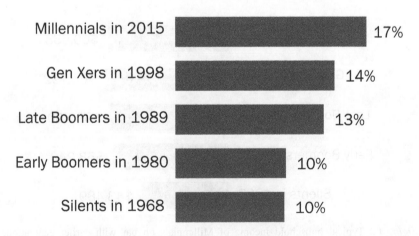

*Figure 4.3* Millennial households more likely be in poverty than earlier generations: households headed by a 25- to 34-year-old in poverty

Source: Tabulations of the 2015, 1998, 1989, 1980, and 1968 Current Population Survey Annual Social and Economic Supplements

The picture emerging from the income and poverty data is that Millennial household incomes tend to be no lower than earlier generations of young adults overall, but there are significantly greater numbers of Millennial households with low incomes than was the case for earlier generations.

Household income is one of the most important measure of economic well-being but it is not the only measure (see also Chapter 5). Income captures what the household brings in over a year. Net worth or wealth captures the amount a household has built up over time, that is its accumulated savings, which in some cases is also income generating. Formally, net worth is the value of what the household owns (assets) minus what it owes (liabilities). Wealth serves as a cushion to tide the household over when incomes are interrupted. In the event of job loss or illness, wealth can be tapped to sustain the household's spending until the household restores its sources of income. Wealth generates financial security. Some forms of wealth can also be used for downpayments on a home purchase.

In 2013 the typical US household (including those headed by older adults as well as Millennials) had a net worth of about $83,000. In 2013 households headed by Millennials had a median net worth amounting to only $9,473. So the typical Millennial household did not have much wealth (Figure 4.4)

Millennial households do not have much wealth because they are young, not because they are Millennial. It takes decades for a household to save and build a nest-egg (households headed by those age 65 to 74 tend to have the greatest wealth) and Millennials are just beginning amassing wealth. No generation has built up much wealth during their young adult years. The typical wealth of Gen X households was only $10,331 in 1998. When the boomers were young the typical net worth of boomer households was only $11,980. These are not statistically different amounts.

Federal Reserve economists have also examined the asset and debt data used for the analysis of wealth here (Dettling & Hsu, 2014). They conclude that on many measures Millennials are faring better than earlier generations of young

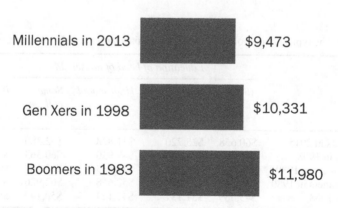

Millennials in 2013       $9,473

Gen Xers in 1998          $10,331

Boomers in 1983           $11,980

*Figure 4.4* The wealth of young generations has been modest at start: median net worth in 2014 dollars

Notes: Household net worth for households headed by 18- to 32-year-olds. Source: Tabulations of the 2013, 1998 and 1983 Survey of Consumer Finances

adults in terms of debt but that Millennials are more likely to be paying off student debt. In general, US Millennials tend not to have much wealth but neither have earlier generations when they were just starting out. The question, of course, is whether or not the same opportunities still exist for Millennials to amass wealth over their lifetime as it did for earlier generations (see Chapter 7).

## The divergence of outcomes among Millennials

Focusing on the outcomes of the typical or median household does not illuminate the greater diversity of outcomes for this generation. A dramatic way to see the diverging fortunes of Millennials is to analyse outcomes by education level. Although the typical Millennial household overall does not have much greater household income than earlier generations when they were young, college-educated Millennials tend to be much better off and less-educated Millennials substantially worse off.

Table 4.1 shows that household incomes of Millennials varies widely by education. Comparing Millennials to the early Boomers in 1980, Millennial households headed by a young adult with at least a bachelor's degree had a typical household income of about $97,000. Households headed by Boomers with at least a bachelor's degree had a household income of only $78,000. For the less-educated, outcomes have deteriorated over time. For example, the median household income of Millennial households headed by a young adult with only high school education was about $42,000 in 2015, 21% lower than the amount of household income of an early Boomer household with similar education in 1980 ($52,658).

It is therefore essential that any appraisal of how Millennials are faring relative to earlier generations should take into account this growing divergence among young adults (also see Chapters 5, 7, 15, 17). Better-educated young adults tend to be faring better but less-educated young adults are faring significantly worse (which contributes to the rise in poverty among Millennials).

*Table 4.1* The typical household income of Millennials varies widely by education

|  | All | Less than high school | High school | Some college | Bachelor's degree or more |
|---|---|---|---|---|---|
|  |  | *Education of head of household* | | | |
| Millennials in 2015 | $60,658 | $25,720 | $41,828 | $52,380 | $96,512 |
| Gen Xers in 1998 | $61,792 | $26,506 | $49,930 | $60,363 | $94,757 |
| Late Boomers in 1989 | $59,560 | $28,855 | $51,633 | $63,639 | $91,432 |
| Early Boomers in 1980 | $58,240 | $33,438 | $52,658 | $62,926 | $77,818 |
| Silents in 1968 | $43,489 | $31,131 | $43,124 | $50,149 | $62,262 |

Notes: Median adjusted household income in 2014 dollars. Household income for households headed by 25- to 34-year-olds. Income standardized to a household size of three. Source: Tabulations of the 2015, 1998, 1989, 1980, and 1968 Current Population Survey Annual Social and Economic Supplements. PEW RESEARCH CENTER

## The fortunes of college-educated Millennials

Households headed by college-educated Millennials have higher household incomes than those of college-educated Boomers back in the 1980s. This is plausible in light of the changes in the labour market and other parts of the economy over the past 35 years. Among young adults with at least a bachelor's degree, those with advanced degrees (master's, doctorate and professional degrees) earn the most. The Census Bureau did not begin to collect information on the attainment of advanced degrees until the 1990s, so the exact nature of the degree attainment of Boomers back in the 1980s is unclear. College-educated Millennials in 2015 were better-educated than college-educated Gen Xers were in 1998. More than 30% of the Millennial household heads with at least bachelor's degree had attained advanced degrees. Among similar Gen-X household heads only 25% had advanced degrees.

At root employers pay significantly higher earnings to college-educated young adults today than they did back in 1980. In other words, the rising household incomes of college-educated young adults are a reflection of the rising returns to education in the labour market that labour economists have copiously documented (Goldin & Katz, 2008). For example, from 1980 to 2012, fully employed males with college degrees saw their real hourly earnings increase between 20 and 56%; and with those who hold advanced degrees seeing the largest increases (Autor, 2014). Earnings are the fundamental source of household income among young adult households (as opposed to earnings generated from wealth). Given that college-educated young adults are the one skill group that employers have paid higher wages over time, one would expect rising household incomes, and wealth accumulation over time, among the college-educated.

Specific trends in the marriage market have also bolstered the household incomes of the college-educated. Millennials are less likely than earlier generations of young adults to be married and hence potentially live in a dual-income household. However the decline in marriage has been more pronounced among less-educated young adults. Among Boomer households in 1980, the college-educated were less likely than their high school educated counterparts to be married. Due to dramatic declines in the likelihood of marriage among less-educated young adults, this pattern is now flipped among Millennial households. College-educated Millennial households are more likely than less-educated Millennial households to involve a married couple.

Patterns of "educational homogamy" (i.e. who marries whom) have also changed. Increasingly, the more educated pair up with the more educated. In 1980 only 51% of married college-educated male 25- to 34 year-olds had a spouse with at least a bachelor's degree. In 2015, 80% of this group had a spouse similarly educated. Married college-educated Boomer heads of households were less likely to be married to a college-educated spouse (who tended to have higher earnings). Married Millennial college-educated households are significantly more likely to be paired up with another college graduate compared to Boomer marriages. These trends will further accentuate differences in earnings and wealth among Millennial households by educational attainment over time.

## The importance of intra-generational diversity in labour market outcomes

Millennials are frequently described as the most diverse generation in history. These claims are made in terms of race or ethnicity, values, culture, and attitudes. The analysis in this chapter adds another dimension. The assessment of economic outcomes of households headed by young adults underscores the diverse fortunes of Millennial young adults (also see Moos, 2014).

Overall, standard economic measures indicate that Millennial households are not faring much differently than Gen X and Boomer households were when they were young adults. All these more recent generations of young adults are doing better than households headed by the Silent generation in 1968 due to the rise of young women in the paid labour market and declines in household size.

However, the overall outcomes of Millennials conceal large differences in how they are faring by educational attainment. Households headed by college-educated Millennials are doing significantly better than college-educated Boomer households were in 1980. Less-educated Millennial households are faring worse than earlier generations of similarly educated households. This reflects the deteriorating wages paid to less-educated young adults in the labour market. In short, the shift toward a knowledge economy that results in the rise in returns to education manifests itself in the diverse economic outcomes of today's young adult households.

## Appendix: Data sources and methods

Most of the analysis in this chapter is based on the Annual Social and Economic Supplement to the Current Population Survey (CPS). The CPS is a monthly labour force survey collected by the US Census Bureau and analysed by the Bureau of Labor Statistics. The CPS is perhaps most widely known as the basis for the nation's monthly unemployment rate trumpeted on the first Friday of each month.

The Annual Social and Economic Supplement is collected in March and features a larger sample than the basic monthly CPS. Since 2002 the Annual Social and Economic Supplement has been based on a sample of about 100,000 households (Pacas & Flood, 2016). The annual income and poverty reports published by the Census Bureau are based on the Annual Social and Economic Supplement. The CPS data files provided by the Integrated Public Use Micro Samples at the University of Minnesota were utilized (Flood et al., 2015).

A household's characteristics are based on the characteristics of the household head. Age and education are based on the year of the survey. Household income and poverty refer to the calendar year before the survey. Household income captures the household's total money income or the before tax earnings and other income sources (for example, rents, dividends and interest) including in-cash government transfers (such as public assistance).

Household wealth is based on the Survey of Consumer Finances (SCF). Collected every three years by the Board of Governors of the Federal Reserve, the

SCF is the most authoritative source on the assets and liabilities of the nation's households (Bricker et al., 2014). Following Census Bureau practice, nominal dollar amounts are adjusted for inflation using the Consumer Price Index for All Urban Consumers Research Series (CPI-U-RS).

In order to better gauge the economic resources of the household it is standard procedure to adjust household incomes for the size of the household. For a given household income, larger households are worse off because there are more mouths to feed and backs to clothe. But simply examining household income per member is not adequate because some items within a household can be shared among the members regardless of size (for example, the furnace, vacuum cleaner, and cable TV subscription). The analysis uses the following common adjustment to capture economies of scale in household well-being (Burkhauser & Larrimore, 2014):

$$\text{Adjusted household income} = \text{household income}/(\text{household size})^{0.5}$$

Under this adjustment, a two-person household does not need twice the household income to be as equally as well off as a one-person household. Rather, the adjustment assumes that a two-person household needs 1.414 times the income of a one-person household to be equally well off.

## References

Aguiar, M., & Hurst, E. (2007). Measuring trends in leisure: The allocation of time over five decades. *Quarterly Journal of Economics, 122*(3), 969–1006.

Autor, D. H. (2014, May 23). Skills, education, and the rise of earnings inequality among the "other 99 percent". *Science, 344*(6186), 843–850.

Autor, D. H., & Wasserman, M. (2013). *Wayward sons: The emerging gender gap in labor markets and education.* Washington, DC: Third Way.

Bricker, J., Dettling, L. J., Henriques, A., Hsu, J. W., Moore, K., Sabelhaus, J., Thompson, J., & Windle, R. (2014). Changes in US family finances from 2010 to 2013: Evidence from the survey of consumer finances. *Federal Reserve Bulletin, 100*(4), 1–41.

Burkhauser, R. V., & Larrimore, J. (2014). Median income and income inequality: From 2000 and beyond. In J. Logan (Ed.), *Diversity and disparities: America enters a new century* (pp. 105–138). New York: Russell Sage Foundation.

Colby, S. L., & Ortman, J. M. (2014). *The baby boom cohort in the United States: 2012 to 2060.* Washington, DC: US Census Bureau.

Dettling, L., & Hsu, J. W. (2014). *The state of young adult's balance sheets: Evidence from the survey of consumer finances.* Retrieved from www.stlouisfed.org/~/media/Files/PDFs/HFS/20140508/papers/Dettling-Hsu_paper.pdf.

Flood, S., King, M., Ruggles, S., & Warren, R. J. (2015). *Integrated public use microdata series, current population survey: Version 4.0* [Machine-readable database]. Minneapolis, MN: University of Minnesota.

Gale, W.G., & Pence, K.M. (2006, Spring). Are successive generations getting wealthier, and if so, why? Evidence from the 1990s. *Brookings Papers on Economic Activity*, 155–213.

Goldin, C., & Katz, L. F. (2008). *The race between education and technology.* Cambridge, MA: Harvard University Press.

Macunovich, D. J. (2010, November). Reversals in the patterns of women's labor supply in the United States, 1977–2009. *Monthly Labor Review*, 16–36.

Moos, M. (2014). Generational dimensions of neoliberal and post-fordist restructuring: The changing characteristics of young adults and growing income inequality in Montreal and Vancouver. *International Journal of Urban and Regional Research, 38*(6), 2078–2102.

Pacas, J., & Flood, S. (2016). *Using the annual social and economic supplement with current population survey panels.* Retrieved from https://cps.ipums.org/cps/resources/linking/4.workingpaper16.pdf.

# 5 Underwater generation?

## Debt and wealth among Millennials

*Alan Walks, Dylan Simone and Emily Hawes*

Of the many trends affecting the Millennial generation, one of the most salient issues discussed by scholars and the mainstream media concerns rising indebtedness and the ability of young adults to build wealth through homeownership. Young adults are somewhat contradictorily seen as, on the one hand, relatively spoiled and highly educated, while on the other hand, relatively disadvantaged in contemporary labour and housing markets (Twenge, 2006; Greenberger, Lessard, Chen, & Farruggia, 2008; Elsby, Hobijn, & Sahin, 2010; Holt, Marques, & Way, 2012; Xu, Johnson, Bartholomae, O'Neill, & Gutter, 2015). Many claim that the Millennial generation will be the first not to do better than their parents in accumulating wealth, due to a relative lack of professional employment opportunities (including permanent positions with retirement pensions), lower real wages, and higher costs of housing (Ferri-Reed, 2013; MacDonald, 2013; Steinberg, 2013; Fry, 2014; Gardiner, 2016). Of key concern is the worry that young adults are taking on generationally unprecedented levels of debt in order to compete for education, jobs and housing.

Of course, these claims and discourses mainly apply to young adults in wealthy developed nations, given that economic growth in many developing nations has raised living standards among many young adults there. And even if young adults in some wealthy developed nations are on average less able to accumulate wealth, it is not clear this applies in all places. One of the main complaints is that housing is more expensive in real terms than it has been for past generations, and that this has required taking on ever-increasing amounts of debt lest it constrains Millennials to a life as "generation rent" (Alakeson, 2011; Blackwell & Park, 2011). Even if true, housing affordability varies considerably from place to place, and it is not clear how this affects young adults as a whole.

There are many challenges in ascertaining the relative wealth of contemporary generations, not least of which is the fact that young adults today have yet to live through most of their life cycle. Thus, even if currently disadvantaged, it is possible that, over time, members of the Millennial generation may be able to surpass the living standards of previous generations. In addition, a greater proportion of young adults today are enrolled in programmes of higher education and, for this reason, fewer participate in labour markets than earlier generations (Statistics Canada, 2013), complicating attempts at multi-year comparisons relating to employment and earnings. Having said that, it is still possible to measure the assets held by young adults, as well as the debts they carry, and to compare these across age cohorts.

## Measuring generational differences in debt and wealth

In this chapter we examine the degree to which the most recent generations of young adults have been able to accumulate wealth – including housing wealth – over time in comparison with previous generations, as well as in relation to other age cohorts. We begin by reviewing the emerging research on the subject, focusing mainly on scholarship from developed English-speaking nations. We then examine these questions via a case study of the Canadian situation, by analysing the data from Statistics Canada's Survey of Financial Security (SFS) across three survey years: 1999, 2005, and 2012. We use the SFS micro-data (raw files) in this chapter, analysed in the Toronto Region Statistics Canada Research Data Centre (RDC). Levels of assets, including housing assets, and debts, are analysed across six age cohorts. Variation in the debts and assets held by young adult households, among different cities, and across socio-economic strata, are examined. The implications of this research for generational inequalities in wealth are discussed in the conclusion.

Given historical, cultural, and institutional similarities to the United States (USA), the United Kingdom (UK), Australia, and New Zealand, Canada makes an excellent case study for examining changes in wealth over time. As in the above nations, the worry that the current crop of young adults will not be able to attain the same standard of living as their parents has grown, with lifetime capacities for labour income and wealth accumulation in doubt (Clark, 2007; MacDonald, 2013; Russell, 2016). Much of the analysis to date, in Canada or other nations, has been conducted at the national scale, implicitly assuming that wealth accumulation among different generational cohorts has advanced similarly regardless of where one lives within a country. In the last section of this chapter, we thus compare net worth for young adults over time in Canada's three largest metropolitan areas: Montreal, Toronto and Vancouver.

To circumvent problems in other studies that make overly limiting assumptions to facilitate comparability between the SFS and survey data from earlier years (Lafrance & LaRochelle-Cote, 2012; Lafrance & Turcotte, 2015), we restrict our analysis to the three waves of SFS data, and use a more standard methodology. To ascertain the changes in net worth and their compositional factors, we classify families and households (not "adults") with major income earners aged 21 or older into six age cohorts. We utilize the major income earner's age to define which age cohort the household belongs to, as this is the most accurate proxy for the primary household maintainer – the standard method of determining household age (Hou, 2010). In our empirical analysis, we use a broad delineation of young adults, defined here as those aged 21 through 34.

We also do not divide the asset and debt levels by the number of adults in a household (as is done in Lafrance & LaRochelle-Cote, 2012 and Lafrance & Turcotte, 2015; see Appendix for further information), but instead base our calculations on the household aggregate values, which in our view is a more accurate method of ascertaining inequalities in wealth, given families and households are the units that consume housing, and often pool incomes for making debt payments. As is true for most studies on this topic, we cannot adequately address the issue of young adults living with their parents.

## Wealth inequalities across generations

The literature suggests that recent generations of young adults may be the first in a number of countries to amass less wealth over their lifetime than their parents. This issue has received increasing attention since the global financial crisis (GFC), and the research on this question has been mainly focused on the recent experience of young adults in Anglo nations.

In the USA, Steuerle, McKernan, Ratcliffe, and Zhang (2013) found that unlike older generations (those aged 47 and higher), who witnessed significant growth in their net worth between 1983 and 2010, those in younger generations experienced limited growth (those aged 38 to 46, and those aged 20 to 28), or in the case of those aged 29 to 37 (the so-called "Generation Y"), a decline in net worth. The GFC had more severe effects on the wealth accumulation capacities of these three younger generations than older generations (see also Chapter 4 by Fry in this volume).

Hood and Joyce (2013) found that in the UK, those born in the 1970s had saved less, and had accumulated less property wealth at equivalent points in their life cycles, than had earlier generations by 2010. This is despite the fact that their incomes had generally been higher than previous generations experienced at a similar time in their life cycles. Furthermore, the 1970s generation was the only one to have their property wealth noticeably affected by the GFC. Unfortunately, this study did not examine the net worth of those born in the 1980s or later (ibid.).

Meanwhile, in Australia, although real incomes have increased across cohorts through the 2000s, the net worth of younger cohorts has generally stagnated. This has occurred even as the wealth of those over 45 years of age (born before 1967) continued to grow rapidly, despite the decrease in wealth caused by the GFC (Daley, Wood, Weidmann, & Harrison, 2014). Indeed, those in the youngest age cohort (25 to 34 years old), representing the Millennial generation in Daley et al.'s (2014) study (for the 2012 data, this pertains to people born between 1978 and 1987), saw their net worth decline by 2.7%. The main cause of these changes in wealth relate to rising housing costs and the declining abilities of successive generations to purchase property in the private market. Homeownership rates in Australia have declined for all cohorts aged 55 and under, particularly among the youngest age cohort (aged 25 to 34). Among the poorest quantiles in each cohort, homeownership rates have declined by between 12% (for those aged 45 to 54) and 31% (for those aged 25 to 34) (ibid.).

Thus, in each of the USA, the United Kingdom and Australia, despite widely varying circumstances with regards to labour-related income, the Millennial generation would appear to be worse off in their abilities to amass wealth, particularly housing wealth, than equivalent generations in the past. Clearly the literature is in its primacy and more research is required on this topic. To date, this subject has attracted insufficient research in the Canadian case. It is to this that we now turn.

## Generational differences in wealth accumulation

Net worth is the standard measurement of household wealth, calculated as total household assets minus total household debts. Pensions and other retirement funds often compose a significant portion of net worth. In the 2012 SFS, pensions and retirement funds comprised approximately 30% of total net worth on average, for all households aged 21 and above (for more information, see our methods appendix). All figures presented in this chapter have been adjusted for inflation, displayed in 2012 constant dollars.

All age cohorts have seen their average net worth rise over the entire study period, by 64% in real terms on average (in 2012 constant dollars). However, average net worth has not advanced equally among household age cohorts. Whereas households with a major income earner aged 21 through 27 largely matched this pattern (net worth increased by 66%), households with their major earner aged 28 to 34 saw their net worth decline between 1999 and 2005, although it increased again between 2005 and 2012 such that their net worth was approximately 46% higher in 2012 than in 1999.

However, these rates of change mask the difference between absolute and relative increases in net worth. For example, although rising at roughly the average rate, households aged 21 to 27 experienced the smallest absolute change between 1999 and 2012 (net worth increased by $39,080). The 55 to 64 cohort, although experiencing the slowest rate of change (40% increase), had the highest absolute level of wealth, and enjoyed a large absolute change in net worth, from $614,930 to $863,660 per household (a $247,730 increase). The oldest cohort (aged 65 and older), meanwhile, had their average net worth rise at the mean rate of all cohorts (64%), but experienced the greatest absolute increase in wealth ($286,122) over the period. As it is the total absolute wealth that matters most when it comes to financial power (e.g. for use as a downpayment for home or vehicle purchase, or starting a business) these patterns present a picture of widening wealth disparities. Indeed, the standard deviation in real net worth across age cohorts increased significantly, from $192,025 to $282,916, over the study period (all in constant 2012 dollars).

These trends highlight the importance of considering both the relative rates of changes and the absolute magnitudes of change. This similarly applies to household debt. While wealthier households often have the largest magnitudes of debt, they reveal the lowest *relative* burden of debt, whether measured in relation to income or assets. Conversely, it is the lowest-income households that often have the highest relative debt burdens, as exemplified through debt-to-income and debt-to-asset ratios (see Hurst, 2011).

Aggregated measures of net worth can mask important differences among classes of assets and debts and their associated change over time. One common discourse in mainstream media and among economists is that growing (unprecedented) levels of household indebtedness in Canada may not be a cause for concern given that asset levels have also been growing, leaving households in better financial positions today than in the past (Cross, 2015). Higher levels of indebtedness, while perhaps a concern for highly-leveraged households, might then be offset in the aggregate economy through rising asset prices.

If, however, we examine how various asset and debt classes have increased (or decreased) *relatively* over time among different age or wealth cohorts, the story looks different. Notably, average absolute magnitude increases in asset prices are significantly influenced by the growth of asset values among wealthier and older households. As such, absolute increases in levels of assets and debts can mask the changing relative rates and burdens of servicing debt among different subgroups. As well, whereas asset prices can change rapidly (witness the bursting of stock market and housing bubbles in the USA, Ireland, and elsewhere), debt obliga- tions typically remain. Indeed, as Irving Fischer long ago noted (1933) declining asset values make it more difficult to pay down debt, and, even when debts are being paid down, overall leverage ratios can rise during a period of asset-value deflation. Households facing asset depreciation can fall "underwater", and can have their debt levels outweigh their asset levels and their relative debt servicing burdens reach unmanageable levels.

Figure 5.1 shows the level (in constant 2012 dollars) and composition of both assets and debts for each age cohort, from 1999 to 2012. Clearly, the total value of assets exceeds total debt levels in all years and for all cohorts, and the big story is the significant increase in total asset values and net worth between 2005 and 2012. However, although average asset values have increased by a greater *dollar*

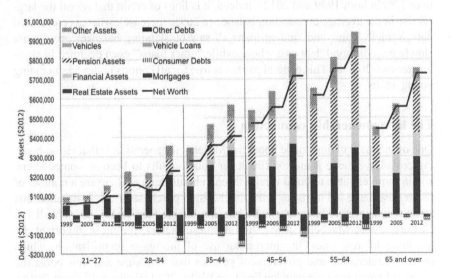

*Figure 5.1* Assets, debts, and net worth, by age cohort, Canada 1999, 2005, 2012

Notes: Age cohorts are classified by the age of the major income earner in each household. "Other" assets include furniture, collectables and business equity, which although part of one's estate is not usually considered part of "marketable wealth" (Wolff, 2002). Source: Calculated by the authors using the raw data from Statistics Canada, Survey of Financial Security (SFS), 1999, 2005, 2012

amount than have average debts, all household age cohorts except the youngest (aged 21 to 27) have seen average total debt levels increase at relatively *faster* rates than average asset levels. For instance, among households aged 28 to 34, total debt levels increased by 90% from 1999 to 2012, while total asset levels increased by only 60%. Among households aged 35 to 44, total debts increased by 133%, while total assets increased by 64%, placing this cohort in a significantly more vulnerable financial position with respect to potential asset value shifts, even if total asset values have outstripped the rise in debt levels in absolute terms.

The result of these changes is a shift in the composition of assets and debts held by each household age cohort. Among asset classes there is a clear and unequivocal shift toward holding larger shares of real estate assets (as a percentage of total assets) across all household age cohorts. The patterns of change for other asset classes, meanwhile, are not as clear-cut. Whereas the value of public and private pensions has increased substantially in real terms for all age cohorts, it is only among the youngest (age 21 to 27) cohort that pensions represented a rising share of total assets over time. All other household age cohorts actually see a decline in pension assets as a share of total assets (mainly due to the rapidly rising share of real estate in their asset portfolios). For financial (non-pension) assets, meanwhile, the youngest cohort (age 21 to 27) saw their level rise faster than other cohorts (a 73.9% increase) but from the smallest base level of any of the cohorts and so ending the period in 2012 with financial asset worth only half as much as the next cohort, and only 11% that of the richest cohort (age 55 to 64).

The increase in real estate holdings is matched by an increase in mortgage balances, despite mortgages maintaining a stable share of overall debt levels (just over 77% in both 1999 and 2012). Indeed, it is lines of credit that reveal the largest rise in the average outstanding balance (a 269% increase), whereas credit card debt, vehicle loans, and student loans all saw outstanding balances grow more slowly than did total debt as a whole, while "other loans" even declined in absolute terms (-20.6%). This general pattern is true for all cohorts under 55, including young adults.

## The homeownership connection

One of the main claims made about the Millennial generation is that rising housing costs have made it more difficult for young adults to become homeowners, and in turn, less able to build wealth through home equity. There are a number of factors that have led to rising Canadian housing purchase prices, a situation that some commentators have labelled a housing bubble (see Walks, 2014; Walks & Clifford, 2015; Macbeth, 2015). Among these factors, the most important are declining interest rates, the increasing use of mortgage securitization, which reduces lender risk, and government policies that subsidize and thus boost the supply of credit for mortgage lending (see Walks, 2014; Walks & Clifford, 2015).

Despite housing cost increases, the rate of homeownership among young adults in Canada has *not* declined from 1971 to 2011, for either those in their twenties or early thirties. This is unlike the situation in the USA, the UK or New Zealand.

Instead, after initially peaking in 1976, the homeownership rates among these cohorts in Canada declined until 1996 but then began rising again. By 2011 the homeownership rate among those in their 20s reached its highest point on record, whereas for those in their early 30s, homeownership rates rose to 59.2% in 2006 and remained unchanged in 2011 (lower than they were in the late 1970s, but higher than at any point since 1986) (calculated by the authors from the 2011 NHS and the data in Hou, 2010). The two oldest cohorts (aged 55 to 64; 65 and over) have experienced virtually monotonic increases in homeownership rates with each passing census, with only 2011 showing a (slightly) lower level (but of course the 2011 National Household Survey, which replaced the long-form census, is neither as reliable as, nor strictly comparable to, the previous censuses). However, the rate of increase in homeownership between 1996 and 2011 among those in their 20s (36.3%) and early 30s (10.7%) far exceeds that of all other age cohorts (average of 2.8%). Thus, in both relative and absolute terms, young adults appear to have been disadvantaged in the 1976 to 1996 period in terms of access to homeownership, but not since 1996. Rising home prices have clearly not hampered the ability of younger adults to become homeowners in the aggregate.

However, increasing rates of homeownership do not necessarily facilitate accumulation of wealth if the levels of debt that young adult households need to take out rise even faster than their income or assets. Indeed, average levels of indebtedness as indicated by debt-to-income ratios have risen substantially, as have average debt-to-asset ratios (Figure 5.2). It is the youngest three age cohorts that have the highest debt ratios in each survey year, both compared to income and in relation to assets. While the third cohort (aged 35 to 44) reveals the highest average debt-to-income ratios in both 2005 and 2012, the two youngest Millennial cohorts are close behind. The average debt-to-disposable (after-tax) income ratios for the two youngest cohorts sit at roughly 250% in 2012, up from 155% (aged 21 to 27) and 188% (aged 28 to 34).

Similarly, while the debt-to-assets ratio has either remained flat or increased slowly (from generally very low levels) for older working-age cohorts, it has grown substantially for the three youngest cohorts. This is particularly true for the youngest cohort (age 21 to 27), due to the coupling of greater homeownership rates with rapidly rising housing prices. Average debt-to-asset ratios are above 100 for the three youngest cohorts in 2012, indicating that, *on average*, households in these cohorts could not pay off all their debts if they sold all their assets, a situation that is new for the second and third cohorts.

The reason this finding contrasts with the net worth values shown in Figure 5.1 is that the latter refer to aggregate totals, which are more influenced by high values (of wealth, debt and income), whereas the average ratios give the same weight to each household regardless of their level of income or wealth. This discrepancy between the two measures indicates that it is lower-wealth households that are more highly leveraged. Thus, while real estate values rose markedly in absolute terms through 2012, if the value of the real estate collateral backing the mortgages were to fall, it would put many households, particularly young adult households who have not built much equity, in a precarious financial position.

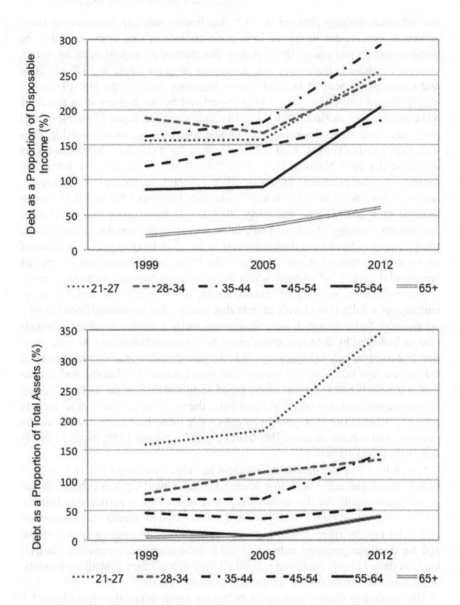

*Figure 5.2* Average debt-to-income and average debt-to-asset ratios, by age cohort, Canada 1999, 2005, 2012

Notes: Graphed are the average debt-to-disposable income and debt-to-assets ratios for households in each age cohort. This method treats all households equally regardless of income, and thus is different from economy-wide ratios that aggregate all households together first before calculating the ratios. Age cohorts are classified by the age of the major income earner. Figure (a) was calculated for only those households reporting positive values of income, while graph (b) was calculated for only those households reporting positive levels of assets. Therefore, our ratios can be viewed as conservative estimates. Source: Calculated by the authors using the raw data from Statistics Canada, Survey of Financial Security (SFS), 1999, 2005, 2012

## Geographies of wealth

Finally, the national picture of assets, debts, and net worth painted above masks significant geographic variation in the ability of young adults to accumulate wealth over time. One important reason for this is that changes in real estate values and, in turn, the mortgage amounts that are needed to facilitate housing purchase, have not continued apace in the same way everywhere. While some cities with healthy job markets for young people have seen relatively rapid increases in housing and land values, other cities have seen slower, or more sporadic, increases. While labour incomes and employment prospects clearly vary among cities, a significant proportion of the variation in wealth accumulation is related to the workings of housing markets and the degree to which young adults feel compelled to purchase housing in a given place.

In cities with high demand and tight housing markets, where rental housing is both expensive and limited, young people and newly-formed households will be under more pressure to purchase owner-occupied housing. A key factor in this is the shift of the federal and provincial governments since the early 1990s away from building social housing, and toward encouraging the private market to provide rental housing (typically, by encouraging would-be landlords to purchase condominium units for renting out: see Walks & Clifford, 2015; Rosen & Walks, 2015). This shift reduced the supply of affordable rental housing, and only partially replaced it with new (mostly high-cost) private rental units, thus driving up average rents. The people who have benefitted from this system are those landlord-speculators who entered this market prior to the bubble; young adults of the Millennial generation are generally too young to have been able to purchase housing during this time.

Here we consider geographic differences among Canada's three largest census metropolitan areas (CMAs): Montreal, Toronto, and Vancouver. First, while all household age cohorts have seen their net worth increase between 1999 and 2012 in all three cities, the rates of such change vary considerably between places (Figure 5.3). The average net worth of all households increased by 61.4%, 52.9%, and 86.0% in Montreal, Toronto and Vancouver, respectively. This translates to an increase in average net worth among all households in Montreal of $184,981, in Toronto of $215,029, and in Vancouver of $366,685. In all three cities, real estate values in the aggregate have risen faster than mortgage debt levels, but each at very different rates (in Montreal, 86 vs. 58%, in Toronto 79.7 vs. 66.9%, and in Vancouver 129.3 vs. 94%).

Similarly, the pattern of asset and debt accumulation differs among younger-adult households depending on where they live. Whereas those with a major earner aged 21 through 34 in Montreal and Toronto experienced minor increases in their average net worth (increasing 8.1% from $117,380 to $126,876, and 5.8% from $182,700 to $193,280, on average, respectively), Vancouver's young adult households saw an increase in average net worth of 105.1%, from $107,988 to $221,430 per household. Thus, the Vancouver CMA's young adult (Millennial generation) households went from having the lowest average net worth among the three CMAs in 1999, to having the highest average net worth among young adult

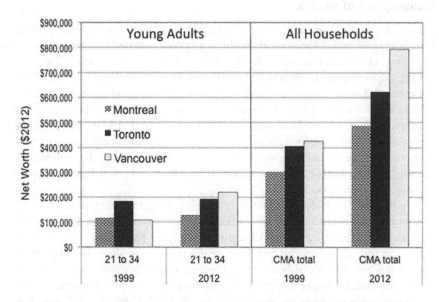

*Figure 5.3* Net worth, all households and young adult households: Montreal, Toronto and Vancouver Metropolitan Areas (1999 and 2012)

Notes: Households are the unit of analysis. Net Worth = Total Assets – Total Debts. The net worth of households was calculated to control for changes in the age composition of primary earners within each cohort in each CMA. Thus, changes in net worth within each cohort are not the result of changing internal age compositions. Average levels of net worth were calculated using all reported values (including those being zero or even negative, where the household has a larger magnitude of total debt when compared to total assets).

Source: Calculated by the authors using the raw data from Statistics Canada, Survey of Financial Security (SFS), 1999 and 2012

households in 2012. In some cases, older adults owning property in Vancouver may have used that as collateral to allow their children to enter the property market, from which much of these gains in net worth derive.

Relative rates of change in assets and debts provide another picture of the changing debtscape in each metro. Debt levels among Montreal's young adult households have increased at the fastest rate compared to total asset levels (43.8% vs. 16.8%, respectively), and the average young adult household mortgage indebtedness similarly increased at a faster rate than their real estate values over time (54.5% vs. 43.9%), although overall debt levels are lowest among young adults in Montreal (at 116.8% of disposable income in 2012). Nearly identical patterns are observed in Toronto. Young adult households exhibit faster rates of change in their total debt levels (39.6%) than in their total asset levels (15.1%), while mortgage indebtedness grew at a faster rate (42.5%) than that of real estate assets (36.5%). Vancouver's young adult households are again the significant outlier here. Not only have their total asset levels (98.3% increase, much of which is real estate) risen at much higher rates than in the other two metropoli, but they rose a faster pace than their total debt levels (85.8%). Indeed, their average real estate values (134.5%) grew faster than outstanding mortgage balances (92.9%). However, the latter rate of change is still very high, with the result that Millennials in Vancouver saw the most rapid increase in average debt-to-income levels (76.3%) and ended the period with much higher levels of debt (310.5% of disposable income in 2012) than their counterparts in either Montreal or Toronto (198.4% of disposable income). These contrasts are directly related to housing market dynamics, which are quite different among the three cities.

Clearly, local context matters considerably. These differences in generational debt, assets and net worth would likely be even more amplified if we compared a broader range of cities, for instance cities with highly variable growth rates or very different population totals (see Walks, 2013, 2016).

## Discussion

Trends in saving, borrowing and wealth accumulation among young adult households reveal both continuities and disruptions. On the one hand, there has been a general continuity over time in the net worth position of each cohort, with young adults understandably amassing less wealth than older cohorts regardless of the decade in question. Furthermore, with the secular increase in homeownership rates over time, there has been a trend toward an increasing weight of real estate assets in each cohort's asset composition, and young adult households are not alone in this. In fact it is solely real estate assets that have consistently increased in proportional share over time across all household age groups. Linked to this, the most recent period from the early 2000s through 2012 is marked by rapid increases in debt levels, particularly mortgage debt. Indeed, at the national scale, total debt levels have increased at a faster (relative) rate than total asset levels for all household age groups except the youngest cohort (aged 21 to 27), and in some cases two to three times faster than net worth.

These trends are disproportionately pronounced among young adult households. If real estate values were to decline, or if a recession were to lead to widespread job losses, this would potentially place many households in a precarious financial situation, particularly young adults who own homes with lower levels of housing equity. While young adults in Canada do not exhibit lower rates of homeownership than older cohorts, higher debt loads (both relative and absolute levels) make young adults vulnerable to interest rate changes or employment struggles, which could put them at risk of not accumulating as much equity in housing as did previous generations. As Lafrance and Turcotte (2015) note, it is possible that today's generation of young adults could be the first to carry high debt loads well into retirement years, worsened by the fact that many will be required to continue working to service their debt. However, the financial position of young adults varies considerably across the country. Even just among the three largest metropolitan areas, there is substantial variation in the levels, and rates of change, of net worth and financial vulnerability. Indeed, there is virtually as much difference in Millennials' financial situations displayed across these metropolitan areas as there is among age cohorts at the national scale. One's chances at wealth accumulation, and therein one's financial well-being, to a large degree depend on where one lives.

This chapter provided an initial look at Millennial wealth and debt in Canada. It demonstrates that the distribution of wealth and its components is far more complex than presented in the mainstream media. Canadian young adults are shown to have quite varied financial situations and, although there are similarities between their experiences and young adults in other nations, the image of "generation rent" that has been applied in other contexts cannot so easily be transferred to the Canadian context. Within Canada, young adult households in different cities exhibit very different financial situations and vulnerabilities. There is a clear need for more research on the diversity of experiences among generations, in Canada and elsewhere.

## Appendix: Methods

The Survey of Financial Security (SFS) is a survey conducted periodically by Statistics Canada with questions pertaining to asset holdings and liabilities. It allows for determination of the degree to which young adults have fared over time. The SFS contains information on all sources of debt and wealth, including pensions and retirement funds. The SFS values pensions and retirement funds from both on a going-concern basis and a termination basis. In our analysis, we value pension funds on a going-concern basis. This values the funds in the pension on the assumption that the pension will continue indefinitely, and so is a more realistic assessment of the actual future value (whereas a termination basis values the funds currently available on the assumption that the pension must be soon terminated and all future benefits paid out, which is rarely a valid assumption but often used to ascertain whether a pension meets strict solvency criteria). Our chapter seeks to answer questions about generational effects on wealth accumulation using data in the SFS across three survey years: 1999, 2005, and 2012.

In our analysis, we do not divide the asset and debt levels by the number of adults in a household as do Lafrance and Turcotte (2015), but instead base our calculations on the household aggregate values, which are the values actually reported in the SFS. This is the standard and more accurate method of ascertaining inequalities in wealth, given families and households are the units that consume housing, and often pool incomes for making debt payments, and we found that how one defines an "adult" has large and problematic implications for the results derived from a per-adult analysis. We compare the average levels of assets, debts, and net worth, for each of the six age cohorts over the period from 1999 to 2012.

Neither our methods, nor those of Lafrance and Turcotte (2015), can adequately deal with situations in which younger adults live at home with their parents. Lafrance and Turcotte's methodology involves some cases in which such households are categorized based on the younger adult's age, and thus incorrectly assigns all the parents' assets and debts to the younger age cohorts of adults (which is one reason their methodology makes it look like younger generations of adults have more wealth than they really do – if young adults become more likely to live at home over time, then the proportion of households mis-classified in this way will increase with each subsequent survey, improperly making it look like younger adults' wealth is increasing). Our methodology, on the other hand, assigns these households to the parent's age cohort, unless the parents' income is lower than that of their younger adult children (which is unlikely, except in the case of adults living with retired parents). Either way, it will be necessary going forward to derive a methodology for ascertaining the wealth and debt levels of young adults who live at home.

# References

Alakeson, V. (2011). *Making a rented house a home: Housing solutions for 'generation rent'*. London: Resolution Foundation.

Blackwell, A., & Park, A. (2011, May). *The reality of generation rent: Perceptions of the first time buyer market*. London: National Centre for Social Research.

Clark, W. (2007). Delayed transitions of young adults. *Canadian Social Trends, 84*, 14–22.

Cross, P. (2015, May). *A longer-term perspective on Canada's household debt*. Fraser Research Bulletin, Fraser Institute. Retrieved from www.fraserinstitute.org/sites/default/files/longer-term-perspective-on-canadas-household-debt.pdf.

Daley, J., Wood, D., Weidmann, B., & Harrison, C. (2014). *The wealth of generations*. Grattan Institute. Retrieved from http://grattan.edu.au/wp-content/uploads/2014/12/820-wealth-of-generations3.pdf

Elsby, M. W., Hobijn, B., & Sahin, A. (2010). The labor market in the great recession. *Brookings Papers on Economics Activity, 41*(1), 1–69. Retrieved from www.nber.org/papers/w15979.pdf.

Ferri-Reed, J. (2013). Millennials – generation "screwed" or generation "shrewd"? *The Journal for Quality and Participation, 36*(1), 22–23.

Fischer, I. (1933). The debt-deflation theory of great depressions. *Econometrica, 1*(4), 337–357.

Fry, R. (2014, May). *Young adults, student debt and economic well-being*. Washington, DC: Pew Research Center's Social and Demographic Trends Project, pp. 1–24. Retrieved from www.pewsocialtrends.org/files/2014/05/ST_2014.05.14_student-debt_complete-report.pdf.

Gardiner, L. (2016, July). *Stagnation generation: The case for renewing the intergenerational contract*. Intergenerational Commission Report, Resolution Foundation. Retrieved from www.intergencommission.org/wp-content/uploads/2016/07/Intergenerational-Commission-launch-document.pdf.

Greenberger, E., Lessard, J., Chen, C., & Farruggia, S. P. (2008). Self-entitled college students: Contributions of personality, parenting, and motivational factors. *Journal of Youth and Adolescence, 37*(10), 1193–1204.

Holt, S., Marques, J., & Way, D. (2012). Bracing for the Millennial workforce: Looking for ways to inspire Generation Y. *Journal of Leadership, Accountability and Ethics, 9*(6), 81–93.

Hood, A., & Joyce, R. (2013). *The economic circumstances of cohorts born between the 1940s and the 1970s*. London: Institute for Fiscal Studies. Retrieved from www.ifs.org.uk/comms/r89.pdf.

Hou, F. (2010). *Homeownership over the life course of Canadians: Evidence from Canadian censuses of population*. Research Paper Cat. No: 11F0019M-No.325, Statistics Canada, Ottawa.

Hurst, M. (2011, April 21). *Debt and family type in Canada*. Component of Cat. no. 11-008-X, Statistics Canada, Ottawa. Retrieved from www.statcan.gc.ca/pub/11-008-x/2011001/article/11430-eng.pdf.

Lafrance, A., & LaRochelle-Cote, S. (2012, June 22). *The evolution of wealth over the life cycle*. Perspectives on Labour and Income, Component of Cat. No. 75-001-X, Statistics Canada, Ottawa.

Lafrance, A., & Turcotte, J. (2015). *Is today's generation of young Canadians saving less than earlier generations? – a cohort analysis of the wealth accumulation process*. Manuscript prepared for the Department of Finance, Government of Canada. Ottawa: Statistics Canada internal memo.

Macbeth, H. (2015). *When the bubble bursts: Surviving the Canadian real estate Crash*. Toronto: Dundurn Press.

MacDonald, D. (2013, November 8). *What the feds should be focusing on*. Canadian Centre for Policy Alternatives: Behind the Numbers. Retrieved from http://behindthenumbers.ca /2013/11/08/what-the-feds-should-be-focusing-on/.

Rosen, G., & Walks, A. (2015). Castles in Toronto's sky: Condoism as urban transformation. *Journal of Urban Affairs*, *85*(1), 39–66.

Russell, A. (2016, January 13). 4 in 10 Canadians believe they are "worse off" than their parents: Ipsos poll. *Global News*. Retrieved from http://globalnews.ca/news/2451943/4-in-10-canadians-believe-they-are-worse-off-than-their-parents-ipsos-poll/.

Statistics Canada (2013, April 7). Changing labour market conditions for young Canadians. *The Daily*. Retrieved from www.statcan.gc.ca/daily-quotidien/130704/dq130704a-eng.htm.

Steinberg, S.A. (2013, June 5). *America's 10 million unemployed youth spell danger for future economic growth*. Center for American Progress. Retrieved from www.americanprogress.org/issues/economy/report/2013/06/05/65373/americas-10-million-unemployed-youth-spell-danger-for-future-economic-growth/.

Steuerle, E., McKernan, S-M., Ratcliffe, C., & Zhang, S. (2013). *Lost generations? Wealth Building among Young Americans*. Washington, DC: Urban Institute. Retrieved from http://inequality.stanford.edu/_media/working_papers/steuerle-et-al_lost-generations.pdf.

Twenge, J. M. (2006). *Generation me: Why today's young Americans are more confident, assertive, entitled and more miserable than ever before*. New York: Free Press.

Walks, A. (2013). Mapping the urban debtscape: The geography of household debt in Canadian cities. *Urban Geography*, *34*(2), 153–187.

Walks, A. (2014). Canada's housing bubble story: Mortgage securitization, the state, and the global financial crisis. *International Journal of Urban and Regional Research*, *38*(1), 256–284.

Walks, A. (2016). Homeownership, asset-based welfare, and the neighbourhood segregation of wealth. *Housing Studies*, *31*(7), 755–784.

Walks, A., & Clifford, B. (2015). The political economy of mortgage securitization and the neoliberalization of housing policy in Canada. *Environment and Planning A*, *47*(8), 1624–1642.

Wolff, E. N. (2002). *Top Heavy: A Study of Increasing Inequality of Wealth in America*. New York: The New Press.

Xu, Y., Johnson, C., Bartholomae, S., O'Neill, B., & Gutter, M. S. (2015). Homeownership among Millennials: The deferred American dream? *Family and Consumer Sciences*, *44*(2), 201–212.

# 6 The Millennial urban space economy

## Dissolving workplaces and the de-localization of economic value-creation

*Richard Shearmur*

There is a long tradition of conceptual and empirical work that attempts to make sense of the way economic activity locates across metropolitan areas and the way the creation of economic value plays out across urban space. Concepts such as the CBD (Central Business District), suburban employment centres, polycentrism, edge cities and – more recently – edgeless cities and scatteration have provided a conceptual vocabulary for planners, residents and businesses to understand their city. These concepts have, in turn, had a powerful effect on how metropolitan areas have developed, since they have been integrated into urban policy thinking, have directed infrastructure development, and have become part of many people's everyday lives. Cities, like national communities, are imagined (Anderson, 2006), and this imagination rests upon shared concepts and vocabulary.

In this chapter I argue that these concepts and vocabulary, and, by extension, the empirical work and planning ideas that rest upon them, espouse a fundamental assumption that has always been an approximation, but that may no longer be a valid one for Millennials, a generation that has grown up with mobile communications technology, internet, and an increasingly precarious and short-term job market (Palfrey & Gasser, 2008; Boltanski & Chiapello, 1999).

The assumption is straightforward – economic activity (and its concomitant value-creation) has a location. Economic activity can be conceptualized in a variety of ways. It is usually understood – at least when urban structure is discussed – as referring to economic establishments and to the workers attached to these establishments. The concepts listed above refer to *where people work*. Thus, for instance, the observation of specialized employment centres (Marshall, 1890; Porter, 2003) led to the idea that the geographic clustering of economic activity is important for productivity. Today, it is similarly believed that localized innovation districts (Katz & Wagner, 2016) or creative neighbourhoods (Spencer, 2016) will, by virtue of the spatial concentration of economic activity, generate interactions that lead to economic growth. Following this line of thinking, city planners have actively promoted the spatial clustering of economic activity on the understanding that workers will interact with each other thereby generating knowledge exchange, innovation, and competitive advantage.

If, however, economic activity and value-creation are no longer tied to particular locations, then we need to re-think our understanding of the spatial-economic processes that occur within cities. There is, of course, considerable inertia in the way cities function. Many jobs remain rooted in space, and the physical reality of past layers of infrastructure and real estate mean that, whatever the trends and concepts discussed in this chapter, traditional concepts remain useful to decipher existing cities. However, at the margin – a margin that I argue is sizeable, growing, and inhabited by many young people today – the question of *where* work actually takes place is increasingly pressing. The assumption that most work takes place in a fixed location needs to be reassessed.

The chapter marshals recent literature, reports and current observations of cities, workplaces and technologies in order to set out a research agenda. There is considerable anecdotal evidence, case-study work and theory that points to the need for this agenda, but little or no systematic research since many of the concepts discussed here have yet to be widely accepted. It is only once the Millennial city has been conceptualized that these concepts can be operationalized and investigated systematically. Millennials are therefore addressed in two distinct ways. First, as workers, the Millennial generation is at the front-line of the changes described. Second, as planners and researchers, Millennials will need to move beyond historic conceptualizations of the space economy in order to understand, plan and manage the cities they inherit.

Following this introduction is a brief review of existing work on spatial structures of the urban economy. In the third section, recent work by sociologists on the changing workplace and nature of work is presented, as is the role played by mobile communication technology. In the fourth section, some consequences of these changes on how the economic structure of metropolitan areas should be conceived are explored. I propose a new way of categorizing employment that complements sectoral and occupational classifications by introducing a typology of spatial attachment. Although many jobs and functions are becoming less attached to particular locations, I also suggest that geographic places and transport infrastructure will remain important, but that their economic role is rapidly evolving.

## The metropolitan space economy: a critique of existing concepts

Although cities have been built, studied and conceptualized since antiquity (Pinol, 2003), current thinking about the urban space economy can be traced back to the Chicago school (Park, Burgess, & McKenzie, 1925). These sociologists explored how neighbourhoods evolved as people, communities and economic activities moved into, out of and around the city.[1] Two types of mobility were considered. First, long-term "ecologic" mobility, as more dynamic communities and economic sectors encroached on older neighbourhoods. Second, short-term, daily, mobility is implicit in Burgess's (1925) concentric model of Chicago. The central business district (CBD) encompasses services, heavy industry and transport, the transition zone encompasses lighter industry and low quality housing, and outer zones encompass suburban residential areas. Workers commute from outer residential zones to inner zones where workplaces are.

This research embodies many of the approaches that influence how metropolitan space economies are understood even today. First, it posits places where people reside, and places where they work. Second, it posits that some zones in the city are characterized primarily by economic function. The argument developed in this chapter is that, whilst metropolitan structures have evolved, the way we think about them has not. Thus, for instance, as the Chicago school's notion of a concentric city was augmented by Hoyt (1939) and Harris and Ullman (1945), it still retained the premise that workers had a place of work. Economic geographers such as Alonso (1964), Vernon (1959), Berry (1959) formalized the Chicago school models and described how the industrial city was evolving whilst retaining this conceptual building block.

As metropolitan areas emerged from the early 1980s recession, it became evident that knowledge and information were the new currency (Daniels, 1985). In addition, the highway system introduced in the 1960s and 1970s had reconfigured the way economic activity was distributed over urban space. A growing number of researchers analysed the polycentric nature of Western cities (McDonald, 1987; Hartshorn & Muller, 1989; Giuliano & Small, 1991). Debates emerged over how sub-centres could be identified (e.g. Coffey & Shearmur, 2001), over distinctions between types of suburban employment centres, suburban downtowns and edge cities (e.g. Garreau, 1991), and over the way different economic functions were distributed across metropolitan areas (e.g. Shearmur & Alvergne, 2002). In the early 2000s it had become evident that economic activities were not only locating in the CBD and suburban employment centres, but were also scattering (Gordon & Richardson, 1996; Lang, 2003; Shearmur, Coffey, Dubé, & Barbonne, 2007). By the mid-2000s it was accepted that metropolitan areas were polycentric, that sub-centres often differed between metropolitan areas, and that transport networks (and highways in particular) were key elements structuring the location of economic activity. The most recent work on metropolitan space economies has focused on refining the way its constitutive elements can be identified and analysed (e.g. Redfearn, 2007; Cladera, Duarte, & Moix, 2009; Shearmur, 2012, Leslie, 2010; Montejano Escamilla, Cos, & Cardenas, 2016). In short, the vocabulary with which metropolitan space economies are described has stabilized, and current studies mostly address methodological and empirical issues. Research in this domain for almost the last century is premised on the idea that economic activity has a location, and that this location is the workplace.

Some exceptions to this premise are of course acknowledged, although these are usually treated as marginal sources of error by analysts. For instance, some employment in the transport (e.g. taxi or truck drivers), construction (e.g. trades people) and service (e.g. salespeople) industries are not performed in fixed places. These jobs are usually assigned to the administrative address of the employer, and this has sometimes merited an explanatory footnote.

In the next section, I suggest that the growth of mobile communications, the internet, and gig work – all of which have become commonplace since the late 1990s and which underpin the work environment with which Millennials have contended since the start of their careers – has altered the way we should think about the geography of economic activity. Researchers and planners can no longer ignore the fact that many jobs are not associated with a specific workplace. Furthermore, even jobs that *are* associated with a workplace are often performed – in part – elsewhere.

## Changes in the workplace, work, and career structure

It is not possible to outline all the changes that have occurred in the workplace and in the nature of work since the early 1990s. Two key changes will be described. First, the types of contract and employment experience faced by Millennials differ from those that prevailed prior to the early 1990s. Second, for many types of worker – particularly "creatives" (Florida, 2002) and "symbolic analysts" (Reich, 1992) – technology allows value-creation to be performed from a wide variety of locations (Schieman & Young, 2010). These changes are connected, and both invite a reappraisal of the connection between economic activity and place of work.

Prior to the 1990s, most young people entering the workforce, especially postsecondary graduates, entertained reasonable expectations of full-time employment, and of a career structured by periodic moves between stable jobs. A quarter of a century later, young people are entering a "gig economy". As Boltanski and Chiapello (1999) described, work is increasingly project based, people are integrated and dropped on the basis of their specific skills, with each person monitoring their surroundings for the next project (or gig) (see also Sennett, 1999). Only a small number of more senior managers, themselves often involved in a variety of projects, benefit from stability, but this stability reflects their relationships and position within professional networks rather than stable salaried employment. As Friedman (2014) shows with respect to the USA:

> A growing number of American workers are no longer employed in "jobs" with a long-term connection with a company but are hired for "gigs" under "flexible" arrangements as "independent contractors" or "consultants," working only to complete a particular task or for defined time and with no more connection with their employer than there might be between a consumer and a particular brand of soap or potato chips
>
> (p. 171)

The prevalence of this type of job has been growing rapidly in the construction, business service and other services sectors (Friedman, 2014, p. 177), but growth has been fast across the whole economy. Approximately 85% of all new jobs created between 2005 and 2013 in the US economy had alternative contractual arrangements, rather than contracts "with fixed hours, location and certain expectations of security" (Friedman, 2014, p. 176).

As a result of this casualization of labour contracts, workers face intense pressure and competition. They are constantly defending their current gig, as well as preparing the next one. Hence, the workday extends well beyond traditional hours (Boltanski & Chiapello, 1999; Friedman, 2014) and work is performed in all types of locations (e.g. homes, offices, cafes, parks, transit). As a consequence, the separation between work and leisure – in terms of time and location – is blurring (Schieman & Young, 2010). This is sometimes described as a positive change – with work presented as play, occasionally ostentatiously so as in Google's offices (AOL.COM, 2016). However, it is also seen by some as corrosive, destroying the possibility of personal and community development outside the economic system (Sennett, 1999; Schieman & Young, 2010).

As Friedman (2014) points out, and as revealed in studies such as Bowlby (2008), ILO (2016) and Worth (2016), the rise in non-standard work started in the early 1990s and accelerated in the mid-2000s, just as Millennials were entering the job market (also see Chapters 3 and 7). Therefore Millennials, whilst not the only ones facing gig-type employment, are the first generation to have entered a labour market where short-term work contracts and work instability have become the norm, and are often a rite of passage before (maybe) acceding to more stable work (Perlin, 2012).

The second factor – linked to, but distinct from, the gig economy – is the revolution in communication technologies, which evolved in the 1970s and 1980s but took off in the mid-1990s as they became ubiquitous, reliable, and increasingly mobile. This has had a variety of consequences. The first is that many alternatives to the traditional workplace have become feasible. At first the principal alternative that was envisaged was working from home. As we saw in the previous section, our understanding of urban structure is premised on two key types of location – place of residence (the space of reproduction) and place of work (the space of production). In the 1990s internet and personal computers made it increasingly feasible for employees and self-employed people to operate, at least part of the time, from home (many studies focused upon tele-commuting, e.g. Handy & Mokhtarian, 1995; Nilles, 1994). Whilst these studies acknowledge some of the impacts of communications technologies, they retained the idea that specific activities occur in specific places – in this case work occurred either at home or in the office.

The advent of mobile phones and other handheld devices has more fundamentally altered contemporary spatial work patterns. Most basically, these devices allow access to social media, to web-based documents, to conference calls from almost any urban location. More subtly, they allow for the real-time coordination of meetings and of other activities. And these two changes are having a major impact on work location, particularly for Millennials, who are more attuned to and at ease with these possibilities than older workers (Deal, Altman, & Rogelberg, 2010; Rainie & Wellman, 2012).

Indeed, with access to the tools and information required for knowledge work at almost all urban locations, the idea that knowledge workers – "creative" workers or "symbolic analysts" – need to be sitting in an office is increasingly outdated. This has been recognized in some workplaces – the workplace has become a place for socializing and for meeting, it being understood that when quiet is required one does not retreat to one's office but to a café, a park or to one's home (Bennet, Owers, Pitt, & Tucker et al., 2010; Waber, Magnolfi, & Lindsay, 2014). Furthermore, transport networks themselves have become places where work occurs. What Augé (1992) characterized as non-spaces – platforms, sidewalks, rails, airports – have now become places from which people phone, respond to e-mail, arrange meetings, write; in short, places from which economic activity can be, and often is, performed.

A more profound effect of mobile technology is the way it replaces geographic location as a coordinating mechanism. Indeed, prior to mobile communications, sitting behind a desk for pre-determined periods of each day was an excellent way of ensuring that communication could occur. The worker could be contacted by fixed-line phone or fax, and colleagues could drop in and discuss things when necessary. There was in fact no practical alternative for coordinating activities within a workplace or team. Thus *fixed workplaces were an essential coordinating mechanism*. Out-of-office meetings of course occurred, but these were arranged in advance and recorded in a diary – again a way of being accessible and of coordinating agents in a work environment. Whilst some types of economic activity did occur outside of the office – reports could be read on trains, academic papers read in cafés – any activity requiring coordination and communication called for a fixed location.

The two changes just described – the gig economy and mobile communications – call for a fundamental reevaluation of the workplace (Chapter 3). The gig economy means not only that individuals often work in many different places in quick succession, but they are also under pressure to continually operate at peak performance levels, work extended hours and blur work and personal time. Mobile communications not only enable and exacerbate the gig economy, they also mean that all workers, not only those caught up in short-term contracts, can also perform their economic activities from a wide variety of locations..

In the next section I draw out the consequences of these changes in the workplace for conceptualizing the urban economy's spatial structure. I propose a typology of jobs according to mobility (the variety of places from which the job can be conducted) and to the control workers have over their mobility. In the light of this typology, approaching the urban space economy as if economic activity occurs at fixed places no longer allows us to understand how the metropolitan economy functions geographically.

## Reconceptualizing the urban space economy

In the previous section, the rationale for questioning the core concept of "place of work" when trying to understand the current urban space economy was elaborated. Whilst there is considerable evidence that this concept should be re-evaluated,[2] it is also important to think through the extent of this re-evaluation.

The dissolution of the workplace is not occurring for all types of jobs nor for all types of worker. In Table 6.1 a classification of jobs is proposed, along two dimensions. The first dimension is mobility, understood as the capacity to perform the job from a variety of locations. Three broad mobility categories are suggested:

• A *hyper-mobile* job is one where many of its components[3] can be performed away from a particular geographic location. For instance, job components that require e-mail correspondence, telephone calls, reading and writing can be performed anywhere that a mobile phone and/or wifi connections are available. Many management, consulting and other high-order service jobs have this type of component. Of course, even the most hyper-mobile jobs sometimes require meetings at fixed places – a university professor has a hyper-mobile job yet still needs to be in front of their class at a fixed time and place (unless of course the class is taught online).

- *Semi-mobile* jobs are not fixed in space, but there is little freedom to roam. For example, truck drivers or airline pilots do not have a fixed place of work, but they cannot perform their economic activity anywhere – their jobs are associated with specific physical capital (trucks, airplanes) and networks (highways, airports).
- *Immobile* jobs are jobs that must be performed in specific locations and at specific times. For example, traditional factory work, fast-food cooking, and janitorial work are immobile occupations, as are air-traffic control and hotel management. This type of job is fixed because it cannot occur away from fixed physical capital, and – particularly for immobile jobs in the service sector – because clients expect to interact with the service provider at a particular place and time.

The second dimension, as alluded to by Bauman (2000) and Urry (2002), is power, understood as economic and social power. In the workplace one consequence of power is the freedom to shape one's own timetable without being subjected to the whims and desires of managers and bosses. Whereas Bauman and Urry suggest a straightforward positive correlation between mobility and power, in the proposed classification – and bearing in mind the preceding discussion of gig work – I suggest that power and mobility are independent of each other. Indeed, although some mobility reflects power, some mobility is imposed and endured. Thus, both a powerful financier and a struggling independent contractor have mobile jobs. But whereas one has *control over* his or her mobility, and can use it to enhance his or her interests, the other is *subjected to* mobility, working in the gig economy with often little alternative but to perform work activities in cafés, metro stations or wherever a space is found, often at the beck and call of whomever is in charge of the project.

Given this classification, the way the spatial structure of metropolitan economies is conceived needs re-thinking. Except for *immobile* jobs, the notion of place of work does not necessarily apply to economic activity. *Semi-mobile* jobs have been recognized by most urban analysts, but usually considered as a source of minor error in the analysis of intra-metropolitan economic structures. *Hyper-mobile* jobs is an entirely new category, which is expanding within some of the fastest growing segments of the economy.

It is difficult to estimate the number of jobs that can be classified as *hyper-mobile*. Whereas data can provide us with some idea of the contractual arrangements under which people work (Friedman, 2014), they provide no information about where jobs' various components are actually performed. Almost all jobs are *assigned* a place of work, but this assignment reflects the administrative unit to which the worker is attached, not where job components are undertaken. Thus, for

*Table 6.1* A classification of jobs by mobility and economic power

|  | Little power | High power |
|---|---|---|
| *Hyper-Mobile* Jobs that can be performed from a wide variety of locations, subject to periodic meetings and presence in specific places at specific times. | • Young "contracted" consultant • Post-doctoral student • Sales representative | • Financier • Tenured university professor • Established consultant |
| *Semi-mobile* Jobs that are performed from a wide variety of locations, but these locations are constrained by the frequent need to be at particular places at particular times, and often by the connection between the job and specific physical capital (e.g. ships, cars, land and building materials) and networks (ports, roads, airports). | • Truck-driver • Construction worker • House cleaner • Taxi driver | • Airline pilot • Ship captain |
| *Immobile* Jobs that are (mostly) performed from a specific location, either because of close connection between the job and fixed machinery, or because clients expect the job holder to be at a particular place. | • Fast food cook • Janitor • Factory worker | • Surgeon • Luxury hotel manager • Air traffic controller |

Note: Examples of occupations are given in each box. The table should be read as follows: "Financier is an occupation where workers have *high power* and are potentially *hyper-mobile* and *high power*"; "Fast-food cook is an occupation where workers are typically *immobile* and have *little power*".

instance, my place of work is McGill University. Yet I regularly work from home, in cafés, from airport lounges, and not infrequently from hotel rooms and lobbies. My economic activity, the value I create,[4] is artificially assigned to a particular address, whereas in reality it is generated from a complex pattern of mobility. There exist virtually no statistics that record the proportion of jobs that are potentially or actually mobile, though there are increasing attempts to gather such data. Kesselring (2015, p. 572), who emphasizes this dearth of evidence, cites a 2008 German survey showing that 37% of workers work – to various degrees – from multiple locations.

Whatever the actual number of mobile workers (which current evidence suggests is large, but which is difficult to ascertain without first operationalizing and then conducting large-scale surveys), the *potential* to work from multiple locations is clearly increasing. Thus, the notion that the urban space economy can be adequately conceptualized on the assumption that economic activity occurs at "places of work" or in establishments is not tenable. By extension, analyses that subdivide metropolitan areas into a CBD, employment sub-centres and corridors on the basis of establishment location and place of work data may not be telling us much about *where* economic value is created, or about *how* metropolitan space is mobilized by economic agents as they produce value.

Figure 6.1 represents a schematic view of the concepts currently mobilized to understand the metropolitan space economy, as outlined in section 2 above. There is a CBD, and there are sub-centres, each of which interacts with its local hinterland. Furthermore, there are axes – both radial and concentric – along which economic activity is also organized. Some of these axes structure corridors of economic activity. Metropolitan areas all differ but are usually described using this type of concept and vocabulary.

In Figure 6.2, I propose a conceptualization of the Millennial city. Many workers and economic agents are no longer assigned fixed places; rather, each has a daily trajectory (Massey, 2005), represented by the arrows and lines going through the nodes. Indeed, the city is still punctuated by fixed places – though these are no longer places where high-value work is visibly performed: They are places where mobile workers meet. These places are characterized by small private offices (meeting places rather than places to spend the working day), but also by cafés, restaurants and leisure activities – face-to-face interaction plays

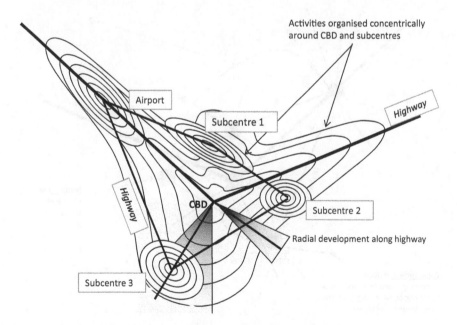

*Figure 6.1* Current conceptualization of the metropolitan space economy: fixed employment centres and places of work

a key social function facilitated by this type of venue. From a planning perspective, from the perspective of the physical city, sub-centres can therefore still be identified and delineated, but their function has changed – sub-centres visibly gather (in terms of place-of-work metrics) *immobile* workers linked to personal services, retail, restaurants, etc . . .). Their economic function is to provide spaces for *hyper-mobile* workers to interact and create value. Hyper-mobile work, even if notionally ascribed to specific locations, in fact occurs across the whole city, in nodes but also along trajectories and at different locations throughout the day. These new spatial configurations and trajectories of economic value-creation have yet to be empirically explored.

If this view of the metropolitan space economy is realistic, then it implies that our understanding of the urban space economy should be re-thought. If the value created by workers in *hyper-mobile* jobs cannot be assigned to any specific place of work, then the economic function of cafés and leisure-related centres may be far higher than the food and retail trade that they appear to generate. They may be providing the context within which the Millennial space economy creates value. It also means that the mere fact of observing the location of creative and high-tech industries at an intra-metropolitan scale may tell us little about the spatial workings of the urban economy (Huber, 2012; Spencer, 2016).

From a planning perspective this raises problems. If cities are increasingly structured by leisure-related centres, acting as meeting places for *hyper-mobile* workers

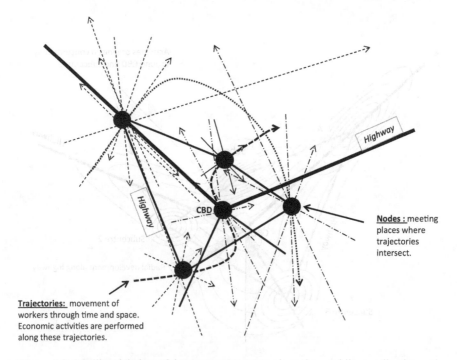

*Figure 6.2* A Millennial view of the space economy: trajectories, mobile coordination and meeting places

following a wide variety of daily trajectories, planning becomes difficult. Indeed, the success or failure of meeting places (or of "cool" neighbourhoods), depends on taste and fashion, which evolve quickly. Whereas economic value-creation attached to fixed places of work possesses some spatial inertia (business parks or office centres can be physically planned), value-creation linked to the trajectories of mobile workers – even if it sometimes focuses on specific neighbourhoods – is spatially volatile. Another, wider, question concerns the distribution of the rewards of value-creation. If cafés, leisure spaces and transport networks are increasingly important to the way value is created in the city, to what extent should they reap some of the rewards (not merely the direct price of the coffee they sell)? This raises far wider questions of inequality, justice, and the distribution of value that is created in the economy (Piketty, 2013). The point being made here is that these questions also have a spatial dimension, and the question of how and by whom this productive (semi) public realm is financed is a key one (Soja, 2010; Shearmur, 2016).

While meeting places remain visible in metropolitan areas and are increasingly structuring their space economy, much urban value-creation is becoming invisible. Rather than occurring in readily observable offices and business parks, zoned and built for high-value-added knowledge intensive activities, the urban economy is seeping into people's homes, onto subways and into parks. The way the urban fabric can accommodate this seepage must be considered, and its desirability should be questioned. Whereas the personal impacts of the economic takeover of time and space have already been identified for workers and families (Sennett, 1999; Boltanski & Chiapello, 1999; Schieman & Young, 2010), the broader impact of the economic takeover of urban spaces and networks has yet to be understood.

## Conclusion

On the basis of a wide-ranging literature review, focused on mobility, communication technology and the workplace, I argue in this chapter that a fundamental change in the geography of economic value-creation in cities is underway. It has been assumed, since the 1920s when the Chicago School analysed the city, that economic value-creation occurs at places of work. This has always been recognized as an approximation, but until the advent of ubiquitous mobile communications and the rise of the gig economy it was reasonably accurate.

However, a growing number of knowledge-related jobs are becoming – or have become – *hyper-mobile*. Whether the worker is a young and struggling independent contractor or artist, or a successful businessperson or financier, they can perform many components of their job from a variety of urban locations. Furthermore, as coordination becomes easier through cell-phone use, social media and instant messaging, more jobs are becoming *semi-mobile* (e.g. dog walkers, hairstylists and fitness trainers who come to clients' homes); these occupations have no fixed place of work, albeit without the locational freedom of *hyper-mobile* work.

In this context, it no longer makes sense to assume that economic value-creation occurs at fixed places of work. By extension, even if employment centres and clusters of economic activity are identified using census data or business registries that record administrative addresses, this may mean very little. It is not because a large number of *hyper-mobile* jobs are administratively assigned to a specific set of co-located addresses that the workers themselves are in spatial proximity. Actual value-creation can be occurring anywhere within the city and even beyond it.

Bearing this in mind, a series of new research questions needs to be articulated in order to understand the metropolitan space economy within which Millennials are operating. First, qualitative studies are necessary to better flesh out the daily, weekly, monthly and yearly trajectories of *semi-mobile* and *hyper-mobile* workers: Where do they perform their economic activities? Which activities are performed where? Why? Similarly, specific urban places should be investigated: To what extent are parks, cafés, transit networks, homes and other urban places becoming sites of economic value-creation? What types of value-creation characterize different urban locations?

Second, once the geography of metropolitan value-creation is better understood from a qualitative perspective, surveys and/or wider data should be collected in order to understand how these micro-processes combine and interact to form an urban economic system. Economic value-creation can no longer be neatly assigned to specific places, yet cities clearly contribute to value-creation. Thus, a systems approach is necessary to understand it. This leads to wider questions of how (and why) value created in a system can be appropriated by specific agents (Piketty, 2013); such questions extend well beyond the metropolitan space economy.

Planners, who deal with the practical world, should take note of these changes, but also push for their thorough empirical investigation. Their implications, in terms of zoning, building use, the economic function of (semi) public spaces, and questions of spatial justice are immense, yet they should not be acted upon before they are better understood, measured and nuanced. Concepts and theories *are* important in a planning context. They are implicit in many "common sense" ideas such as employment zones, clusters and innovation districts (all of which take for granted the idea of place of work). If, however, there are changes occurring in cities that call for a re-evaluation of current paradigms – changes outlined in this chapter, but which call for detailed empirical evaluation – then many standard planning tools and solutions will need re-thinking.

In sum, if one accepts this chapter's basic point that economic value-creation no longer occurs at specific places of work, then the conceptual underpinnings of the way metropolitan space economies have been understood, studied and planned need to be re-thought.

## Notes

1 The ecological model has been criticized for its social Darwinism. However, in a market system – with market competition determining which people and communities live where – the model reflects some on-going processes (such as current 21st century gentrification), whether these are desirable or not.

2 There is a growing literature on mobility – much of it in the wake of Urry's seminal book (2000) – some of it looking specifically at mobility and work. It establishes that worker

behaviour is changing: mobile communications are enabling work to occur in multiple locations (Hislop & Axtell, 2009; Kesselring, 2015), such as trains (Berry & Hamilton, 2010), cars (Hislop, 2013), and in a variety of "third-places" or "hybrid spaces" (Frith, 2012) that mobile workers frequent. Real estate agents are reassessing office require-ments in the light of "hot-desking" – Cushman and Wakefield (2016) speak of "digital disruption" – and in Sweden the concept of Hoffices – individuals who open their homes to independent workers seeking a more sociable work environment – is taking off (see http://hoffice.nu/en/).

3 Most jobs involve a variety of quite distinct activities (or components), each of which can be undertaken in a different set of locations. Thus, for example, writing this chapter, a component of my own job, could occur anywhere that affords enough quiet and elec-tricity to charge my laptop computer. Meeting my grad students requires an office or a public place where one has some privacy, such as a café. Teaching requires my presence in the classroom. Answering my dean's phone call can occur anywhere that I have a phone connection. Each component of my job has different locational requirements.

4 I kindly request that readers suspend their disbelief, and accept, for the sake of argument, that a university professor creates value.

# References

Alonso, W. (1964). *Location and land use: Toward a general theory of land rent*. Cam-bridge, MA: Harvard University Press.

Anderson, B. (2006). *Imagined communities* (2nd ed.). London: Verso.

AOL.COM (2016). *Take a peek inside Google's unbelievable headquarters*. AOL. Finance. Retrieved from www.aol.com/article/2016/08/16/take-a-peek-inside-google-s-unbelievable-headquarters/21452955/.

Augé, M. (1992 [1995]). *Non-places: Introduction to an anthropology of supermodernity*. London: Verso Books.

Bauman, Z. (2000). *Liquid modernity*. Cambridge: Polity Press.

Bennet, J., Owers, M., Pitt, M., & Tucker, M. (2010). Workplace impact of social network-ing. *Property Management, 28*(3), 138–148.

Berry, B. (1959). Ribbon developments in the urban business pattern. *Annals of the Asso-ciation of American Geographers, 49*(2), 145–155.

Berry, M., & Hamilton, M. (2010). Changing urban spaces: Mobile phones on trains. *Mobilities, 5*(1), 111–129.

Boltanski, L., & Chiapello, E. (1999). *Le Nouvel Esprit du Capitalisme*. Paris: Gallimard.

Bowlby, G. (2008). *Studies in "non-standard" employment in Canada*. Ottawa: Statistics Canada. Retrieved from http://wiego.org/sites/wiego.org/files/resources/files/bowlby_presentation_2008_non-standard_employment_Canada.pdf.

Burgess, E. (1925). The growth of the city: An introduction to a research project. In R. Park, E. Burgess & R. McKenzie (Eds.), *The city* (pp. 47–62). Chicago: University of Chicago Press.

Cladera, J., Duarte, C., & Moix, M. (2009). Urban structure and polycentrism: Towards a redefinition of the sub-centre concept. *Urban Studies, 46*(13), 2841–2868.

Coffey, W., & Shearmur, R. (2001). The identification of employment centers in Canadian metropolitan areas: The example of Montreal, 1996. *The Canadian Geographer, 45*(3), 371–386.

Cushman and Wakefield Inc. (2016). *Digital disruption in the workplace*. Retrieved from www.cushmanwakefield.ca/en/research-and-insight/2016/digital-disruption-in-the-workplace-2016/.

Daniels, P. (1985). *Service industries: A geographical appraisal*. London: Methuen.

Deal, J., Altman, D., & Rogelberg, S. (2010). Millennials at work: What we know and what we need to do (if anything). *Journal of Business Psychology, 25,* 191–199.

Escamilla, J. M., Cos, C. C., & Cardenas, J. S. (2016). Contesting Mexico City's alleged polycentric condition through a centrality-mixed land-use composite index. *Urban Studies, 53*(11), 2380–2396.

Florida, R. (2002). *The rise of the creative class: And how it's transforming work, community, and everyday life.* New York: Basic Books.

Friedman, G. (2014). Workers without employers: Shadow corporations and the rise of the gig economy. *Review of Keynesian Economics, 2*(2), 171–178.

Frith, J. (2012). Splintered space: Hybrid spaces and differential mobility. *Mobilities, 7*(1), 131–149.

Garreau, J. (1991). *Edge city.* New York: Doubleday.

Giuliano, G., & Small, K. (1991). Subcentres in the Los Angeles region. *Regional Science and Urban Economics, 21,* 163–182.

Gordon, P., & Richardson, H. (1996). Beyond polycentricity: The dispersed metropolis, Los Angeles, 1970–1990. *Journal of the American Planning Association, 62,* 289–295.

Handy, S., & Mokhtarian, P. (1995). Planning for telecommuting measurement and policy issues. *Journal of the American Planning Association, 61*(1), 99–111.

Harris, C., & Ullman, E. (1945). The nature of cities. *Annals of the America Academy of Political and Social Science, 242,* 7–17.

Hartshorn, T., & Muller, P. (1989). Suburban downtowns and the transformation of metropolitan Atlanta's business landscape. *Urban Geography, 10,* 375–395.

Hislop, D. (2013). Driving, communicating and working: Understanding the work-related communication behaviours of business travellers on work-related car journeys. *Mobilities, 8*(2), 220–237.

Hislop, D., & Axtell, C. (2009). To infinity and beyond? Workspace and the multi-location worker. *New Technology, Work and Employment, 24*(1), 60–75.

Hoyt, H. (1939). *The structure and growth of residential neighbourhoods in American cities.* Washington, DC: Federal Housing Administration.

Huber, F. (2012). Do clusters really matter for innovation practices in information technology? Questioning the significance of technological knowledge spillovers. *Journal of Economic Geography, 12*(1), 107–126.

ILO (2016). *Non-standard work around the world.* Geneva: International Labour Organization.

Katz, B., & Wagner, J. (2016). *The rise of innovation districts: A new geography of innovation in America.* Washington, DC: Brookings Institution.

Kesselring, S. (2015). Corporate mobilities regimes: Mobility, power and the socio-geographical structurations of mobile work, *Mobilities, 10*(4), 571–591.

Lang, R. (2003). *Edgeless cities: Exploring the elusive metropolis.* Washington, DC: The Brookings Institution.

Leslie, T. (2010). Identification and differentiation of urban centers in Phoenix through a multi-criteria kernel-density approach. *International Regional Science Review, 33*(2), 205–235.

Marshall, A. (1890). *Principles of economics* (8th ed., 1920). London: MacMillan's.

Massey, D. (2005). *For space.* London: Routledge.

McDonald, J. F. (1987). The identification of urban employment subcentres. *Journal of Urban Economics, 21,* 242–258.

Nilles, J. (1994). *Making telecommuting happen: A guide for telemanagers and telecommuters.* New York: Van Nostrand Reinhold.

Palfrey, J., & Gasser, U. (2008). *Born digital: Understanding the first generation of digital natives*. New York: Perseus Books.

Park, R., Burgess, E., and McKenzie, R. (Eds.). (1925). *The city*. Chicago: University of Chicago Press.

Perlin, R. (2012). *Intern nation: How to earn nothing and learn little in the brave new economy*. London: Verso Books.

Piketty, T. (2013). *Le Capital au XXIe Siècle*. Paris: Seuil.

Pinol, J.-L. (Ed.). (2003). *Histoire de l'Europe urbaine* (Vol. 1). Paris: Seuil.

Porter, M. (2003). The economic performance of regions. *Regional Studies, 37*, 549–578.

Rainie, H., & Wellman, B. (2012). *Networked: The new social operating system*. Cambridge, MA: MIT Press.

Redfearn, C. L. (2007). The topography of metropolitan employment: Identifying centers of employment in a polycentric urban area. *Journal of Urban Economics, 61*, 519–561.

Reich, R. (1992). *The work of nations*. New York: Basic Books.

Schieman, S., & Young, M. (2010). The demands of creative work: Implications for stress in the work – family interface. *Social Science Research, 39*, 246–259.

Sennett, R. (1999). *The corrosion of character: The personal consequences of work in the new capitalism*. London: Norton.

Shearmur, R. (2012). The geography of intra-metropolitan KIBS innovation: Distinguishing agglomeration economies from innovation dynamics. *Urban Studies, 49*(11), 2331–2356.

Shearmur, R. (2016). Debating urban technology: Technophiles, luddites and citizens. *Urban Geography, 37*(6), 807–809.

Shearmur, R., & Alvergne, C. (2002). Intra-metropolitan patterns of high-order business service location: A comparative study of seventeen sectors in Ile-de-France. *Urban Studies, 39*(7), 1143–1164.

Shearmur, R., Coffey, W., Dubé, C., & Barbonne, R. (2007). Intrametropolitan employment structure: Polycentricity, scatteration, dispersal and chaos in Toronto, Montreal and Vancouver, 1996–2001. *Urban Studies, 44*(9), 1713–1738.

Soja, E. (2010). *Seeking spatial justice*. Minneapolis, MN: Minnesota University Press.

Spencer, G. (2016). Knowledge neighbourhoods: Urban form and evolutionary economic geography. *Regional Studies, 49*(5), 883–898.

Urry, J. (2000). *Sociology beyond societies: Mobilities for the twenty-first century*. London: Routledge.

Urry, J. (2002). Mobility and proximity. *Sociology, 36*(2), 255–274.

Vernon, R. (1959). *The changing economic function of the central city*. New York: Committee for Economic Development.

Waber, B., Magnolfi, J., & Lindsay, G. (2014). Workspaces that move people. *Harvard Business Review, 92*(10), 68–77.

Worth, N. (2016). Feeling precarious: Millennial women and work. *Environment & Planning D, 34*(4), 601–616.

# 7 The privilege of a parental safety net

## Millennials and the intergenerational transfer of wealth and resources

*Nancy Worth*

This chapter examines the intergenerational transfer of wealth and resources from the parents of Millennials to their children as a mechanism for coping with labour and housing insecurity. As Millennials finish education and enter the world of work, many experience labour market precarity in the form of under-employment and unemployment, or find jobs with little security or potential for upward mobility. For many Millennials who entered the labour market during or after the economic crisis of 2008–2009, short-term contracts, freelancing and multiple part-time jobs are a reality of working life (Foster, 2013; Lyons, Ng, & Schweitzer, 2012; Ross, 2009). Rather than high unemployment, this age cohort has high rates of underemployment, where young people are over-skilled for the jobs they have or cannot get enough hours and are unwillingly part time. In Canada, the Parliamentary Budget Office considers 40% of university graduates age 25–29 to be underemployed (Lao & Scholz, 2015).

In order to manage this insecurity, Millennials often turn to their parents for various forms of support, including money for school or day-to-day living, housing, job connections, as well as care and emotional support. This intergenerational transfer of resources is a form of privilege – not all parents can financially support their children through their 20s, have the space for their adult children to live at home, or have the social capital to get their children meaningful work. As a result, an *intra*generational divide emerges as some Millennials have a parental safety net that helps them cope with insecure work while other struggle to make it on their own.

This chapter draws on two projects, *Working Lives: Millennial Women and Work*, and *GenY at Home: Well-being, Autonomy and Co-residence with Parents* to explore intergenerational wealth and resource transfers. Data from the *Working Lives* project includes a survey with over 600 women in Canada, born in the 1980s – the leading edge of the Millennial age cohort, as well in-depth interviews with 33 women and seven of their mothers. *GenY at Home* follows a similar structure with more than 800 survey responses from Millennials living in the Greater Toronto Area (GTA), with in-depth interviews with 35 men and women from diverse backgrounds. This chapter uses indicative examples from both projects to tease out how different forms of intergenerational transfer impact young people's possibilities and choices in the labour and housing markets of Canadian cities.

In order to examine the use of intergenerational transfers, the chapter sets out the concept of intergenerationality and the practice of transferring resources across generations. The chapter then moves on to explore four forms of inter-generational transfer: money, professional connections, co-residence in the parental home, and care and emotional support. The chapter concludes by framing these transfers as a form of privilege, only available to some Millennials (Pease, 2010).

## Intergenerationality and the Millennial city

Thinking about the city through a generational lens is useful as it scales up inquiry beyond the individual and adds an important temporal element. In *Intergenerational Space*, Robert Vanderbeck and I highlight "a three-fold understanding of the concept of generation in terms of life stages (e.g. "child", "teenager", "middle-aged adult", or "pensioner"/"retiree"), membership of a birth cohort (which is often ascribed particular characteristics and dispositions based on shared historical position and experience) and positions within a family structure (Hagestad & Uhlenberg, 2005; R. M. Vanderbeck, 2007)" (R. Vanderbeck & Worth, 2014, p. 2). As a result, a term like "intergenerational transfer" has some flexibility – for the purposes of this chapter, definitions of generation overlap – but I am interested in examining the resources that young adults (Millennials) receive from their parents (Baby Boomers). This attention to parent–child transfers sits within an interest on families, and viewing families as economic actors (Hallman, 2011; Holdsworth, 2013).

Beyond the increasing unaffordability of cities in Canada, the wider context around intergenerational relations (and the need for transfers of resources) is the changing relationship between families and the state. Kasearu and Kutsar (2012) offer the useful framework of *crowding out* and *crowding in* of family support. One might assume that the rise of the welfare state would lead to a *crowding out* of family support, as the state takes over roles traditionally assumed by families, including education and care. Yet in practice, there is a *crowding in*, as families and the state share responsibilities. In the European context, in southern Europe in-kind transfers dominate (co-residence, care) while in northern Europe financial support is more common (Albertini & Kohli, 2012). In East Asia, norms of reciprocity and obligation are shifting as care for older people moves outside the family to become a state responsibility; yet, economic woes mean adult children are relying longer on their parents (Izuhara, 2010). In Canada, with a welfare regime that sits between systems found in Nordic countries and southern Europe, with a diverse population (especially in the Greater Toronto Area), the complex blend of class position, cultural background and state provision mean that a diverse range of intergenerational transfers are common. This chapter aims to demonstrate how rich members of a generation can compound the privilege of their children. In this way, intergenerational transfers can contribute to increases in social inequality (Szydlik, 2012).

## *Money: transfers of wealth across generations*

The most direct form of wealth transfer is money – given to Millennials for education, a downpayment for a house, or to support day-to-day life. One of the most common (and expected) forms of support that Millennials identify is money to partly or fully support college, undergraduate or graduate education. For middle-class parents, paying for school is often an expectation, tied up with definitions of being a "good parent", and a sense of family capital (Holmstrom, Karp, & Gray, 2011). An important aspect of family capital involves using parental resources to invest and maintain children's class position, and therefore the position of the family as a unit. In many cases, financial support from parents was not what made education possible, but it meant that young people did not need a job while studying or accumulate debt.

An intergenerational transfer of wealth is often thought of as inheritance. However, Baby Boomer parents are increasingly giving "inter-vivos gifts", or money gifted during a person's lifetime. For Albertini and Radl (2012), these gifts function as a form of status reproduction, allowing children to achieve the same milestones as their parents and allaying fears that their children are falling behind. Gwen told me about the importance of her parents' support when she bought a house at 22:

> When I was engaged and we were getting married I just assumed we would buy a house. [. . .] I had it all figured out and he was just like, "I don't know about this. This is scary." And I was like, "So, let's go talk to my parents." Because they know this stuff. And then my parents, unbeknownst to us had this plan that they were going to help all of us so they had actually put aside a bit of money in an RSP [an investment account in her name, which could be used to buy a first home]

And reflecting further on her approach to money,

> I think partly the way I was raised, but I've always sort of felt like more in the driver's seat when it came to those types of things. I've always been way more confident about those types of things, more willing to take a bit of risk. I was nervous but I was also like – I had that sort of confidence of the young. I was like, who knows? If worse comes to worse, my parents will help me out –, if it comes to that.

The privilege Gwen possesses as a result of her parents subsidizing her downpayment also extends to her overall financial confidence – knowing her parents can step in and help if needed. Beyond large transfers of wealth through one-off gifts, it is common for parental support to be incremental and ongoing. According to a recent CIBC/Angus Reid survey of parents who are financially supporting adult children, provision of accommodation is widespread, followed by assistance with day-to-day expenses (see Table 7.1 and discussion below).

For many young adults, debt is a growing and inevitable part of life (Horton, 2015). For those with parents with available financial resources and who

*Table 7.1* CIBC/Angus Reid poll of Canadian parents who are financially supporting adult children (CIBC, 2015)

| How parents are financially assisting their adult children: | |
| --- | --- |
| Free room & board in my home | 71% |
| Help out with groceries/household expenses | 47% |
| Pay cell phone bills (voice & data plans) | 35% |
| Car payments or vehicle-related expenses (full or partial contribution) | 23% |
| Rent (full or partial contribution) | 17% |
| Debt repayments | 12% |
| Contribution toward a downpayment for home purchase | 5% |
| Other | 21% |
| **Amount of financial support or assistance parents provide to their adult children:** | |
| Less than $100 | 10% |
| $100 – $249 | 24% |
| $250 – $499 | 31% |
| $500 – $999 | 16% |
| $1,000 or more | 7% |
| I don't know | 11% |

are willing to ease this burden, the impact can be powerful and cumulative. For example, if parents subsidize living costs, young people can take on low-paid or unpaid work in their desired field. This gives some Millennials a coveted foot in the door, the "experience" that is key to the job market. Yet, although money is the most obvious form of intergenerational transfer, many Millennials benefit from transfers of resources that are more difficult to measure but critical to their success.

## *Professional connections: networks of privilege*

In contrast to money, one of the more ephemeral forms of intergenerational transfer is the use of parental networks to find work. To use Bourdieu's (1986) framing, some parents are able to translate economic capital into social and cultural capital to give their children an advantage. This form of transfer is becoming increasingly important as precarity in the labour market invades the working lives of many Canadian Millennials – and even jobs that are objectively secure still often *feel* precarious (Worth, 2016). As a result, Millennials with parents who find them first jobs or provide a personal connection to gain experience in a chosen field find themselves in a privileged position compared to their peers. I asked Louise, the mother of three Millennials, about other resources she and her husband provided their kids beyond money. She told me:

> So resumes, cover letters, and I taught them to customize everything. [. . .] Our kids lucked out too, my husband was with the [City] at that time so they all got early jobs working in the concession at the arena or the reception at

the pool until the City . . . wouldn't let him do that anymore. [laughs] But he would still know – gee, they are hiring for this one so get your resume in like, now.

Probably the one way I think I helped my kids the most was my daughter, [Kelly]. So she went through a crisis of "what do I do?" She was going to quit [university] and she was actually working for my firm at the time. So yeah, I guess I gave her that job too – we were working that early summer and I said, "[Kelly], the thing you loved about music as much as the music was the camaraderie, the working towards a goal. To me, that sounds like business." She eventually applied [to college]. I helped her with resumes, again. And she went into that and she is now a small business advisor at [CIBC] so it worked out well for her.

For Millennials like Kelly and her siblings, having an easy pathway into employment as teenagers and young adults – and parents who not only have the time and experience to support job-searching but that can also create jobs for their children – is a significant advantage in a climate of precarious work and underemployment. Not only did Kelly get a first job from her parents, she also worked for her mother during university, where her mother also helped with applications and resumes. The classic sociological text *Learning to Labour: How Working Class Kids Get Working Class Jobs* traced class-based social capital, including how working class young people found work through the pub, through the factory their father worked at, or from the local neighbourhood (Willis, 1977). Critically, the support allowed young people to find a job similar to their parents, thus maintaining rather than destabilizing class positions (Devine, 2004). In our current neoliberal "economy of experience" (Heath, 2007), professional connections are a powerful form of intergenerational transfer that can solidify social inequalities. Bank of Canada governor Stephen Poloz (2014) advocated that young people work for free – "If your parents are letting you live in the basement, you might as well go out and do something for free to put the experience on your CV." Yet, this is only feasible for Millennials who have financial support during their time of unpaid work.

### Co-residence: the boomerang kids

Besides money and family connections, co-residence is a significant form of intergenerational transfer for Canadian Millennials. Co-residence is often referenced in the media as a proxy for the narrative that Millennials as a generation have failed to achieve key markers of adulthood. However, co-residence is an increasingly common experience that deserves careful attention. In Ontario, 50% of twenty-somethings live at home (Statistics Canada, 2011), rising to 57% in the Toronto CMA – with the rate of co-residence for 25–29-year-olds doubling in just 30 years (Milan, 2016). Moving beyond stereotypes is important, as Millennials report diverse push and pull factors that bring them (or keep them) at home. At its most basic level a free (or subsidized) place to stay is an opportunity for some Millennials to save on housing costs, especially in expensive cities like Toronto and Vancouver. For many of the young adults involved in the *GenY at Home* project, saving money (79%) was the top survey answer for "why do you live with your parents", with looking for work (25%), paying off debt (33%) and unable to afford rent (42%) as other (overlapping) financial reasons.

In Toronto, rent accounts for a growing portion of income and the market for buying a home in Toronto is increasingly challenging for young adults; saving for a downpayment now takes three times as long (15 years) as it did 30 years ago (see Figure 7.1). Millennials who are lucky enough to live with parents can often enter the housing market much sooner than their peers (Clapham, Mackie, Thomas, & Buckley, 2014). One respondent from the *GenY at Home* survey told me:

> Living with my parents has been a huge help. It has allowed me and my partner to save our money and invest it as a down payment towards a [. . .] condo unit. Living with my parents has also allowed both of us to focus on our education and career-building. For example, living at home has allowed me to focus all of my energies in my [grad] studies. As you know, grad students do not make enough money as TAs to live a sustainable life (i.e. after monthly tuition reduction we have $1300 for our rent, transportation, groceries, social life etc.). By living at home and not paying rent, I haven't had to take on a second job or take out loans. Similarly, my partner returned to post-secondary to complete a Bachelor's of Education to become an Ontario Certified Teacher. He was living with me and my family during this time and he also was able to focus all of his energies on his school work and placement without the external stress of paying rent. We both also feel extremely well supported by my parents and feel welcomed and accepted as a common-law couple. [. . .]. I believe living at home has helped me set-up a strong future in terms of both of our careers and future housing.

For this respondent, co-residence was what allowed her and her partner to save for a home of their own while also pursuing graduate education. According to the Toronto Real Estate Board (2016), the average monthly rent for a one-bedroom apartment in Toronto is $1,662, meaning that living at home for a year can generate a $20,000 down payment. As housing continues to rise as a percentage of income, an *intra*generational divide is emerging between Millennials who can afford to buy a house due to parental support and those who may never be able to enter the housing market. In the *Working Lives* project, many women could not imagine ever being able to afford a house. Moreover, the family money that allows some young adults to enter home ownership may be contributing to the housing vulnerability of those without this intergenerational support. According to a Toronto real estate agent, family money has increased the number of first-time buyers by almost one-third, which in turn increases housing prices and shuts some new buyers out of the market (McLaren, 2014). Statistics Canada is currently tracking how an estimated 1 trillion dollars is transferred from the Greatest Generation to their Baby Boomer kids. The growing wealth inequality connected to inheritance is poised to continue, as some Baby Boomers are or will soon be able to support their Millennial children in a challenging housing market.

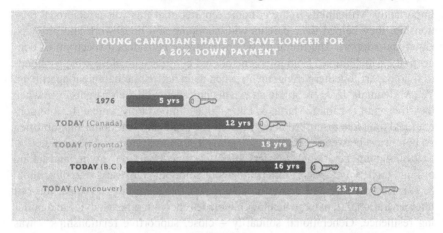

*Figure 7.1* Young Canadians have to save longer for 20% downpayment
Source: www.gensqueeze.ca (Kershaw & Minh, 2016)

A final notable characteristic about co-residence is that it goes against the stereotype of the neoliberal individual, who conquers the housing market independently. Millennials with parents in large urban centres are strategically choosing to move home to save money, not in response to financial stress. A common refrain was that participants did not want to "waste money on rent". This choice not only gives Millennials an advantage when they leave home again compared to peers stuck with pricy rent but also demonstrates a complex interplay of changing expectations around the meaning of (independent) adulthood, housing affordability and the retrenchment of state support (Arundel & Ronald, 2016).

### Care and emotional support: moving beyond the material

This final section aims to capture an under-examined form of intergenerational transfer: care and emotional support. In terms of care, some Millennials with young children have parents who are willing and able to care for their grandchildren (Kohli & Albertini, 2008). The gendered division of labour within the household continues within the Millennial generation, coupled with a lack of affordable childcare. Having parents that can and are willing to look after their grandchildren is transformative for Millennial women. For Millennials living at home with parents, support can extend to domestic tasks, including shopping, cleaning and making meals (Gorman-Murray, 2015). The *GenY at Home* survey asked more than 800 young adults who live at home about the support they receive from parents:

- 85% of respondents reported receiving help with shopping, running errands, or transportation;
- 87% reported receiving support with housework and chores inside the home; and
- 84% reported receiving advice and emotional support.

Importantly, Millennials living at home reported that this was a reciprocal relationship; they helped their parents with shopping and housework and provided emotional support. The literature on co-residence considers a spectrum of connectivity, which moves from intergenerational families with close affective bonds to living apart together arrangements, such as in a separate basement apartment. Yet, Millennials' lives are still to an extent enmeshed with their parents even when they live apart (Scabini, Marta, & Lanz, 2006). Importantly, unlike the intergenerational transfers discussed above, transfers of care and emotional support often go both ways between generations, creating a support system and a safety net.

While intergenerational transfers are often functional, the "giving and taking of time, money and space", others are about emotional closeness (Szydlik, 2008, p. 98). Beyond care and domestic labour, affective support is a critical form of intergenerational transfer, which often includes encouragement, advice and building resilience. Generational solidarity – close, supportive relationships – was common in both the *Working Lives* and *GenY at Home* studies, echoing the (limited) literature on this form of intergenerationality (Schwarz, 2013). For example, Sara told me that her parents help her

> with the emotional side of things. I talk to my Mom every single day for like, at least 10 minutes if not an hour [laughs]. I see her a couple of times a week and [my parents] are INCREDIBLY supportive. Everything I've done they have been rah, rah, rah, good for you. And great advice and thoughtful feedback. Yeah, just really supportive. Like, do whatever you want to do.

Sara's mother told me she relished this close relationship with her daughter – something that she did not have with her own mother. For Waithaka (2014, p. 476), this kind of social support underlies other intergenerational transfers, noting that it is important to "tease apart significant support that facilitate[s] transmission of family advantages".

Yet, co-residence can also be stressful, because it threatens young adults' autonomy. One respondent wrote in the *GenY at Home* survey:

> I find it challenging to express and maintain an adult relationship with my partner's parents, as our living with them reinforces their idea of us being children. It is also difficult as there are issues around personal space and living space boundaries, and interpersonal pressures and expectations beyond the expected personality differences (e.g. eating together, or not; being obliged to spend time with them or at least not keep to myself). I also do not feel as if this is my "home", in the sense of being comfortable and able to fully be myself, except for when they are on holiday away from the house. Finally, it is sometimes difficult to assert and feel like an independent adult, in terms of resources (financial and otherwise).

Intergenerational transfers allow many Millennials to mitigate labour market and housing vulnerability in the short term and save for future goals. While intergenerational transfers are positioned as a form of privilege in this chapter, Kins and

Beyers (2010) are concerned that parental support can damage the well-being of young adults, as it goes against the expectation of autonomy that some Millennials expect (or hope for)(Nairn & Higgins, 2007). The power dynamics connected to intergenerational transfers – who gets to be the decision maker around Millennial finances, employment and housing – are complex, as parents can have their own expectations and plans for their adult children.

## Conclusion

This chapter has examined different forms of intergenerational transfers received by Millennials in Canada. The goal of the chapter has been to mark the diversity of resources beyond inheritance that some parents are able to provide during their lifetimes to their children. In practice, some Millennials are able to have a smoother transition through youth, young adulthood and beyond, mitigating risks from the labour and housing markets with help from their parents. As feeling precarious becomes a more generalized feeling for many Millennials (Worth, 2016), knowing that a parental safety net is there is a powerful advantage, allowing some young adults to take risks, get ahead and benefit from their parents' support in a myriad of ways. From a policy lens, concerns about growing wealth inequality within the Millennial generation could suggest a need for a gift tax on inter-vivos gifts, but this would need to be quite high to be effective, and money is only one form of intergenerational transfer that contributes to inequality. In fact, care, co-residence and job connections are familial resources that have stepped in to replace services of a welfare state. To truly confront the privilege that some parents are able to create for their Millennial children, we need to develop policies that remove housing and care from market forces to make both more accessible – a challenging and highly aspirational goal.

Intergenerational transfers in Canada will increase in the future. Millennials will increasingly rely on their parents' wealth as housing becomes less affordable (especially in large urban centres), work becomes more precarious, state support weakens and parents receive their own inheritances. Moving beyond simply measuring the capability of parents to support their Millennial children, Albertini and Radl (2012) suggest that we conceptualize intergenerational transfers as a dualism between reciprocity and altruism. As support for their children's higher education becomes almost a social requirement of parents, while the gift of money for a downpayment remains optional, evolving *parental motivations* around the type and amount of intergenerational transfer are worthy of future study – especially as their children move further into adulthood. Moreover, *Millennials' expectations and negotiations* with parents about the various forms of intergenerational transfer discussed in this chapter is another fruitful avenue for future research.

This chapter aimed to position intergenerational transfers as a form of privilege, with significant intragenerational consequences for Millennials. The precariousness of those young adults without this safety net needs further attention. Do intergenerational transfers perpetuate a lifestyle (homeownership) that would otherwise no longer be the norm? How much do transfers contribute to social stratification relative to other factors? Finally, as statistics are published about rates of intergenerational transfer, including the prevalence of co-residence with parents, higher rates of financial support into the early 30s, hours of grandparent daycare, etc., more qualitative research is needed to contextualize these trends. It is important to look deeper into the data to examine how Millennials experience growing wealth inequality in their everyday lives, including how it changes their social relationships and expectations for the future.

## Acknowledgements

The Social Sciences and Humanities Research Council and a Banting Fellowship funded the research presented in this chapter.

## References

Albertini, M., & Kohli, M. (2012). The generational contract in the family: An analysis of transfer regimes in Europe. *European Sociological Review, 29*(4), 828–840. doi:10.1093/esr/jcs061.

Albertini, M., & Radl, J. (2012). Intergenerational transfers and social class: Inter-vivos transfers as means of status reproduction? *Acta Sociologica, 55*(2), 107–123. doi:10.1177/0001699311431596.

Arundel, R., & Ronald, R. (2016). Parental co-residence, shared living and emerging adulthood in Europe: Semi-dependent housing across welfare regime and housing system contexts. *Journal of Youth Studies, 19*(7), 885–905. doi:10.1080/13676261.2015.1112884.

Bourdieu, P. (1986). The forms of capital. In J. Richardson (Ed.), *Handbook of theory and research for the sociology of education*. New York: Greenwood Press.

CIBC. (2015). *Parents say their adult kids are draining their nest egg: CIBC poll [Press release]*. Retrieved from www.newswire.ca/news-releases/parents-say-their-adult-kids-are-draining-their-nest-egg-cibc-poll-523820651.html.

Clapham, D., Mackie, P., S., O., Thomas, I., & Buckley, K. (2014). The housing pathways of young people in the UK. *Environment and Planning A, 46*, 2016–2031.

Devine, F. (2004). *Class practices: How parents help their children get good jobs*. Cambridge: Cambridge University Press.

Foster, K. R. (2013). Disaffection rising? Generations and the personal consequences of paid work in contemporary Canada. *Current Sociology, 61*, 931–948. doi:10.1177/0011392113501820.

Gorman-Murray, A. (2015). Twentysomethings and twentagers: Subjectivities, spaces and young men at home. *Gender, Place & Culture, 22*(3), 422–439.

Hagestad, G. O., & Uhlenberg, P. (2005). The social separation of old and young: A root of ageism. *Journal of Social Issues, 61*(2), 343–360. doi:10.1111/j.1540-4560.2005.00409.x.

Hallman, B. (Ed.). (2011). *Family geographies: The spatiality of families and family life*. Toronto: Oxford University Press.

Heath, S. (2007). Widening the gap: Pre-university gap years and the "economy of experience". *British Journal of Sociology of Education, 28*(1), 89–103. doi:10.1080/01425690600996717.

Holdsworth, C. (2013). *Family and intimate mobilities*. Basingstoke: Palgrave Connect.

Holmstrom, L. L., Karp, D. A., & Gray, P. S. (2011). Why parents pay for college: The good parent, perceptions of advantage, and the intergenerational transfer of opportunity. *Symbolic Interaction, 34*(2), 265–289. doi:10.1525/si.2011.34.2.265.

Horton, J. (2015). Young people and debt: Getting on with austerities. *Area*. doi:10.1111/area.12224

Izuhara, M. (2010). Housing wealth and family reciprocity in East Asia. In M. Izuhara (Ed.), *Ageing and intergenerational relations* (pp. 77–94). Bristol: Policy Press.

Kasearu, K., & Kutsar, D. (2012). Intergenerational solidarity in families: Interplay between the family and the state. In I. Albert & D. Ferring (Eds.), *Intergenerational relations: European perspectives on family and society* (pp. 25–38). Bristol: Policy Press.

Kershaw, P., & Minh, A. (2016). *Code Red: Rethinking Canadian Housing Policy*. Vancouver, BC: Generation Squeeze. Retrieved from http://bit.ly/GSCodeRed.

Kins, E., & Beyers, W. (2010). Failure to launch, failure to achieve criteria for adulthood. *Journal of Adolescent Research, 25*(5), 743–777.

Kohli, M., & Albertini, M. (2008). The family as a source of support for adult children's own family projects: European varieties. In C. Saraceno (Ed.), *Families, ageing and social policy: Intergenerational solidarity in European welfare states* (pp. 38–58). Cheltenham: Edward Elgar Publishing.

Lao, H., & Scholz, T. (2015). Labour market assessment 2015. *Parliamentary Budget Officer Blog*. Retrieved from www.pbo-dpb.gc.ca/en/blog/news/Labour_Market_Assessment_2015.

Lyons, S. T., Ng, E. S., & Schweitzer, L. (2012). Generational career shift: Millennials and the changing nature of careers in Canada. In E. Ng, S. T. Lyons, & L. Schweitzer (Eds.), *Managing the new workforce: International perspectives on the Millennial generation* (pp. 64–85). Northampton, MA: Edward Elgar.

McLaren, L. (2014). The bank of mom and dad: Confessions of a propped up generation. *Toronto Life*. Retrieved from http://torontolife.com/city/life/the-bank-of-mom-and-dad/.

Milan, A. (2016). *Diversity of young adults living with their parents*. Insights on Canadian Society. Retrieved from www.statcan.gc.ca/pub/75-006-x/2016001/article/14639-eng.pdf.

Nairn, K., & Higgins, J. (2007). New Zealand's neoliberal generation: Tracing discourses of economic (ir)rationality. *International Journal of Qualitative Studies in Education, 20*(3), 261–281.

Pease, B. (2010). *Undoing privilege: Unearned advantage in a divided world*. London: Zed Books.

Poloz, S. (2014). *Stephen Poloz on youth unemployment* [Press release]. Retrieved from www.theglobeandmail.com/report-on-business/stephen-poloz-on-youth-unemployment/article21448687/.

Ross, A. (2009). *Nice work if you can get it: Life and labor in precarious times*. New York: NYU Press.

Scabini, E., Marta, E., & Lanz, M. (Eds.). (2006). *The transition to adulthood and family relations*. New York: Psychology Press.

Schwarz, B. (2013). Intergenerational conflict: The case of adult children and their parents. In I. Albert & D. Ferring (Eds.), *Intergenerational relations: European perspectives on family and society* (pp. 131–146). Bristol: Policy Press.

Statistics Canada. (2011). *2011 Census: Living arrangements of young adults aged 20 to 29* (Catalogue CS98-312/2011003-3E-PDF). Retrieved from http://publications.gc.ca/site/eng/429527/publication.html#.

Szydlik, M. (2008). Intergenerational solidarity and conflict. *Journal of Comparative Family Studies, 39*(1), 97–114. doi:10.2307/41604202.

Szydlik, M. (2012). Generations: Connections across the life course. *Advances in Life Course Research, 17*(3), 100–111. doi:http://dx.doi.org/10.1016/j.alcr.2012.03.002.

Toronto Real Estate Board. (2016). *Rental market report.* Retrieved from www.trebhome. com/market_news/rental_reports/pdf/rental_report_Q1-2016.pdf.

Vanderbeck, R. M. (2007). Intergenerational geographies: Age relations, segregation and re-engagements. *Geography Compass, 1*(2), 200–221.

Vanderbeck, R. M., & Worth, N. (Eds.). (2014). *Intergenerational space.* London: Routledge.

Waithaka, E. N. (2014). Family capital: Conceptual model to unpack the intergenerational transfer of advantage in transitions to adulthood. *Journal of Research on Adolescence, 24*(3), 471–484. doi:10.1111/jora.12119

Willis, P. E. (1977). *Learning to labour: How working class kids get working class jobs.* Westmead, Hants: Saxon House.

Worth, N. (2016). Feeling precarious: Millennial women and work. *Environment and Planning D, 34*(4), 601–616.

# 8 Planning for the sharing economy

*Sean Geobey*

Millennials are abandoning ownership, or so the headlines in publications like *Fast Company*, *Forbes*, and *The New York Times* suggest. The shift from the 20th century model of owning major capital goods such as houses and cars to a greater reliance on renting and access to transportation needs to be accompanied by changes in planning policy. This shift in consumption patterns is likely to be a permanent one, though how great the shift from ownership to access will be is still uncertain. What is clear is that the sharing economy has emerged as a new set of industrial technologies and consumption patterns that are often radically different from the institutional equilibrium that planners are used to working alongside. The kinds of public regulations that will co-evolve alongside the institutional experimentation of new industrial technologies is also unclear, and many sharing economy proponents have argued that in the absence of a clear regulatory model, platform self-regulation is sufficient.

In this chapter I argue that planners need to view industries being disrupted by new industrial technologies as undergoing a process of institutional experimentation. As new technologies develop and allow new forms of online-enabled production, consumption, and coordination of goods and services – the sharing economy platforms – they can render the existing institutional equilibrium governing the industry obsolete. During the period of transformation it will be unclear as to what role planners should play in an industry as it enters a phase of institutional experimentation. During this phase, many organizations advancing new industrial technologies will not view profit maximization as a core motivator. This may suggest that leaving the industry to self-regulation will be sufficient to ensure socially beneficial outcomes. However, once this institutional experimentation produces a viable profitable business model, the industry will quickly reorganize around a new institutional equilibrium. This new institutional equilibrium may resemble a decentralized market, a centralized monopoly, or some industrial structure in-between, and there will be a role for planners in supporting this structure and mitigating its destructive excesses. The key, ongoing challenges for planners will be in encouraging institutional experimentation and avoiding developing premature planning frameworks for industries undergoing these transformations.

## The sharing economy

While a single definition of the sharing economy does not exist, there are a number of examples of platforms and services that are often considered part of the sharing economy and are useful for illustrative purposes.

- Uber, often characterized as a ride-sharing service, allows people to use their smartphones to hail a point-to-point ride from a driver much as one would a taxi, but with the entire transaction conducted through a mobile app.
- Airbnb allows people to rent out space in their homes or entire houses as short-term accommodations, effectively creating a network of small hotels.

While Uber and Airbnb are the two most prominent sharing economy companies – leveraging physical capital, i.e. cars and real estate, respectively – there are also numerous platforms that enable more generalized commodities such as short-term labour and financial capital.

- TaskRabbit, oDesk and other flexible workforce platforms enable short-term employment either virtually or in-person.
- Lending Club, Zopa, OnDeck, and other marketplace lending platforms allow retail investors to provide small loans directly to individuals or firms.

For the purposes of this discussion it is unclear whether the term sharing economy is the most appropriate, so a few of the alternative and related terms will be outlined here. The term sharing economy itself has been defined a number of different ways. Rachel Botsman provides useful definitions and examples of a number of terms that are related to the sharing economy (see Table 8.1).

*Table 8.1* Sharing economy definitions

| Term | Definition | Examples |
| --- | --- | --- |
| Sharing Economy | An economic system based on sharing underused assets or services, for free or for a fee, directly from individuals | Airbnb, Cohealo, BlaBlaCar, JustPark, Skillshare |
| Collaborative Economy | An economic system of decentralized networks and marketplaces that unlocks the value of underused assets by matching needs and haves, in ways that bypass traditional middlemen | Etsy, Kickstarter, Vandebron, Lending Club, Quirky, TransferWise, TaskRabbit |
| Collaborative Consumption | The reinvention of traditional market behaviours – renting, lending, swapping, sharing, bartering, gifting – through technology, taking place in ways and on a scale not possible before the internet | Zopa, Zipcar, Yerdle, Getable, thredUP, Freecycle, eBay |
| On-Demand Services | Platforms that directly match customer needs with providers to immediately deliver goods and services | Instacart, Uber, Washio, Shuttlecook, DeskBeers, WunWun |

(adapted from Botsman, 2015)

While all these definitions have value and apply a different lens to this related set of concepts, *sharing economy* will be used here as it is the most commonly used of these terms, according to a Google Trends search as of July 11, 2016, and has been so since 2013. This chapter will not seek to settle the question of how to define the sharing economy. Rather, it will ask the question: *How should planners understand their role in the governance of the sharing economy?*

## Millennial attitudes toward and adoption of sharing economy models

As Millennials enter the phase of their lives in which people generally make major capital purchases such as cars and homes, there has been a trend in which they have delayed or avoided many of these in favour of paying for "access to" rather than "ownership of". According to a study on the sharing economy by Pricewater-houseCoopers (2015), Millennials aged 18 to 34 comprise 38% of the providers on sharing economy platforms, and people between the ages of 18 and 24 have been the most enthusiastic adopters of these platforms. Millennials' adoption of sharing economy tools comes from both the supply side and the demand side. As suppliers of goods and services on sharing economy platforms, Millennials may be displaying a preference for sharing economy values and a "sharing ethos"; they are searching for alternative revenue streams or consumer activities that previous generations would have preferred if the barriers to engaging in them were lower.

As consumers of sharing economy services, Millennials might be demonstrating a strong preference for variety in their consumption experiences. Millennials seek access to lower-priced goods and services or demand quicker access to goods and services than would otherwise have been available to previous generations. For both supply-side and demand-side drivers, what is unclear is whether Millennials are coming to the sharing economy by choice or are forced into it by circumstance. While Millennials have grown up with access to these technologies, they have also entered the job market at a time when access to mid-20th century permanent employment opportunities with strong wages and benefits are few and far between (also see Chapters 4, 5 and 7). Furthermore, many have acquired student loan debts that are larger than those of previous generations, particularly when compared to the lack of available work at wages that could quickly repay these loans. The sharing economy may offer benefits like choice and convenience for many Millennials. Yet, the sharing economy is also a rational response to greater degrees of income instability facing the Millennial generation. Millennials are both a driving force behind, and the earliest people to experience, the impact of the sharing economy.

## Managing free-riders

As sharing economy platforms expand and move into new industries, they disrupt those industries. At any given time an industry will have access to a particular set of technologies that are used for the production, consumption, and distribution of goods and services. The emergence of new industrial technologies allows for new ways of producing, consuming, or distributing goods in a more cost-effective way, or in a manner that can increase the gains to consumers. However, the existence of new industrial technologies does not itself lead to their implementation. Organizations in an industry must see it as being to their benefit to implement these new technologies. Here the concept of an institutional equilibrium is valuable, and it is defined as a state where "given the bargaining strength of players, no player would find it advantageous to spend the resources to renegotiate the framework" (North, 1990, p. 86).

When the institutional equilibrium is disrupted, it is not clear to new or incumbent players what a viable new institutional equilibrium will look like. What is clear is that there will be a variety of different possible institutional equilibria, and each of these equilibria will contain different players with different bargaining strengths and a different share of the spoils from the gains the new industrial technologies provide. The exact structure of any of these equilibria can only be discovered through experimentation. A process of inductive, trial-and-error experimentation is needed to identify what viable new ways of implementing new technologies in an industry are possible (Arthur, 1994). Until a viable approach to implementation is discovered, the new technology will not have the traction needed to have a scalable impact that disrupts and transforms the industry.

The knowledge of what models are viable is difficult to contain and effectively is a public good, knowable to all players in the industry. Incumbents are unlikely to invest extensive resources in experimenting with new business models that are costly to implement and whose benefits can be exploited by competitors. Similarly, well-financed potential new market entrants will often hold off on market entry until an early mover expends the resources to identify a new viable business model, rather than undertake that expense themselves. This leads to a free-rider problem when a new industrial technology emerges. However, there is often still demand for what the new industrial technology could allow.

Into this gap can emerge a variety of alternative governance structures built around managing free-riding. These institutional experiments are not explicitly profit-seeking in their organizational structure and are largely self-governed by the participants, be they consumers, producers, volunteers or a combination thereof. This state of institutional experimentation is largely characterized by commons-led structures that are similar to those seen in stable non-profit, non-governmental sectors of the economy such as social services, environmental management and basic research. Such commons-led approaches have theoretical and empirical strengths, as Axelrod (2006) shows that when participants repeatedly interact over time around a common pool of resources, self-regulating governance without the collapse of the commons can be achieved. Coase (1974) connects these experiments to reductions in transaction costs arguing they can help market participants negotiate self-regulating solutions to common pool resource problems. Ostrom's (1990) studies of how real-world common resource pools are managed demonstrate that polyarchic approaches, in which a variety of institutions with diverse interests jointly manage common pool resources, can show a remarkable degree of resilience. Yet, as these institutional experiments uncover new knowledge of viable business models, they create a momentum that pushes the industry towards

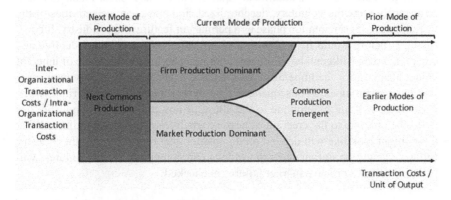

*Figure 8.1* Transaction cost model of industrial succession

institutional equilibria that are not commons-led structures. Institutional experimentation can be modelled as a shift from one institutional equilibrium to another as transaction costs per unit of output fall (see Figure 8.1).

What this new institutional equilibrium looks like will depend on what the transaction costs of this new structure look like. Depending on the governance structure within organizations that make decisions collectively, different types of firm ownership structures – investor-owned, worker-owned, other producer-owned, consumer-owned, unowned or nonprofit, and publicly owned – will tend to dominate the production of different types of goods (Hansmann, 1996). Furthermore, if the cost of transacting between organizations is lower than that of transacting within organizations, the institutional equilibrium will tend towards decentralization. Alternately, if the cost of transacting within organizations is lower than between organizations, it will tend towards hierarchical organization (Williamson, 1996).

In this chapter, I will refer to these as market institutional equilibria and monopolistic equilibria, respectively. However, it is worth noting that industrial organization theory posits a number of further distinctions and hybrids such as natural or artificial monopoly, dominant firms, oligopolistic structures, monopolistic competition, perfect competition, or a common pool resource organization. Amongst these organizations are a variety of possible ownership models that could dominate, including privately held for-profit corporations, publicly traded for-profit corporations, worker cooperatives, consumer cooperatives, producer cooperatives, non-profit corporations, or publicly owned corporations. Furthermore, within the full ecosystem of an institutional equilibrium are public regulators, industry associations, volunteer networks, organized labour, and a whole host of other complementary, competitive, and regulatory bodies that coexist in an ecosystem with markets or monopolies.

In the sharing economy industries also move from one institutional equilibrium to another as platforms emerge that adopt new industrial technologies. However, there is often a period of activity that looks like noncommercial, commons-style production before a new institutional equilibrium is established. Planners might erroneously view the voluntary, decentralized, and non-coercive activities of the institutional experimentation phase as a permanent feature of the industry. Indeed, such a structure would fit well with a vision of governance being adaptive and co-created with Millennials. While this commons-style model may be of inherent value, planning for the hope of its permanence when it is likely only temporary will have unanticipated consequences. The new institutional equilibrium, when it emerges, will not be immune to the abuse of power if it is monopolistic, nor will it be immune to the creation of new externalities if it is decentralized. What these might look like will depend on the institutional transition path the industry follows. Acting on unfounded expectations can compound market failures with planning failures or leave market failures unchecked.

## Alternative institutional equilibria in institutional transition

Figure 8.2 shows the transition between institutional equilibria as a function of transaction costs per unit of output and plots the ongoing transformation of three different industries on it: Point-to-point transportation, short-term accommodations, and consumer finance. I will analyse these disruptions by applying the institutional experimentation model developed here to these three industries.

## Point-to-point transportation, taxis and Uber

Perhaps the most visible and controversial sharing economy platforms are those for point-to-point transportation. The most prominent is Uber, which, although it explicitly resists the sharing economy label, is often cited as the primary example of it. A similar service, Lyft, was a precursor to Uber but has since adopted a similar user interface and service delivery model. Both of these services allow customers to use their mobile devices to order a vehicle to their location to provide transportation to some other location, much like taxis have traditionally done. Through mobile apps, Uber, Lyft, and similar services manage the payments

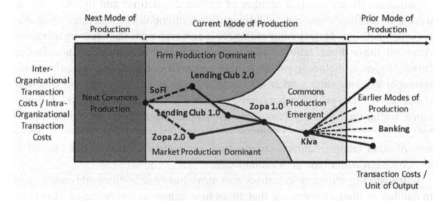

*Figure 8.2* Development of sharing economy platforms

between the passenger and driver, manage routing, and provide a ratings system for drivers and passengers. These services largely operate outside the regulatory regime that currently applies to taxis, providing an alternative governance system for the delivery of point-to-point transportation services. Point-to-point transportation services tend towards natural local monopolization, as dispatch services benefit from strong local economies of scale, and like many other natural monopolies, they have attracted public regulation. What this means for the new sharing economy versions of point-to-point transportation is that the same platform in different jurisdictions can be legal, illegal, or rest in a regulatory grey zone.

The development of point-to-point transportation services has a long history. Coach services, services similar to taxis, existed before the advent of the automobile. Alongside the development of the modern automobile-based taxi industry has grown a variety of regulatory and ownership models, which vary from jurisdiction to jurisdiction. For-profit, cooperative, and public sector ownership of taxi companies are all common models, as are regulatory environments that prescribe various health and safety requirements, insurance protocols, pricing, and supply quantity. In each jurisdiction, the relationships between companies and regulators co-evolved for decades to reach a stable equilibrium that allows for a balance among returns to cab owners, wages for drivers, pricing for customers, and infrastructure for a variety of modes of transportation. The equilibria reached in each jurisdiction provides reasonable safety and security for drivers, passengers, and other road users in a particular context. The balance among all of these objectives is different in different jurisdictions. There is no reason to believe that in any jurisdiction the balance that has occurred is optimal, but until recently they were in equilibria. However, the governance arrangement that evolved in each jurisdiction with a viable taxi industry was relatively stable until recently.

The first internet-enabled point-to-point transportation services existed at the margins of this industry. Online carpooling bulletin boards were precursors. Lyft produced the earliest platform that resembles the current point-to-point transportation model. However, originally Lyft had a business model different than it does today. Though drivers and passengers were matched using Lyft's monopolistic system, the payment for these services followed a voluntary, commons-style donation system. Lyft also offered a ratings system for both drivers and passengers on this platform, a commons-style governance system that is ubiquitous amongst sharing economy platforms. We can think of this development as "Lyft 1.0", an early experiment in how to develop a business model on top of an online point-to-point transportation service.

The entry of Uber into this market dramatically changed the scale at which online-enabled point-to-point transportation systems worked. Uber had a similar model to Lyft but added algorithmically generated pricing that responds to local supply of and demand for point-to-point transportation services. Uber raised prices when demand outstripped the available supply of drivers and lowered prices when the supply of drivers exceeded demand for rides at the current prices. Contrast this with the more common standard pricing of taxi services and this model can be seen as a simple shift along a trade-off curve with Uber offering greater price variance in exchange for more predictable wait times, as taxis are more prone to shortages and gluts. Lyft quickly adopted a similar pricing model, a "Lyft 2.0". A number of similar services were launched in different jurisdictions, such as China's Didi Chuxing. These centralized sharing economy platforms have continued the strong natural monopolization that has existed in the point-to-point transportation system using prior industrial technologies, though the local regulatory environment has not necessarily caught up to the implications of this new industrial technology in most jurisdictions. However, the trajectory of this industry's institutional transition is such that the institutional equilibrium appears to be monopolistically organized.

## Short-term accommodations, hotels and Airbnb

Just as dramatic in its impact on cities has been the growth of a market in online-enabled short-term accommodations. By providing accommodations for travelers and others on a night-by-night basis, short-term accommodation platforms such as Airbnb are competing head-to-head with the hotel industry. Much like taxis, hotels and similar services like bed and breakfasts, hostels, and boarding houses have a long history and have co-evolved with local public regulations around the world. However, services like Airbnb could potentially impact more than just tourist accommodation. Not only has this new model led to competition between Airbnb and hotels, but there has also been a conversion of long-term rental accommodations to Airbnb units (see Chapter 13). The rental housing market and the broader real estate industry to which it is connected are also highly regulated environments that have traditionally been kept separate and distinct from travel-oriented accommodations. However, Airbnb and similar platforms have brought residential space into the short-term accommodations market and as a consequence are blurring the lines between these two previously distinct markets with strong disruptive potential in both.

The first major development in the online-enabled short-term accommodation market came not from Airbnb, but from the volunteer-driven website Couchsurfing. Whereas Airbnb facilitates paid transactions between guests and hosts, Couchsurfing only facilitates unpaid connections between guests and hosts. However, as a precursor to Airbnb, Couchsurfing developed many of the social networking, search, and peer-review functions that would later be used on the Airbnb platform to facilitate exchange. Couchsurfing was an institutional experiment, which demonstrated that people are willing to provide accommodation in their own home to strangers and that trust-based mechanisms could allow this to occur safely enough for people to use and grow this service. Couchsurfing established the basic social infrastructure, a commons good, that Airbnb used to provide a similar service on a transactional basis.

Although Airbnb dominates the short-term rental market, it is not a straightforward analogue to Uber's role in the point-to-point transportation market. The Airbnb platform functions much more like a market than Uber's platform. Uber's service model offers much more standardization in both pricing and in the type of point-to-point transportation being offered. There are a number of Uber services that are offered depending on jurisdiction, but each uses a narrow range of vehicles. Further, the Uber platform dictates the route a driver will take. To participate as an Uber driver or passenger is to be involved in a service that is highly standardized, with tightly internally regulated service qualities and internally set pricing.

Airbnb's platform has moved more in the direction of standardizing service delivery over time, but this has been quite relaxed and mostly focused on standardizing booking and insurance coverage. Beyond these loose requirements, Airbnb hosts have a wide range of accommodations they can offer and no restrictions on their pricing. While Uber has largely centralized decision making beyond the choice of whether or not to participate on the platform, the Airbnb platform appears to largely be a monopolistically competitive market. This is not simply a matter of corporate philosophies; it arises from differences between the two industries they serve. Uber is a commodity service offering transportation from one place to another. Airbnb, on the other hand, enables a range of accommodations from turnkey room rentals to luxury exotic villas, and a range of guest–host relationships, from those in which guests stay cheek-to-jowl with the hosts and their families to ones where the guest never sees another human being at the accommodation site. Add to this that Airbnb's geographic scope ranges from the densest of urban areas to the most remote communities and it is clear that service standardization is much more challenging on the Airbnb platform, and it is likely the case that standardized services are easier to centralize than nonstandard ones. Variety imposes intraorganizational transaction costs.

Yet for both Uber and Airbnb these features have long been the case. Taxi services have long tended towards monopoly, and accommodation services have long tended towards monopolistic competition. Despite the institutional experimentation in their interregnum periods, the sharing economy platforms in each industry did not change these core features of the services provided in either market. What these new platforms did was shift the balance of who governs those markets away from local governments and towards privately owned, multinational online platforms.

## Consumer finance, banks and peer-to-peer lending

Shifts in the finance world are less noticeable than shifts in accommodation and transportation; however, they are just as impactful in their impact on people and the economy while also deeply intertwined with public regulation. The types of innovations here are quite varied, from the development of blockchain alternative currencies, such as Bitcoin, to automated investment advisors, such as Wealthsimple. One of the most dramatic industries that emerged in finance is crowdfunding. Crowdfunding platforms allow a large number of funders to place small amounts of money into a single project. Funders can make pure donations, be offered rewards such as branded merchandise, pre-purchase a product being offered, or have their funding structured as an investment.

The most common investment structures are a share of future company royalties, share equity in a company, and loans. The crowdfunding sector has been dominated by peer-to-peer lending, which structures their funding arrangements as debt instruments. In 2015, of the $34 billion (US) processed through crowd-funding platforms globally, $25 billion was through peer-to-peer lenders (Mas-solution, 2015, p. 23). These platforms allow a borrower to request funds from a large number of lenders, each of whom pledges part or all of the value of the request. If the amount requested is pledged, then the money will be packaged as a loan and transferred from the lenders to the borrowers, with the borrowers then given terms for the repayment of the loan with interest. Unlike a standard loan administered through a bank, the lender has a specific person or set of people to whom they have provided funds, and the borrowers have specific people to whom they owe money. With a standard loan product, a financier such as a bank receives deposits from lenders and provides loans to borrowers. The peer-to-peer lending platform makes the loan a person-to-person affair rather than a person-to-bank-to-person affair, disaggregating the entire set of transactions.

Although the peer-to-peer lending market represents only a small share of the global financial marketplace, it is a new form of financing that is growing at a rapid pace. What peer-to-peer lending platforms displace is unclear, but there are a variety of financial institutions that they could be displacing, including tradi-tional banks and credit unions, payday loan services, informal networks of fami-lies and friends, and black market loan sharking. Regulated financial institutions tend to operate at a relatively large scale, be it regional, national, or international, and the regulatory environments they operate under vary considerably. This leads to a different balance between for-profit, cooperative, non-profit, and publicly owned banks in each jurisdiction, as well as a range of industry concentrations. As peer-to-peer lending platforms have emerged in this wide variety of regulatory environments, they have evolved in different ways.

The earliest platforms had strong elements of commons production. Two of the earliest peer-to-peer lending platforms, Kiva and Zopa, are good examples of the commons-led institutional experimentation that existed during institutional transition. Kiva, launched in the USA in 2005, is a non-profit platform that allows lenders in the developed world to provide microloans to people in the develop-ing world. Non-profit microfinanciers based in the developing world administer these loans. In the United Kingdom, Zopa, founded in 2004 but launched in 2005, initially modelled itself on the non-profit friendly societies that proliferated in the country more than a century earlier (Hulme & Wright, 2006). Based on this prec-edent, they developed a working hypothesis that an internet-enabled version of the friendly society model could fill a gap in managing economic instability in the household that could generate substantial social gains. Though founded as a for-profit platform, Zopa was originally designed to be highly reliant on an existing infrastructure of social capital held between the users of its platform. The Zopa platform focused on the stories of individuals and lending between identifiable people, a commons-style feature.

The first American lending platforms, Lending Club and Prosper, were similar to Zopa but shifted their model substantially after the 2008 global financial crisis. Whereas Zopa focused on lenders making assessments of the borrowers, Prosper and Lending Club quickly pivoted to a greater reliance on the use of algorithmic screens to support and eventually replace decision making by individual lenders.

Zopa eventually integrated more of these tools into its platform, reducing transaction costs, but has largely remained a person-to-person platform. Lending Club and Prosper, on the other hand, have used their reliance on algorithms to attract institutional investors as lenders. With a person-to-person platform, institutional investors placing millions of dollars would quickly find that the expense involved in manually deciding to whom lend money would eat into any profits they could make from entering this new industry. However, algorithmic decision-making tools allow the rapid deployment of millions of dollars onto a platform. Thus these platforms developed from a person-to-person model to a person-algorithm-person model, with retail lenders now being displaced by institutional investors and leading to an institution-algorithm-person model. With this shift, there is a sense that the old person-to-bank-to-person model is reemerging in a much more dynamic way.

The hybridization of traditional finance and peer-to-peer lending platforms is triggering a new wave of consumer finance institutions. These experiments are not entirely new, as in 2007 Zopa attempted to enter the US market and struck a short-lived partnership with a handful of credit unions to do so; but this model quickly fell apart and Zopa reversed its US expansion. Furthermore, Kiva, which has a model of raising capital primarily from developed countries and distributing capital through microfinance institutions in the developing world, has always operated as a pseudo-hybrid of online and brick-and-mortar operations, albeit not one in which those services are tightly integrated. The launch of SoFi in the USA in 2011 represented a shift towards greater true hybridization as their model is built on university alumni networks to attract lenders to finance the student loans of current students from their alma mater. SoFi has since expanded this service to graduate student, small business, and home loans. SoFi's model has also inspired other online–offline partnerships, such as a small business loan programme co-developed by Lending Club and Wal-Mart's Sam's Club.

It is unclear how the combination of online lending platforms with the infrastructure of existing brick-and-mortar organizations will evolve. The networks being leveraged include non-profits like SoFi's reliance on universities, Lending Club's programme with the for-profit Sam's Club, and the experiments Zopa and others have conducted in partnership with cooperatively owned credit unions. However, it is likely that the variety of regulatory and ownership models found in the lending models of earlier industrial technologies will be repeated in different jurisdictions in the online peer-to-peer lending arena.

## Five lenses for policymakers to adopt

There are a number of implications for policymakers suggested by the institutional experimentation model that we are seeing play out on the sharing economy platforms. However, the complexity of each industry and each jurisdiction, as well as the unpredictable interactions between them, means that a prescriptive set of policies would be ill-advised. Instead the lessons are best viewed as analytical lenses and critical questions through which planners can view these emerging platforms.

First, policymakers should recognize that although it might seem clear when industrial technology is changing in a sector, it will be less clear what the final institutional equilibrium will be. This means that when there is a period of technological experimentation there will also be institutional experimentation alongside it, and the regulatory environment that best serves consumers, producers, and citizens may have to co-evolve alongside this experimentation. Often, policymakers focus only on technological experimentation and overlook institutional experimentation. Early lock-in to regulating for one particular institutional equilibrium, whether as a consequence of public policy or the aggressive advances of a monopolistic firm, can short-circuit the process of experimentation that might lead to an alternative institutional equilibrium that would be preferable to planners and to Millennials as they integrate into broader society. For example, premature regulatory efforts to block the entry of Uber often have the effect of further entrenching the institutional power of taxi companies with little incentive to provide mobile services as convenient or easy to use as Uber. However, waiting too long to wade into the regulatory waters has also led to undermining other regulatory regimes, in particular those related to labour standards and health and safety standards. As the market dominance of platforms like Uber and Lyft has grown, the shape of the challenges presented by point-to-point transportation organized through these platforms becomes more apparent, and alternative regulatory regimes such as those that could harmonize the regulatory burdens on online platforms with those of the taxi industry can now be developed.

Second, policymakers seeking to foster the creation of new sharing economy platforms in their own jurisdictions as a local economic development strategy need to do so from an institutional experimentation lens, not just a technological experimentation lens. Business development, university–business technology transfer, and new enterprise financing infrastructure can be geared towards the creation of new sharing economy startups. However, this infrastructure is often built on the assumption that startups will be for-profit, investor-owned firms. This assumption prematurely channels startups into one particular governance model. Sole proprietorships, partnerships, cooperatives, non-profits, trade unions, public agencies, and hybrid joint ventures could all be viable legal forms for sharing economy startups. Such experimentation may even be achieved if existing organizations with alternative structures to for-profit corporations launch their own sharing economy platforms, such as through freewheeling makerspaces or skunkworks housed in libraries or schools. These models could be used to encourage alternatives to investor-owned platforms or to add additional checks on abuses of power. This may also promote additional experimentation in intellectual property governance, such as those seen in open-source software development, internet protocols, and blockchain algorithms. Even if these legal forms only find traction during the institutional experimentation phase, they can push forward business models and new technology experimentation.

Third, policymakers should keep an eye on the scale of the governance tools being attached to the industry, both from a public policy regulatory perspective and a sharing economy platform perspective. The impact of institutional experimentation is often felt at a local level, as is the case with point-to-point transportation and short-term accommodations. Although the public regulatory bodies are oriented to this level of an industry, the platforms themselves can be national or international in scope. The standardization of service delivery models that sharing economy platforms provide may not always be appropriate to the jurisdictions in which they operate. As a consequence, local governments may be squeezed out as players in an industrial ecology, which could lead to the collapse of either that industry or a closely related industry. For example, there is a legitimate fear in jurisdictions already facing affordable housing shortages that short-term accommodation sharing economy platforms may be creating a crisis in markets for rental housing (Chapter 13).

Fourth, there is a democratic lens that can be applied to the governance of industries shifting towards a sharing economy industrial technology. Although the previous institutional equilibrium was not necessarily subject to deep democratic oversight, it had achieved a degree of institutional equilibrium alongside government. Many sharing economy platforms, whether dominated by market or monopolistic governance, are largely structured as investor-owned, for-profit firms. The lack of democratic participation in this governance model is not unique to sharing economy platforms, but prior institutional transitions co-evolved with public regulations, voluntary norms, organized labour, and the development of civil society organizations while moving towards the formation of a new institutional equilibrium. There is still the possibility that these other components of an industrial ecosystem can emerge, but this is part of the commons-phase experimentation that is needed to produce a new industrial ecosystem. Without these components of the ecosystem the industry may become brittle and greatly reduced in its capacity for innovation and adaptation. For example, a for-profit firm may have a natural monopoly in a sharing economy platform and may underinvest in service quality once it has become entrenched. However, an alternative sharing economy platform that dominates a different jurisdiction but is structured as a consumer cooperative may see its members replace those on its sclerotic board of directors through electoral means until the platform reinvests in service quality. With these democratic governance checks in place, monopolistic platforms can target a wider range of objectives.

Finally, the scale of public regulation may actually be too large to allow for public regulatory co-evolution alongside institutional experimentation. Tightly regulated industries, especially those where public regulations operate at the national and international scale, can impose barriers to entry for alternative governance models that actually prevent a range of possible equilibria to emerge. This is likely the case with sharing economy platforms entering the tightly regulated financial sector. Multiple expensive, overlapping, and often contradictory regulatory regimes exist in the financial sector. A major cost is navigating this environment. With the strong economies of scale, including from the network effects in peer-to-peer-lending, the possibility of achieving monopolistic control by attracting investor capital quickly to overcome regulatory barriers to entry is high.

# References

Akerlof, G. A. (1970). The market for "lemons": Quality uncertainty and the market mechanism. *Quarterly Journal of Economics*, *84*(3), 488–500. Retrieved from www.jstor.org. proxy.lib.uwaterloo.ca/stable/1879431.

Arthur, W. B. (1994). Inductive reasoning and bounded rationality. *The American Economic Review*, *84*(2), 406–411. Retrieved from www.jstor.org/stable/2117868.

Axelrod, R. M. (2006). *The evolution of cooperation* (Rev. ed.). New York: Basic Books.

Botsman, R. (2015). Defining the sharing economy: What is collaborative consumption – and what isn't. *Fast Coexist*. Retrieved from www.fastcoexist.com/3046119/defining-the-sharing-economy-what-is-collaborative-consumption-and-what-isnt.

Coase, R. H. (1974). The lighthouse in economics. *The Journal of Law & Economics*, *17*(2), 357–376. Retrieved from www.jstor.org/stable/724895.

Hansmann, H. B. (1996). *The ownership of enterprise*. Cambridge, MA: Harvard University Press.

Hulme, M. K., & Wright, C. (2006, October). Internet based social lending: Past, present and future. Social futures observatory. *Citeseer*. Retrieved from http://citeseerx.ist.psu.edu/viewdoc/download?doi=10.1.1.130.3274&rep=rep1&type=pdf.

Massolution (2015). *2015CF: The crowdfunding industry report*. Retrieved from Massolutions.com.

North, D. C. (1990). *Institutions, institutional change and economic performance*. Cambridge: Cambridge University Press.

Ostrom, E. (1990). *Governing the commons*. Cambridge: Cambridge University Press.

PricewaterhouseCoopers (2015). *The sharing economy: Consumer intelligence series*. Retrieved from www.pwc.com/us/en/industry/entertainment-media/publications/consumer-intelligence-series/sharing-economy.html.

Williamson, O. E. (1996). *The mechanisms of governance*. New York: Oxford University Press.

# Part III
# Housing the next generation

# 9  Generationing housing

## The role of intergenerational wealth transfer in young adults' housing outcomes

*Markus Moos*

Although public discourse in North America still often invokes a narrative of equal opportunity, there is clear evidence that intergenerational wealth transfer shapes opportunity structures. As Pikkety (2000) notes, "[t]he most obvious channel explaining why inequality can persist across generations is the transmission of wealth from parents to children through inheritance" (p. 436). In previous work I have built upon McDaniels' (2004) work on "generationed" social structures to consider generationed spaces (i.e. generational geographies) and their housing implications (Moos, 2014, 2015). Although there has been growing interest in generational dimensions of housing, it is essential that this line of inquiry be expanded to include intergenerational wealth transfers more explicitly as this is a critical factor in shaping socio-spatial structures (Hochstenbach & Boterman, 2015): That is my intent here.

The transmission of wealth from one generation to the next does not have to occur solely through inheritance after a parent's or guardian's passing. Gifts and loans, and other forms of assistance to adult children, can of course occur when parents are still alive (see Chapter 7). In the present context where there is increasing concern that young adults, particularly the Millennial generation, cannot enter home ownership markets with the same ease as previous generations (see Chapter 11), and some Baby Boomers have seen substantial gains in their housing values, transmission of wealth from one generation to another could become a more important factor in shaping inequalities.

Therefore, in this chapter I examine the ways housing is generationed, and the intergenerational processes that stratify housing, by way of an analysis of the wealth transfer from parents to their children in the homeownership market in the USA and Canada. In this analysis I aim to make a conceptual and empirical contribution to the literature in housing studies and planning that is increasingly paying closer attention to the intergenerational linkages in housing pathways and trajectories (Clark & Mulder, 2000; Skaburskis, 2002; Forrest & Yip, 2013; Moos, 2015).

In Marxist scholarship, intergenerational wealth transfer has traditionally been thought to be a critical factor in reproducing capitalist class structures (Pikkety, 2000; Clark, 2014). By focusing on generationing housing rather than class *per se*, this chapter aims to empirically test the potential role and impact of generational transfers in the housing market in generating and reinforcing inequalities.

The analysis provides an understanding of how social structures (and class) are created, reproduced and/or altered over time.

In the analysis I draw on primary survey data that asked people aged 18 to 40 in the USA and Canada about their demographic, employment and housing characteristics. The analysis focuses on several specific survey questions that asked respondents about the current value of homes, monthly housing payments, affordability concerns, whether and how much parents contributed to their mortgage downpayment, and residential location characteristics, as well as value statements on how respondents would make housing decisions in the case of financial hardship. Details about the survey and its methodology are further explained in an appendix.

Prior research shows that homeownership is linked to the socio-economic and housing wealth of the parents but that these links are more acute in expensive housing markets in large metropolitan areas (Clark & Mulder, 2000; Skaburskis, 2002; Gathergood, 2011). The analysis here cannot compare the impact of intergenerational wealth transfer on the likelihood of owning. This is because respondents that provided information on parental contributions are already owners. Instead, I provide an empirical test of the relationship between having received help with a downpayment from parents and their children's housing demand and affordability.

## Housing as a dimension of intergenerational wealth transfer

Housing has always had a generational dimension with people of different ages living under the same roof, and with intergenerational wealth transfer playing an important role in reinforcing housing inequalities. Historically, it was not uncommon for three or four generations of household members to live, and in many cases even work, together, particularly among lower income households. Moreover, inheritance was one of the principal ways that wealth was passed from one generation to the next (Clark, 2014). Housing outcomes became relatively less linked to parental wealth with the expansion of the middle-class and the increasing provisions of the welfare state during the 1950s and 1960s, though inequalities remained.

There is growing concern that the neoliberalization of housing policy and labour markets is reversing progress made in evening out housing conditions during the 1950s and 1960s (Gathergood, 2011). There are growing affordability concerns associated with stagnating wages, increasingly insecure employment conditions, and declining state involvement in housing provision. These trends have contributed to speculations, and case-specific evidence, that parental housing wealth is once again playing a more important role in shaping future generations' housing outcomes (Forrest & Yip, 2013).

The impact of intergenerational wealth transfers on homeownership has received attention in sociology, family studies and economics for some time. Given heightened concern over growing inequality, there is clearly a need for an expanded research agenda in planning to consider intergenerational wealth transfers in housing markets, and to add a more explicit spatial dimension (Hochstenbach & Boterman, 2015).

Prior studies have found that parents' homeownership status increases that of the children through access to equity but also plausibly "a socialization process in which the child's standard of living . . . in his or her parents' home affects the child's expectations" (Henretta, 1984, p. 138). Mulder and Smits (1999) found that parents in the Netherlands who own also have greater financial ability to help their children acquire housing, and they help more frequently than parents who rent.

More recent research has found a heightened impact of parents' status in the housing market on their children's status (Ost, 2011), but that there is national variation in this effect by welfare regime and homeownership accessibility (Mulder et al., 2015), among other contextual factors (Mulder & Smits, 1999). Elsewhere, Mayer and Engelhardt's (1996) study of Boston showed that financial constraints among young adults were also an important explanation of an increase in financial gifts in housing markets. Parents' home equity was found to impact the timing of ownership with parents' wealth positively correlated with earlier entry into ownership markets (Ma & Kang, 2015). Interestingly, receiving parental support was found to increase the value of a home purchase (Engelhardt & Mayer, 1998; Yukutake, Iwata, & Idee, 2015), as opposed to increasing the size of a downpayment on a home.

Hochstenbach and Boterman (2015) add a novel spatial dimension. In their case study of young adults leaving home in Amsterdam, they found that those who were able to draw on parental wealth predominantly moved into gentrifying or gentrified neighbourhoods. They conclude that "[d]rawing on parental support, young people may be able to outbid other households and hence exclude them from gentrifying neighbourhoods" (p. 1).

Evidently, the transfer of wealth from one generation to the next has implications for the social geography of cities, and planning. Gift giving could increase demand in some parts of the market but also increase need for others who are not so fortunate to have been born to parents' with equity or gift-giving abilities in general. If gift giving results in higher priced housing purchases than otherwise would be the case, parental support can also contribute to inflated housing markets, reducing ownership prospects for those without parental support in the long run.

Intergenerational transfers in the housing market may take the form of money, information, or maintenance (also see Worth, 2009; Boterman, 2012; Hochstenbach & Boterman, 2015; see Figure 9.1). Financial transfers are perhaps the most studied, and include gifts or loans (one time or ongoing), as well as low rent or rent-free living when adult children return to live with parents. Information transfers relate to the social networks and financial literacy of the parents that may allow children of existing homeowners (who in the North American context also tend to be more highly educated) access to preferable rates and generally enhanced insight into the workings of mortgage markets. Maintenance refers to help parents sometimes provide to adult children with ongoing household operations such as repairs, yard work or even cleaning that contribute positively to the state of repair and financial value of a home. Maintenance is perhaps the one intergenerational transfer that relies less on parental financial wealth, as skills training in trades or construction occupation could potentially contribute to cost savings. Overall, it is hoped that this conceptualization of wealth transfers will lead to potential fruitful future areas of research on intergenerational transfers that go beyond the transfer of money.

In this chapter I focus on financial transfers in the form of one-time gifts or loans from parents to their children, used to help with a downpayment on a mortgage. I first compare the demographic, housing and location characteristics of homeowners who received help from their parents with those owners who did not receive parental assistance. The purpose of the comparison is to find out whether there is a particular kind of young adult household receiving help from their parents. Second, I test the impact of receiving parental help with a downpayment on a number of key variable related to housing affordability and location:

- The value of the mortgage
- Number of years to pay off the mortgage

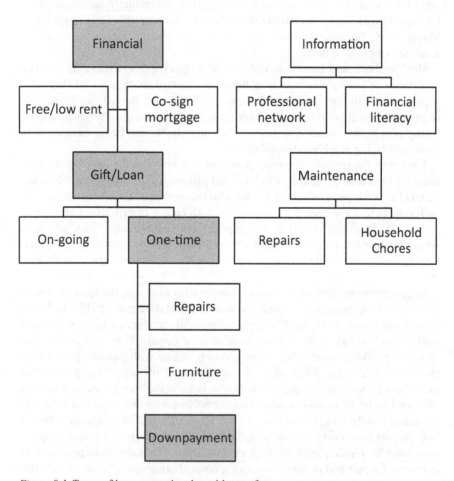

*Figure 9.1* Types of intergenerational wealth transfers

- Monthly mortgage payments
- Self-declared affordability concerns
- Residing in more urban locations versus suburban ones (and housing types)

Based on the literature, we could expect that young adults receiving help with a downpayment would hold higher value mortgages and be more likely to reside in urban (and potentially gentrifying) areas. However, the impact on the other three variables is less certain. If young adults use a parental gift or loan to purchase a higher priced house than they otherwise could afford, the number of years it takes to pay off a mortgage and the monthly payments should not be different from other owners with equivalent mortgages. Affordability concerns should in theory be lower but again if parental help actually leads to the purchase of a more expensive home, it could even potentially negatively affect affordability.

## Measuring intergenerational wealth transfers using survey data

Intergenerational wealth transfers are inherently difficult to study at an aggregate scale due to severe data limitations. Few, if any, datasets collect information on how wealth is transferred from one generation to another, and the amounts transferred. Data are usually not available at the individual or household level, which would permit further analysis.

The data used in this analysis come from an online survey of 1,413 young adults aged 18 to 40 in the USA and Canada. Definitions of young adults vary but generally do not include those over 35 years old (Moos, 2015). Here I expand that definition slightly since first-time home purchase decisions are starting to occur much later, on average, extending into the late 30s.

The survey was available from February 2015 to April 2016. A variety of methods were used to recruit participants including social media, online adverts, traditional media coverage, and posters (see appendix section). The survey received significant exposure at the national level in the USA and Canada through traditional and social media. But given the nature of the data-collection methods, the survey results should not be used to make claims about aggregate trends (e.g. the percentage receiving help from parents, or the amount of help received). However, we still can study the relationship among variables in the survey (e.g. the impact of receiving parental help on mortgage payments), and, with caution and in concert with prior study, generalize the nature of the relationship (e.g. in what ways does parental support impact housing consumption).

Respondents were asked 80 questions regarding their demography, employment, current housing, residential and work location, transportation patterns, and residential preferences, as well as value statements regarding housing and transportation decisions. 863 surveys had complete answers for the variables considered in this paper, and 304 of these are homeowners. In attempting to generalize findings, it needs to be kept in mind that survey respondents are much more highly educated (83 and 78% with one or more university degrees in the USA and Canada respectively) than the national average for the same age range (29 and 25%

in the USA and Canada). In the USA, survey respondents are overrepresented in the southwest and northeast and underrepresented in the west and southeast. In Canada, Ontario is overrepresented and Quebec is underrepresented in the survey data.

The survey asked respondents to indicate the percentage of a downpayment obtained from each of the following sources: Savings, loan from parents, gift from parents, inheritance, government loan, government grant, and other sources. Savings are, on average, the largest source of a downpayment among respondents (61% of total downpayment) and also the most common source. On average, gifts from parents constitute 14% of a total downpayment and loans from parents constitute 8% of total downpayment among respondents.

I compare the characteristics of survey respondents who received more than the average amount from parents with those who received less than the average. I use Chi-squared tests to examine whether the differences between the two groups are different from zero at a statistically significant level in the case of categorical variables and t-tests in the case of continuous variables. I use one-way anova to test whether the average proportion paid by parents differs by demographic and housing characteristics.

The variable measuring parental contribution was modified for the purposes of a multivariate test. The variable is not normally distributed as required by regression assumptions. The distribution of the proportion of a downpayment contributed by parents (either through a loan or a gift) is heavily right skewed as a substantial share of survey respondents does not receive any help from parents at all. In fact, the median is zero with half of respondents receiving no parental help. Due to the large number of zeros, it was difficult to find an appropriate transformation that would normalize the distribution. Instead, the variable was turned into a categorical variable with three categories: those receiving no parental assistance, those receiving parental assistance below the mean, and those receiving parental assistance above the mean. The multivariate analysis uses the value of the mortgage debt as the dependent variable and household income, gender, education and share of parental contribution to downpayment as independent variables. It tests the impact of parental contributions on housing demand, holding other factors that impact demand constant.

## The birth lottery

In Western countries, and especially in the USA, a high degree of social mobility is often seen as the cornerstone of a free society; we often hear the mantra that one's financial and social status derives from one's own work efforts (Clark, 2014). In practice, this does not pan out. For instance, as Clark notes, the USA has a much lower degree of social mobility and higher inequality than more regulated northern European countries. Despite being a "free" country that supposedly rewards individual effort, Clark explains, the birth lottery remains an important factor in determining socioe-conomic outcomes in the USA.

Table 9.1 shows the outcome of this birth lottery by comparing age, gender, income, educational attainment and country of residence between the respondents

*Table 9.1* Respondent characteristics by share of parental contribution to mortgage downpayment

| | Parental contribution | | Chi-Squared | Downpayment | One-Way Anova | |
|---|---|---|---|---|---|---|
| | Below mean | Above mean | p-value | % paid by parent | F | Prob > F |
| **% by age group** | | | | | | |
| Under 29 | 33 | 18 | * | 16 | 4.12 | * |
| 29–40 | 67 | 82 | | 24 | | |
| **% by gender** | | | | | | |
| Female | 55 | 58 | | 24 | 1.65 | |
| Male | 45 | 42 | | 19 | | |
| **% by household income** | | | | | | |
| Under $30,000 | 4 | 1 | | 16 | 0.97 | |
| $30–$59,999 | 8 | 14 | | 31 | | |
| $60–$89,999 | 17 | 18 | | 24 | | |
| $90–$119,999 | 20 | 24 | | 26 | | |
| $120–$149,999 | 18 | 19 | | 20 | | |
| $150–$179,999 | 14 | 11 | | 19 | | |
| $180–$209,999 | 9 | 6 | | 12 | | |
| $210 and above | 10 | 7 | | 16 | | |
| **% by highest level of education** | | | | | | |
| High school diploma or less | 9 | 2 | | 8 | 1.96 | |
| Trades certificate/ college | 11 | 8 | | 19 | | |
| One or more university degrees | 80 | 90 | | 23 | | |
| **% by country** | | | | | | |
| USA | 43 | 38 | | 22 | 0.16 | |
| Canada | 57 | 62 | | 21 | | |
| **% by housing type** | | | | | | |
| Single-family | 61 | 57 | | 19 | 0.54 | |
| Row-house/ townhouse | 20 | 24 | | 24 | | |
| Basement apartment | 1 | 0 | | 0 | | |
| Low-rise apartment, < 5 stories | 6 | 7 | | 23 | | |
| Mid-rise apartment, 5–10 stories | 6 | 5 | | 19 | | |
| High-rise apartment, >10 stories | 6 | 7 | | 28 | | |

(*Continued*)

*Table 9.1* (Continued)

| | Parental contribution | | Chi-Squared | Downpayment | One-Way Anova | |
|---|---|---|---|---|---|---|
| | Below mean | Above mean | p-value | % paid by parent | F | Prob > F |
| **% by monthly mortgage payments** | | | | | | |
| Under $1,000 | 17 | 21 | | 31 | 0.73 | |
| $1,000–$1,999 | 59 | 60 | | 25 | | |
| $2,000–$2,999 | 18 | 16 | | 22 | | |
| $3,000 and above | 6 | 3 | | 17 | | |
| **% by affordability issues** | | | | | | |
| Declared affordability concerns | 14 | 15 | | 20 | 0.01 | |
| No declared affordability concerns | 86 | 85 | | 21 | | |
| **% by neighbourhood type** | | | | | | |
| Urban, single-family | 41 | 32 | | 19 | 0.65 | |
| Urban, low-rise apartments | 18 | 28 | | 25 | | |
| Urban, high-rise apartments | 8 | 9 | | 28 | | |
| Suburban, single-family | 29 | 26 | | 21 | | |
| Suburban, low-rise apartments | 4 | 5 | | 26 | | |
| Suburban, high-rise apartments | 0 | 0 | | 0 | | |
| **n-cases** | 212 | 92 | | | | |

Notes: ***p<0.0001, **p<0.01, *p<05

who received parental contributions below the mean (including those that received no contribution) and those who received parental contributions above the mean. The key finding here is that there are very few differences in the characteristics of young adults by the degree to which parents contributed to a downpayment. There is no statistically significant difference when comparing the number of years respondents have left to pay their mortgage (21 and 20 years for those receiving below and above the median amount of parental support).

The results show that without information on the parents, we cannot predict (among these respondents who are already owners) who is more or less likely to receive parental help. Hence, the utility of characterizing intergenerational wealth transfer in the housing market as akin to a lottery. A comparison using three categories, to separate out those who received no contributions from their parents, displays the same findings (not shown for brevity) as does regressing each of the demographic characteristics (separately) against the share of parental contribution to the downpayment (as a continuous dependent variable).

One variable in the model that is different from zero at a statistically significant level is age. Respondents who received help with their downpayment above the mean are more likely to be older; and the average share of the downpayment paid by parents is higher for those 29 and over (24% paid by parents) than for those under 29 years of age (14% paid by parents). This finding may seem counterintuitive because older respondents should have had more time to save. However, older respondents are also more likely to have completed post-secondary education, which means they are entering the housing market, generally speaking, having had less time to save and potentially with student debt. It could also be the case that the older the respondent the older the parent, and thus the latter are more likely to be in a position to provide assistance either from savings or as a result of downsizing in the housing market.

Also compared in Table 9.1 are the two groups by housing type, monthly mortgage payments, self-declared affordability concerns and neighbourhood/housing type. Again, no differences are found that differ from zero at a statistically significant level between those who received more versus less from their parents toward a mortgage downpayment.

## Mortgage debt and housing demand

At least since the onset of the Great Recession starting in 2008, debt and housing debt specifically have become of greater interest, particularly in the context of historically low interest rates that entice borrowing. Walks (2013) has explored the socio-spatial dimensions of debt, finding highest occurrences in the suburbs, in areas dominated by young, first-time home buyers (also see Chapter 5). Mortgage debt is expected to be higher among the young, as they just entered housing markets. More troublesome is the increasing debt burden, which creates concerns about plausible market collapse should economic shocks occur that impede households' ability to serve the debt (e.g. sudden unemployment).

The ability to borrow using mortgage instruments is critical to success in the ownership market. The more one can borrow, the potentially larger and more luxurious a home one can afford. As shown in the regression results (Table 9.2), mortgage debt is positively correlated with age, household income and educational attainment. Mortgage debt is closely related to income in that higher income earners can borrow more, in part due to their ability to save more for a downpayment and in part due to their higher ability to service a larger mortgage over time.

Older households would generally have lower mortgage debt having had more time to pay off a mortgage. However, the relationship between age and mortgage debt is positive in this analysis because the sample is restricted to those under 40 years of age. For a sample with only those 18 to 40 years of age, it is not surprising that mortgage debt increases with age. The positive relationship between mortgage debt and educational attainment, holding income constant, is of interest

*Table 9.2* Mortgage debt ($) as a function of parental contribution to mortgage downpayment

|  | Coefficient | Standard Error | t | p > \|t\| | Beta |  |
|---|---|---|---|---|---|---|
| Parental contribution = 0 | BASE |  |  |  |  |  |
| Parental contribution < mean | 52359 | 45159 | 1.16 | 0.247 | 0.090 |  |
| Parental contribution > mean | 50678 | 22613 | 2.24 | 0.026 | 0.134 | * |
| Male (vs Female) | −20136 | 19775 | −1.02 | 0.309 | −0.058 |  |
| Age | 6561 | 2235 | 2.94 | 0.004 | 0.189 | ** |
| Household Income | 0.585 | 0.259 | 2.26 | 0.025 | 0.187 | * |
| Highschool or less | BASE |  |  |  |  |  |
| Trades/College | 143857 | 38843 | 3.7 | 0.000 | 0.243 | *** |
| University Degree | 102031 | 31681 | 3.22 | 0.001 | 0.213 | ** |
| Constant | −168035 | 64615 | −2.6 | 0.010 |  | * |

Notes: ***$p<0.0001$, **$p<0.01$, *$p<05$. N-Cases = 272, R-Squared = 0.151, $F_{(7, 264)}$ = 9.9, Prob>F = 0.000

as it suggests there are likely aspects to education that translate into more permanent jobs, which might reflect more positively in the eyes of lenders.

The variable measuring parental contribution to the downpayment is showing results that are statistically significant. Respondents whose parents contributed more than the average hold higher mortgage debt, holding age, income and education constant. The comparison of the distributions shows that among those who received parental contributions above the mean, the value of median mortgage debt is $235,000 versus $159,000 for those with parental contributions below the mean (including those receiving nothing from their parents). The results point to an increase in housing demand as a result of parental support toward mortgage downpayments.

## Dealing with affordability concerns

Rising housing costs, and increasing income inequality, in major cities have contributed to increasing affordability concerns (Moore & Skaburskis, 2004). Affordability, often measured as spending no more than 30% of one's income on housing expenditures, is also a spatial concept (Bunting, Walks, & Filion, 2004). One may be able to afford housing in general but not in a desired location. As discussed before, survey respondents did not have differing rates of affordability concerns (self-declared) by the degree to which parents contributed to a mortgage downpayment. Furthermore, comparison of the share of income spent on monthly mortgage costs did not produce differences that are statistically significant.

But survey respondents were also asked a hypothetical affordability question. They were asked to indicate how likely they would be to pursue one of several strategies if the neighbourhood they wanted to live in was too expensive for them

*Table 9.3* Strategies respondents would pursue if they could not afford desired neighbourhood, by parental contribution to downpayment

| Percent of respondents (%) | Parental contribution to downpayment | | |
| --- | --- | --- | --- |
| | Below mean | Above mean | p-value |
| **Move in with roommates** | | | |
| Not likely | 72 | 86 | * |
| Somewhat likely | 16 | 8 | |
| Very likely | 12 | 7 | |
| **Accept lower quality housing** | | | |
| Not likely | 35 | 28 | * |
| Somewhat likely | 42 | 59 | |
| Very likely | 23 | 13 | |
| **Cut back on expenses** | | | |
| Not likely | 16 | 21 | |
| Somewhat likely | 52 | 40 | |
| Very likely | 32 | 39 | |
| **Move somewhere cheaper close by** | | | |
| Not likely | 10 | 4 | |
| Somewhat likely | 44 | 52 | |
| Very likely | 46 | 44 | |
| **Move to a different neighbourhood altogether** | | | |
| Not likely | 6 | 5 | |
| Somewhat likely | 33 | 44 | |
| Very likely | 62 | 51 | |
| **Ask parents for financial help** | | | |
| Not likely | 81 | 66 | * |
| Somewhat likely | 13 | 22 | |
| Very likely | 6 | 12 | |
| **Find additional work opportunities** | | | |
| Not likely | 55 | 61 | |
| Somewhat likely | 33 | 30 | |
| Very likely | 12 | 9 | |
| **Find a higher paying job** | | | |
| Not likely | 32 | 36 | |
| Somewhat likely | 48 | 46 | |
| Very likely | 20 | 18 | |
| **Save money** | | | |
| Not likely | 17 | 21 | |
| Somewhat likely | 42 | 52 | |
| Very likely | 41 | 27 | |
| **Move in and figure out the finances later** | | | |
| Not likely | 84 | 82 | |
| Somewhat likely | 13 | 15 | |
| Very likely | 3 | 3 | |

Note: ***p<0.0001, **p<0.01, *p<05

(Table 9.3). The differences by parental support for a downpayment are different from zero at a statistically significant level for several questions: Those who received parental support above the mean are less likely to indicate that they would "move in with roommates" or "accept lower quality housing", but they are more likely "to ask parents for financial help".

## Intergenerational dimensions of housing markets

This chapter helps to reveal the intergenerational dimensions of housing markets. First, I summarized literature mainly from sociology and economics on how parents' housing wealth impacts children for a planning audience. I then outlined a conceptual model illustrating the different types of housing wealth transfers from parents to their children, noting that these need not necessarily be financial. I provided empirical evidence of intergenerational wealth transfers in the form of one-time financial gifts/loans from parents toward their children's mortgage downpayment using unique survey data of young adults ages 18 to 40 in the USA and Canada.

The analysis reveals few to no differences in demographic characteristics between those receiving parental support for a downpayment above the mean and those receiving less or no support. This finding is important because it shows that benefiting from parental wealth is more a product of luck than inherited or acquired characteristics.

These results point to existing and reinforcing structures that inhibit social mobility in the housing market and beyond, since housing market success in North America is also closely linked with overall wealth accumulation. Help from parents appears to allow otherwise equivalent owners to acquire more mortgage debt and thus increase their housing demand. Parental support does not seem prompted by affordability concerns, since owners receiving support are purchasing more expensive homes. If parental support was intended to aid in purchasing an existing home, mortgage debt would be similar among respondents regardless of parental support. Affordability concerns also do not differ by level of parental support, although arguably affordability issues among those receiving support would likely have been higher had they still purchased their current home without parental support.

Although Hochstenbach and Boterman (2015) find a link with gentrification in Amsterdam, in this sample here there is no evidence of a statistically significant relationship between parental contributions and social geography. Respondents are just as likely to reside in single-family homes versus other housing types in urban versus suburban locations regardless of the degree of contribution from parents toward a downpayment. There is a larger percentage of those receiving parental support above the mean living in urban high-rise units, pointing to some plausible relationship with parental support and expensive inner city condominium markets; however, the difference is not statistically significant. This requires further study, especially in light of the limitations of the survey data used here.

Although those receiving more parental support are more likely to indicate that they would ask parents for financial help if they could not afford their desired neighbourhood, this does not have a particular spatial or housing type dimension among survey respondents. Nonetheless, I certainly agree with Hochstenbach and Boterman (2015) that "it is imperative for gentrification scholars to take [inter-generational] wealth [transfer] into account as a useful additional capital form to acquire housing in gentrifying neighbourhoods" (p. 19) as these trends would further exacerbate inequalities and affordability of inner city neighbourhoods.

## Appendix

### *Survey recruitment methods*

Participants were recruited using a variety of methods:

- Social media (Twitter and Facebook). Dedicated accounts were used to promote the survey to the target audience, and to organizations, government departments and individuals who further shared the survey invitations amongst their networks
- Social media adverts: Facebook ads were purchased for two consecutive weeks, targeting profiles of users in the USA and Canada between the ages of 18 and 40
- Media coverage: Radio, newspaper and online coverage in national and regional outlets helped bring substantial traffic to the survey site
- Posters: Distributed in neighbourhoods with high shares of young adults and at organizations working with young adults in Austin, Calgary, Chicago, Houston, New York, Philadelphia, San Francisco, Toronto, Vancouver, Washington DC, and Waterloo

### *Respondent characteristics*

Tables 9.4 and 9.5 show the educational attainment and geography of survey respondents. The sample includes a much higher share of 18- to 40-year-olds with a university degree than is the case for this population as a whole. The geographic distribution of respondents in the USA shows an overrepresentation of respondents in the southwest and northeast. In Canada, there is an overrepresentation of Ontarians and an underrepresentation of Quebecois, while the share of respondents from other regions are close to the overall population distribution of 18- to 40-year-olds.

Additional demographic variables are shown in Table 9.6. Over three-quarters of respondents reside in an urban neighbourhood. The gender distribution in the sample is almost even. The majority are renters. The median household size is 2 (range 1–7) and children are present in 13% of respondents' households. Almost 40% of respondents are in quaternary service sector occupations.

*Table 9.4* Educational attainment of survey respondents and population 18–40 in the USA and Canada

|  | Survey | National data |
|---|---|---|
| **USA** | | |
| Less than high school | <1% | 11% |
| High school | 9% | 26% |
| College, diploma, certificate, trades | 10% | 34% |
| University degree | 81% | 29% |
| Total (N cases) | 263 | 65,174,105 |
| **CANADA** | | |
| Less than high school | 1% | 11% |
| High school | 14% | 30% |
| College, diploma, certificate, trades | 12% | 34% |
| University degree | 73% | 25% |
| Total (N cases) | 298 | 9,364,443 |

Notes: Calculated using data from the US Census Bureau (American Community Survey, 5-Year Estimates, 2006–2010) and Statistics Canada (National Household Survey, 2011, Public Use Microdata files, Canadian Census Analyzer, University of Toronto).

*Table 9.5* Share of respondents and population 18–40 by country and region of residence

|  | Survey | Census |
|---|---|---|
| **USA** | | |
| West | 5% | 21% |
| Southwest | 37% | 13% |
| Southeast | 8% | 25% |
| Midwest | 12% | 21% |
| Northeast | 38% | 20% |
| Alaska/Hawaii | 0% | 1% |
| Total (N cases) | 275 | 190,701,204 |
| **CANADA** | | |
| Atlantic | 5% | 6% |
| Quebec | 4% | 23% |
| Ontario | 68% | 38% |
| West | 11% | 19% |
| British Columbia | 13% | 13% |
| North | 0% | 0% |
| Total (N cases) | 320 | 9,421,070 |

Notes: Calculated using census data from the US Census Bureau (State Characteristics Datasets: Annual Estimates of the Civilian Population by Single Year of Age and Sex for the United States and States, 2010) and Statistics Canada (2011 Census data, Public Use Microdata files, Canadian Census Analyzer, University of Toronto).

*Table 9.6* Select characteristics of survey respondents

| Variable | Percent |
| --- | --- |
| Resides in urban neighbourhood | 77 |
| Resides in Canada (versus USA) | 54 |
| Visible/racial minority | 16 |
| Female | 54 |
| Renters | 69 |
| Children present | 13 |
| Quaternary sector occupations | 38 |

Note: N=516

# References

Boterman, W. (2012). Deconstructing coincidence: How middle-class households use various forms of capital to find a home. *Housing, Theory, and Society, 29*(3), 321–338.

Bunting, T., Walks, A., & Filion, P. (2004). The uneven geography of housing affordability stress in Canadian metropolitan areas. *Housing Studies, 19*(3), 361–393.

Clark, G. (2014). *The son also rises: Surnames and the history of social mobility*. Princeton, NJ: Princeton University Press.

Clark, W., & Mulder, C. (2000). Leaving home and entering the housing market. *Environment and Planning A, 32*, 1657–1671.

Engelhardt, G., & Mayer, C. (1998). Intergenerational transfers, borrowing constraints, and savings behavior: Evidence from the housing market. *Journal of Urban Economics, 44*, 135–157.

Forrest, R., & Yip, N. M. (2013). *Young people and housing: Transitions, trajectories and generational fractures*. New York: Routledge.

Gathergood, J. (2011). Unemployment risk, house price risk and the transition into home ownership in the United Kingdom. *Journal of Housing Economics, 20*, 200–209.

Henretta, J. (1984). Parental status and child's home ownership. *American Sociological Association, 49*(1), 131–140.

Hochstenbach, C., & Boterman, W. (2015). Intergenerational support shaping residential trajectories: Young people leaving home in a gentrifying city. *Urban Studies*, early view. doi:10.1177/0042098015613254.

Ma, K. R., & Kang, E. T. (2015). Intergenerational effects of parental wealth on children's housing wealth. *Environment and Planning A, 47*(8), 1756–1775.

Mayer, C., & Engelhardt, G. (1996). Gifts, down payments, and housing affordability. *Journal of Housing Research, 7*(1), 59–77.

McDaniel, S. (2004). Generationing gender: Justice and the division of welfare. *Journal of Aging Studies, 18*(1), 27–44.

Moore, E., & Skaburskis, A. (2004). Canada's increasing housing affordability burdens. *Housing Studies, 19*(3), 395–413.

Moos, M. (2014). "Generationed" space: Societal restructuring and young adults' changing residential location patterns. *The Canadian Geographer, 58*, 11–33.

Moos, M. (2015). From gentrification to youthification? The increasing importance of young age in delineating high-density living. *Urban Studies*, Early view. Retrieved from http://usj.sagepub.com/content/early/2015/09/15/0042098015603292.abstract.

Mulder, C., et al. (2015). The association between parents' and adult children's homeownership: A comparative analysis. *European Journal of Population*, *31*(5), 495–527.

Mulder, C., & Smits, J. (1999). First-time home-ownership of couples: The effect of intergenerational transmission. *European Sociological Review*, *15*(3), 323–337.

Ost, C. (2011). Parental wealth and first-time homeownership: A cohort study of family background and young adults' housing situation in Sweden. *Urban Studies*, *49*(10), 2137–2152.

Pikkety, T. (2000). Theories of persistent inequality and intergenerational mobility. In A. Atkinson & F. Bourguignon (Eds.), *Handbook of income distribution*, chapter 8. Elsevier: New York.

Skaburskis, A. (2002). Generational differences and future housing markets. *Canadian Journal of Regional Science*, *25*(3), 377–404.

Walks, A. (2013). Mapping the urban debtscape: The geography of household debt in Canadian cities. *Urban Geography*, *34*(2), 153–187.

Worth, N. (2009). Understanding youth transition as becoming: Identity, time and futurity. *Geoforum*, *40*(6), 1050–1060.

Yukutake, N., Iwata, S., & Idee, T. (2015). Strategic interaction between inter vivos gifts and housing acquisition. *Journal of the Japanese and International Economies*, *35*, 62–77.

# 10 Is the real estate industry cementing Millennials' residence in urban cores and central cities?

## Insights from Phoenix and Houston

*Deirdre Pfeiffer and Genevieve Pearthree*

Millennials are living differently than previous generations. Instead of transitioning from adolescence to adulthood, Millennials linger in the stage of emerging adulthood, which is defined by having new experiences, moving to new places, and meeting new people (Arnett, 2004). Millennials are less likely to be married and more likely to live alone or with their parents (Fry, 2014). Millennials also express interest in renting in walkable central city neighbourhoods, although homeownership is still a part of their American Dream (Nelson, 2013; Leinberger, 2011; Kolko, 2012). A theory is that these differences will transform urban housing markets by spurring innovative development and increasing demand for dense rental housing in amenity-rich communities in the urban core (Leinberger, 2011).

This chapter adds a supply-side perspective to this discussion. It explores how the real estate industry is cementing Millennials' residence in urban cores and central cities in two overlooked US regions during the 2000s and 2010s – Phoenix and Houston. These are young and sprawling Sunbelt cities with historically weak urban cores that had divergent experiences during the recent Great Recession and foreclosure crisis. We triangulate data from the US Census and local media to examine trends in Millennials' location relative to new housing construction, and the effect that new housing construction had on urban core and central city neighbourhoods from 2000 to 2013. We weave together these trends to assess 1) how the real estate industry affected where and how Millennials live in Phoenix and Houston, and 2) how these dynamics reshaped urban housing markets, if at all. We conclude by drawing lessons for planners.

## Millennials' migration to US urban cores and central cities

A curious trend emerged in US regional housing markets in the first decade of the 21st century: a resurgence of infill housing construction in urban cores and central cities. 70% of the 51 regions with a population of one million or more experienced an increased share of infill housing development from 2005 to 2009 compared to 2000 to 2004 (Ramsey, 2012). This infill development is primarily large multi-family rental construction (Huso, 2015; Joint Center for Housing Studies, 2013). Some of these projects have "concierge style living", with amenities such as pet grooming, movie theatres, and electric car plug-ins (Huso, 2015).

The resurgence of multifamily infill rental housing occurred for several reasons. Starting in the 1950s and 1960s, some cities, such as New York, Philadelphia, Boston, and San Francisco, used federal urban renewal funds to build middle income housing in their downtowns, which primed the pump for future development (Birch, 2005). Apartment ownership became a more attractive investment in the wake of the Great Recession in the late 2000s, as demand for rental housing climbed (Warren, Kramer, Biank, & Shari, 2013). Real estate developers also initiated projects in downtowns to capture emerging preferences in the housing market. In the early 2010s, 60% of Americans preferred to live in a small single-family or multifamily home and have a short commute than live in a larger home with a long commute. In turn, 50% of Americans preferred to live in a walkable, mixed-use community and/or a community with transit (Nelson, 2013).

Millennials, adults who turned 18 years old after 2000, are at the nexus of these trends (Nelson, 2013). In a 2013 survey, Warren et al. found Millennials overwhelmingly preferred to live in compact, mixed use, walkable and transit accessible communities – conditions that define the urban core more than other areas. In their analysis, about 40% of Millennials surveyed lived in a medium to big city at the time, and an equal proportion wanted to live in such a place in five years. A similar proportion of Millennials anticipated they would be in an apartment, duplex, townhome, or row house in the next five years – a rate higher than that of the general population (29%). These trends led Warren et al. (2013) to predict that Millennials will "affect not only how space will be used, but also where it will be used" (p. 29).

Other survey research, however, shows Millennials will gravitate to the suburbs as they age (Kotkin, 2013). Homeownership and parenthood remain goals for most Millennials; yet much urban core housing lacks features to accommodate growing Millennial families (Kotkin, 2013). Further, there is a spatial mismatch between regions where Millennials are concentrated and regions where they can afford to buy urban core homes. Student loan debt, unemployment, and stagnant incomes limit Millennials' ability to make downpayments and afford homeownership, particularly in expensive urban cores in coastal cities like New York, Los Angeles, San Francisco and Seattle. Thus, to live in the urban core, most Millennials make a tradeoff between owning and renting, and between accessibility and space.

An understudied aspect of the potential back-to-the-city phenomenon is how the combination of real estate developers' reinvestment in the urban core and Millennials' preference for living there may change the demographic, socio-economic, and environmental characteristics of these communities and their larger regions. There are concerns that Millennials are gentrifying urban core communities, contributing to higher rents and the displacement of low-income renters (Baum-Snow & Hartley, forthcoming; Birch, 2005). Millennials' crowding of the rental market, particularly in high cost coastal cities, may contribute to a growing rental housing "availability gap" faced by low-income renters (Joint Center for Housing Studies, 2013).

Millennials also may help to economically revitalize urban core communities. Growing demand for urban core living may raise property values and incentives for existing homeowners to maintain and rehabilitate their properties. For instance, the urban cores of Millennial destinations like Washington, DC and Denver now boast higher home prices than formerly high-end suburbs in these regions (Doherty & Leinberger, 2010). Shifts in the regional social structure may also arise from the symbiotic choices of Millennials and downtown real estate developers. A consensus exists among most urban and planning scholars that a "Great Inversion", or a movement towards regional recentralization rather than dispersion, is not only probable but also desirable (Ehrenhalt, 2012).

Developers' and homeowners' reinvestment in walkable and transit-oriented urban core communities may alter housing demand and travel behaviours in the broader region. Reinvestment in the urban core may lead people to walk and take transit more, and drive less (Ewing & Cervero, 2010), which may improve residents' health, the neighbourhood and regional environment, and jurisdictions' fiscal stability.

Millennials' residential decisions also may shift housing demand away from the suburbs, however, potentially leading these areas to depopulate (Doherty & Leinberger, 2010). Eventually, the movement of Millennials to the urban core may contribute to further shifts in poverty from the urban core to the suburbs (Kneebone & Lou, 2014; Suro, Wilson, & Singer, 2011). Yet, these projections are widely debated, as housing and population growth in US suburbs continue to outpace growth in urban cores (Kotkin, 2013; Ramsey, 2012).

## Approach

This chapter helps to illustrate how the interdependent choices of Millennials and real estate developers are changing US regions. We use a comparative approach to uncover trends and their implications.

### Case study sites

We chose two US Sunbelt regions for comparison: Phoenix and Houston. The regions are fast growing, newly built, sprawling, and home to an ample single-family detached housing stock (see Table 10.1). The regions also were home to a sizable Millennial population in 2013: 14% of the population in Phoenix and 15% of the population in Houston. From 2000 to 2013, the Millennial population grew 21% in Phoenix and 29% in Houston. From 2014 to 2018, the Millennial population is expected to grow by about 7% in Houston and about 11% in Phoenix – rates higher than that predicted for the USA (about 4%) (Warren et al., 2013).

Four time periods were selected for the analysis: 2000, 2006, 2010, and 2013. These years cover distinct stages of the recent US housing market cycle. The housing market was stable and modestly growing in 2000, while 2006 captures the housing boom, a time of increasing housing production and demand. 2010 captures the recession and foreclosure crisis, a time of falling housing production and demand, and increasing vacancies. Finally, 2013 captures the recovery period, a time of declining foreclosures and vacancies, and changing property ownership and tenure composition.

The regions experienced similar levels of housing growth during these stages (see Table 10.1). However, Phoenix was harder hit by the recession and fore-closure crisis. Phoenix's economy relied heavily on housing construction, while Houston's more diverse economy was also bolstered by high oil prices. At the height of the foreclosure crisis in 2010, the typical neighbourhood in Phoenix had 11% of loans in foreclosure compared to 6% of loans in Houston, and 3% among the top 50 US regions. About 9% of homes were vacant in Phoenix in 2013 com-pared to about 8% in Houston (see Table 10.1).

Proactive policy making helped to promote infill during the study period. Ari-zona's ongoing Government Property Lease Excise Tax (GPLET) programme eliminates or reduces property values for businesses on government-owned land within the urban core, while Houston's 2013 Downtown Living Initiative offered developers up to $15,000 in tax breaks for each multifamily unit built within the Downtown neighbourhood.

*Table 10.1* Case study characteristics, 2013

| Conditions | Phoenix | Houston |
|---|---|---|
| *Total Population* | | |
| Population | 4,398,762 | 6,313,158 |
| % Population Growth (2000–2013) | 35 | 35 |
| % Latino | 30 | 36 |
| % African American | 5 | 17 |
| % Non-Hispanic White | 58 | 38 |
| % Asian | 3 | 7 |
| Median Household Income | $51,847 | $57,366 |
| % Unemployed | 8 | 7 |
| % Renters | 39 | 40 |
| % College Educated | 29 | 31 |
| *Millennials* | | |
| % Millennial | 14 | 15 |
| % Population Growth (2000–2013) | 21 | 29 |
| *Housing Conditions* | | |
| Homes | 1,832,488 | 2,387,114 |
| % Housing Growth | | |
| 2000–2006 (Boom) | 22 | 18 |
| 2006–2010 (Recession) | 11 | 10 |
| 2010–2013 (Recovery) | 2 | 3 |
| Average neighbourhood Foreclosure Rate, 2010[+] | 11 | 6 |
| % of Homes Vacant[*] | 9 | 8 |
| *Environmental Conditions* | | |
| % of Homes Single-Family Detached | 65 | 62 |
| Median Year Built | 1990 | 1986 |
| % of Workers Commuting by Car | 77 | 80 |

Source: US Census; JCHS (2011)

[*] Excludes second homes and homes for migrant workers.
[+] The number of auctions in 2010 divided by the number of outstanding first lien mortgages in December 2009.

**Data and method**

We triangulated US Census Public Use Microdata Area (PUMA) data and local media publications to answer our research questions. We started with PUMA data to analyse differences in housing and population trends among the urban cores, central cities, and suburbs of the regions. PUMAs in the Phoenix and Houston regions were delineated along two lines: primarily located in the central city or not and, among those located in the central city, those located in the urban core or not. Urban core PUMAs were then excluded from the central city geography, and were based on the City of Phoenix and Houston's definitions of their downtowns (City of Phoenix, 2015; City of Houston, 2015).

We constructed location quotients to assess how the concentration of new housing and Millennials in Phoenix and Houston's urban core and central city compared to their suburbs. We defined new housing as housing built since the previous time period. For example, new housing in 2013 was housing built between 2010 and 2013. We defined Millennials as individuals age 25 to 34, choosing age 25 as the starting point to capture young adults that were independent and out of college. We computed location quotients using the following ratio:

1) $(G_i/G_t)/(R_i/R_t)_{year}$

where $G_i$ is the number of new housing units constructed since the previous time period or the number of Millennials living in the geography, $G_t$ is the total number of housing units or people living in the geography, $R_i$ is the number of new housing units constructed since the previous time period or the number of Millennials living in the region, $R_t$ is the total number of housing units or people living in the region, and $_{year}$ is the current time period. Values greater than 1 mean that the geography has a greater concentration of new housing or Millennials than the region; values less than 1 mean that the geography has a lower concentration of new housing or Millennials than the region.

We constructed an index to compare the socio-economics of the regions' urban cores and central cities to their suburbs. We analysed two characteristics: the percentage of adults age 25 and older with at least a bachelor's degree, and the percentage of households earning above the region's median income for the year (about $44,000 to $57,000, depending on the region and year). First, we calculated z-scores for each PUMA by subtracting the value of the characteristic from the mean value of the characteristic across the PUMAs for the region, and dividing this value by the standard deviation of the characteristic for the PUMAs in the region. Then, we constructed an index by taking the average of the PUMAs' z-scores for each geography. Negative values indicate lower socio-economic status relative to the region; positive values indicate higher socio-economic status relative to the region.

We also conducted a content analysis of popular press news articles, real estate reports, and blog reports within each region in the 2000s and early 2010s to triangulate evidence from the US Census. We constructed and inputted a list of keywords into Google web and scholar search engines to identify relevant sources.

We coded these materials to extract information on real estate trends and developers' intentions, Millennials' housing preferences and residential mobility, and changes to the demographic, socio-economic, and housing conditions in urban core and central city communities.

## Cementing Millennials? Evidence from Phoenix and Houston, 2000–2013

This section describes how the concentration of new housing and Millennials changed in Phoenix and Houston's urban core, central city, and suburbs from 2000 to 2013. We also address how the characteristics of housing, Millennials, and neighbourhoods in these areas changed over time.

### *Spatial trends in new housing and Millennials*

We first compared location quotients for new housing and Millennials in the regions to understand how their spatial patterns changed over time (Table 10.2). Location quotients above 1 (and that are increasing over time) would provide supportive evidence of a housing renaissance in the urban core and central city, and the migration of Millennials to these places.

The results show new housing is far less concentrated in the regions' central cities and urban cores than in their suburbs, and this trend persisted for the entirety of the study period. For instance, the proportion of newly constructed homes in the Phoenix and Houston urban cores and central cities in 2013 was about 20% to 40% less than in the region; in contrast, the concentration in the suburbs was about 15% to 30% more than in the region.

However, the relative concentration of new housing in the urban core *increased* over time. In Phoenix, the urban core new housing location quotient increased from 0.30 to 0.71 from 2000 to 2013. In Houston, the increase in the urban core was greater (0.32 to 0.81). Interestingly, new housing concentration in the Houston central city and suburbs follow opposite patterns. The concentration of new housing in the central city rose from 2000 to 2010, but fell from 2010 to 2013, while it fell in the suburbs from 2000 to 2010, but rose from 2010 to 2013. This evidence suggests a shift in real estate activity to the urban core but not central city.

The strong, post-recession multifamily housing market in both regions' urban cores helps explain this trend. In both Phoenix and Houston in the early 2010s, developers flocked to infill, including large companies that previously developed in the suburbs (Haldiman, 2015; Sarnoff, 2014; Sunnucks, 2014a). In Phoenix, infill grew to 13% of all new home permits in 2013 (The Associated Press, 2013; Reagor & Morrison, 2013). In Houston, the condo and townhome market grew in late 2013/early 2014, while the single family home market stayed flat, and land prices for infill sites around the light rail in Houston's urban core and central city increased up to 40% during this time period (Sarnoff, 2014). Building permits grew by 23% in one year within Beltway 8 (which encompasses the City of Houston but not its suburbs), compared to 9% in the region (Sarnoff, 2014).

Like trends for new housing construction, the relative concentration of Millennials in Phoenix and Houston's urban cores increased over time, although Millennials have long been more concentrated in the urban core and central city than in the suburbs (see Table 10.2). For instance, in 2000 and 2013, Millennials' proportion of the population in the regions' urban cores and central cities was between 7% and 26% more than in the broader region, and, from 2000 to 2013, the Millennial location quotient rose from 1.12 to 1.25 in Phoenix's urban core and from 1.08 to 1.26 in Houston's urban core. The relative concentration of Millennials modestly increased in Phoenix's central city (1.07 to 1.12) but largely remained the same in Houston's central city (about 1.14).

Local media reveal several reasons for Millennial population growth in the regions' urban cores. Sources in both regions cited infill and adaptive reuse projects as Millennial draws (Holeywell, 2013; Reagor, 2015). Phoenix's Roosevelt Row neighbourhood, with its historic buildings and art galleries, has been at the epicentre of the multifamily boom for several years and is also home to rapid infill development (Goth, 2015). In Houston, the long-neglected East Downtown (located outside of the urban core in Houston's central city) experienced a resurgence of infill and adaptive reuse projects starting in the early 2010s, including a pedestrian "esplanade" with retail and restaurants, a cultural centre, public art installations, and upgraded parks and trails (Holeywell, 2013). Additional Houston-specific reasons include diverse housing and amenities, and a strong local economy (Sarnoff, 2011; Downtown Houston, 2010), as well as Millennials entering the workforce and housing market (Dixon, 2015; Mulvaney, 2014; Sarnoff, 2005). In Phoenix, reasons also include a sense of community and proximity to amenities (Reagor, 2015; Tempest, n.d.).

Light rail also drew Millennials seeking a public transit-oriented lifestyle to both regions' urban cores, and catalyzed public and private investment in

*Table 10.2* Spatial trends in new housing and Millennials

| Location Quotients | Phoenix | | | Houston | | |
|---|---|---|---|---|---|---|
| | Urban Core | Central City | Suburbs | Urban Core | Central City | Suburbs |
| New Housing | | | | | | |
| 2000 | 0.30 | 0.69 | 1.20 | 0.32 | 0.64 | 1.30 |
| 2006 | 0.68 | 0.70 | 1.16 | 0.32 | 0.64 | 1.27 |
| 2010 | 0.72 | 0.77 | 1.12 | 0.64 | 0.69 | 1.21 |
| 2013 | 0.71 | 0.68 | 1.15 | 0.81 | 0.57 | 1.28 |
| Millennials | | | | | | |
| 2000 | 1.12 | 1.07 | 0.95 | 1.08 | 1.14 | 0.90 |
| 2006 | 1.18 | 1.08 | 0.95 | 1.02 | 1.09 | 0.95 |
| 2010 | 1.22 | 1.08 | 0.95 | 1.12 | 1.14 | 0.91 |
| 2013 | 1.25 | 1.12 | 0.93 | 1.26 | 1.14 | 0.91 |

Source: US Census

Note: New housing is housing that was built in between the current and previous time period. Millennials are aged 25 to 34.

housing and retail (Brodle, 2014; Downtown Houston, 2010; Goth, 2015; Haldiman, 2015; Holeywell, 2013; Karaim, 2014; Montgomery, n.d.) (also see Chapter 15). However, each region's light rail alignment may impact Millennials' ultimate housing location, since both central cities compete with their suburbs for Millennials. While the city of Houston has a comparative advantage over its urbanizing suburbs (Houston's light rail stays in its central city), mixed-use town centres in suburban The Woodlands and Sugar Land may be attractive to Millennials (Holeywell, 2013). In contrast, Phoenix's satellite cities of Tempe and Mesa are growing urban centres connected to each other and Phoenix via light rail. Millennials who want light rail access do not have to live within the Phoenix city limits. Indeed, multifamily development in Tempe and Mesa grew during the early 2010s (Haldiman, 2015; Karaim, 2014; Reagor, 2015; Sunnucks, 2014a).

### Characteristics of new housing

We next examined how the nature of recently built housing compares to the past, which offers insight into how recentralization trends may reshape urban housing markets. Table 10.3 shows the characteristics of housing built in 2000 and after, and before 2000 ("newer" and "older" housing respectively) using five-year data from the US Census' 2013 American Community Survey. Table 10.3 also offers further evidence of whether the real estate industry is cementing Millennials' residence in central cities by showing how the proportions of Millennial households in newer and older housing differ.

The data show several important features of recent housing construction. By 2013, rental housing comprised the majority of newer homes in both regions' urban cores. This is a departure from the tenure of housing built prior to 2000. In Phoenix's urban core, 35% of older homes were owner-occupied compared to only 16% of newer homes. A similar, though less dramatic trend, was apparent in Houston's urban core (45% vs. 31% respectively). Notably, the tenure of newer and older housing in the central city was similar (majority owner-occupied). In the suburbs, newer housing was more likely to be owner-occupied.

Newer housing in the urban core was also more diverse than older housing in the urban core, as well as newer and older housing in the central city and suburbs. No single housing type comprised the majority of newer housing in the regions' urban cores. Phoenix's urban core exhibited tremendous heterogeneity among its newer homes in particular. Large multifamily buildings (greater than 20 units) comprised the largest proportion of newer homes (36%). Remaining homes were fairly evenly distributed among other housing types (27% medium multifamily, 20% single family, and 17% small multifamily). Among older housing in the urban core, and among newer and older housing in the central city and suburbs, single-family housing was most common.

Newer housing seems to be more expensive than older housing over the study period. These values are not exact, since they are based on average median values (see footnote on Table 10.3). One exception was in Phoenix's urban core, where newer for-sale housing was similarly valued to older for-sale housing. Home values and gross rents in the urban cores also were often less than those in the central city and suburbs, regardless of the age of the housing, although this may change if demand for urban core living continues to grow.

A Millennial was more likely to occupy newer housing than older housing in both regions. Yet, this difference was especially pronounced in the urban core. In Phoenix, 50% of newer homes had a Millennial householder compared to only 30% of older homes. In Houston, 44% of newer homes had a Millennial householder compared to only 23% of older homes. The disproportionate residence of Millennials in newer housing suggests recent real estate activity played a role in attracting them to the urban core.

Some evidence from the regions' local media supports this relationship. Developers in both regions built multifamily rental housing to attract Millennials preferring an urban or "live–work–play" lifestyle, building units that are smaller (Sunnucks, 2014a), require minimal maintenance (Mulvaney, 2015; Sunnucks, 2014b), and provide ample access to cultural amenities, public transit, and walkable neighbourhoods (Brodle, 2014; Haldiman, 2015; Reagor & Morrison, 2013; Reagor, 2015; Sunnucks, 2014a; Sunnucks, 2014b; The Associated Press, 2013). Developers also tailored projects to meet perceived Millennial preferences for 1) unique and eco-friendly properties (Reagor and Morrison, 2013; The Associated Press, 2013), 2) having few material possessions (Sunnucks, 2014a), and 3) renting (Conley, 2015; Mulvaney, 2015).

However, local media also suggest Millennials may soon become owners in both regions, meaning that their residence in new multifamily buildings in the urban core may be fleeting. Reasons pushing Millennials to homeownership include 1) a shortage of good rental properties and rising rents (Conley, 2015; McGraw, 2013; Takahashi, 2015), 2) growing post-recession incomes (McGraw, 2013), 3) family formation (Bradley-Waddell, 2015; Takahashi, 2015), and 4) affordable homeownership opportunities in the suburbs (Feser, 2015; McGraw, 2013; Takahashi, 2015). Homeownership will likely cost more for Millennials who stay in the urban core (Reagor & Morrison, 2013; Reagor, 2015), however, and Millennials unwilling to pay this premium will likely move elsewhere within the central city or to the suburbs (Braddley-Waddell, 2015; Guillen, 2015; Takahashi, 2015; Waddell, 2015).

This trend may already be occurring in Houston. In the mid-2000s, the Millennial homeownership market was strong inside the Loop 610 Freeway (which includes Houston's central city), because Millennials without children wanted to live near cultural amenities and mass transit (Sarnoff, 2005). However, some sources suggested that by the early 2010s, Millennials who preferred to live in the Loop were moving outside to sprawling, suburban areas in search of affordable homeownership (Holeywell, 2013; Levin, 2015; McGraw, 2013; Guillen, 2015). Other sources suggested that growth in the central city multifamily housing market (including 1,400 sub-market rate units) offered sufficient affordable housing for prospective Millennial homeowners who wanted to stay (Downtown Houston, 2010).

Table 10.3 Characteristics of new housing

| Characteristics | Phoenix | | | | | | Houston | | | | | |
|---|---|---|---|---|---|---|---|---|---|---|---|---|
| | Urban core | | Central city | | Suburbs | | Urban core | | Central city | | Suburbs | |
| | Built 2000+ | Built Pre-2000 | Built 2000+ | Built Pre-2000 | Built 2000+ | Built Pre-2000 | Built 2000+ | Built Pre-2000 | Built 2000+ | Built Pre-2000 | Built 2000+ | Built Pre-2000 |
| % Owner occupied | 16 | 35 | 60 | 58 | 69 | 66 | 31 | 45 | 50 | 51 | 73 | 68 |
| Housing type | | | | | | | | | | | | |
| % Single-family | 20 | 51 | 76 | 69 | 82 | 70 | 42 | 65 | 55 | 58 | 77 | 75 |
| % Small multi-family | 17 | 15 | 4 | 5 | 2 | 5 | 10 | 10 | 3 | 5 | 2 | 3 |
| % Medium multi-family | 27 | 15 | 10 | 13 | 7 | 10 | 15 | 11 | 13 | 19 | 9 | 9 |
| % Large multi-family | 36 | 16 | 8 | 9 | 5 | 6 | 32 | 13 | 27 | 16 | 7 | 5 |
| Average median value ($1,000) | 129 | 131 | 214 | 145 | 238 | 161 | 224 | 86 | 333 | 165 | 188 | 117 |
| Average median gross rent ($) | 816 | 709 | 957 | 864 | 1,147 | 915 | 718 | 671 | 1,041 | 840 | 1,057 | 910 |
| % Millennial householder | 50 | 30 | 32 | 23 | 25 | 19 | 44 | 23 | 37 | 24 | 26 | 18 |

Source: US Census

Note: Millennials are aged 15 to 34. Small multi-family housing has 2 to 4 units, medium multifamily housing has 5 to 19 units, large multifamily has 20 or more units. The base data for average median value and gross rent was median value or gross rent by housing age (decade) in each PUMA. The average of these values was taken across the PUMAs in the geography for the time period to arrive at the average median value.

There were a small amount of PUMAs missing values for these variables.

## Characteristics of Millennials

We also explored how the demographic and socio-economic characteristics of Millennials compared to people who were not Millennials among the geographies in 2013 (see Table 10.4). This information lends further insight into the types of Millennials that migrated to Phoenix and Houston's urban core and central city, and how they may be changing the characteristics of communities within these geographies.

With one exception (Houston's urban core), Millennials were less likely to be non-Hispanic White and more likely to be Latino than not Millennials across the geographies. For instance, although 58% of Millennials in the Phoenix suburbs were non-Hispanic White, only 43% and 36% were non-Hispanic White in the central city and urban core respectively. However, urban core residents were less likely to be non-Hispanic White than central city and suburban residents, regardless of age. Differences in racial and ethnic identity among Millennials and not Millennials were narrower in the urban core than in the central cities and suburbs.

Millennials were much more likely than not Millennials to have never married in both regions, particularly those living in the urban cores. For instance, 66% and 55% of Millennials in the Phoenix and Houston urban cores had never married respectively, compared to only 45% and 39% of Millennials in their suburbs.

Millennials in the urban core were more likely to hold a bachelor's degree than not Millennials in the urban core. Millennials also were more likely to be enrolled in school, but rates were relatively similar across the geographies (12% to 15%). This is surprising in Phoenix, given that Arizona State University opened its Downtown Campus in the urban core in 2006. Some local media explain this trend by crediting the campus with activating public spaces and the adaptive reuse of historic buildings in the urban core, thereby catalyzing investment in retail and multifamily housing, and attracting non-student Millennials (Gilger & Jeffrey, 2015; Montgomery, n.d.; Tempest, n.d.).

Millennials across the geographies were more likely to live in poverty than not Millennials, with the exception of those in Houston's urban core. However, differences in poverty rates by age were much narrower in the urban cores than central cities and suburbs. Interestingly, Millennials were slightly less likely to be unemployed than not Millennials in the urban cores of both regions (9% vs. 12% respectively in Phoenix and 8% vs. 9% respectively in Houston).

Millennials were more likely than not Millennials to be born out of state. Millennials had the highest rates of being born out of state in Phoenix's urban core (70%). Millennials were much more transient than not Millennials, particularly in the urban core and central city. About 31% of Millennials in Houston's urban core had moved within the past year, compared to 16% of not Millennials in the urban core and 25% of Millennials in the suburbs. Transient Millennials were slightly more likely to be from out of state in Houston's urban core, but not in any of the other geographies. Millennials' greater mobility is also mentioned in the local media (e.g. Bradley-Waddell, 2015).

Table 10.4 Characteristics of Millennials

| Characteristics (%) | Phoenix | | | | | | Houston | | | | | |
|---|---|---|---|---|---|---|---|---|---|---|---|---|
| | Urban core | | Central city | | Suburbs | | Urban core | | Central city | | Suburbs | |
| | Mill. | Not Mill. | Mill. | Not Mill. | Mill. | Not Mill. | Mill. | Not Mill. | Mill. | Not Mill. | Mill. | Not Mill. |
| Non-Hispanic White | 36 | 44 | 43 | 61 | 58 | 75 | 13 | 13 | 27 | 39 | 40 | 55 |
| Black | 10 | 11 | 7 | 6 | 5 | 4 | 13 | 12 | 20 | 21 | 16 | 14 |
| Latino | 47 | 40 | 42 | 28 | 28 | 16 | 72 | 74 | 44 | 31 | 37 | 23 |
| Asian | 3 | 2 | 5 | 3 | 4 | 3 | 2 | 2 | 9 | 7 | 7 | 7 |
| Never married | 66 | 33 | 50 | 15 | 45 | 10 | 55 | 21 | 50 | 16 | 39 | 8 |
| Enrolled in school | 12 | 3 | 12 | 3 | 15 | 3 | 8 | 2 | 12 | 3 | 11 | 3 |
| College educated | 26 | 21 | 25 | 27 | 29 | 30 | 17 | 9 | 33 | 31 | 27 | 29 |
| Poor | 33 | 31 | 20 | 15 | 14 | 9 | 23 | 24 | 19 | 15 | 14 | 9 |
| Unemployed | 9 | 12 | 10 | 8 | 9 | 7 | 8 | 9 | 9 | 7 | 8 | 6 |
| Born outside state | 70 | 77 | 68 | 82 | 67 | 85 | 51 | 63 | 56 | 63 | 47 | 57 |
| Moved within the year | 34 | 18 | 32 | 14 | 29 | 14 | 31 | 16 | 30 | 12 | 25 | 10 |
| Moved from out of state | 17 | 20 | 16 | 18 | 22 | 30 | 12 | 10 | 18 | 18 | 15 | 18 |

Source: US Census

Note: Millennials are aged 25 to 34. Not Millennials are aged 35 and older.

## Changes in socioeconomic status

Finally, we explored whether real estate developers' and Millennials' activities within Phoenix and Houston's urban cores and central cities changed the character of these places during the 2000s and early 2010s. Table 10.5 reports changes in the socio-economic index (see Data and Methods section) for the geographies from 2000 to 2013. Positive values mean that a geography has higher socio-economic status than others in the region on average; negative values mean that a geography has lower socio-economic status than others the region. The more negative the value, the relatively lower the socio-economic status.

Two trends are evident. First, Phoenix and Houston's urban cores were of much lower socio-economic status than their suburbs, and this trend has persisted over time. Their central cities also had a lower socio-economic status, but the difference is less pronounced. For instance, the SES index value of –2.06 for Houston's urban core means that the geography's rates of college educated adults and affluent households were on average over 2 standard deviations less than the region's rates. Second, the socio-economic statuses of the urban cores have increased markedly over time. The greatest increases occurred from 2010 to 2013 (–1.35 to –1.05 in Phoenix and –1.77 and –1.18 in Houston). No such consistent increase in relative socio-economic status occurred for the central cities, although both registered smaller but not insignificant increases in socio-economic status from 2010 to 2013 (–0.31 to –0.25 in Phoenix and –0.32 to –0.18 in Houston). The suburbs, in contrast, experienced a small decline in relative socio-economic status between 2010 and 2013 (0.22 to 0.18 in Phoenix and 0.24 to 0.14 in Houston).

In short, Phoenix and Houston's suburbs and central cities became more similar in their socio-economic status over time. Their urban cores are still relatively disadvantaged but appear to be catching up. This is evidence that new housing investments that attract Millennials to the urban core and central city may be changing longstanding patterns of socio-economic inequality within these regions. For instance, poor people may be moving from the urban core to elsewhere in the central city or suburbs (Kneebone & Lou, 2014) However, there was little information on this theme in local media, so further confirmation is still warranted.

*Table 10.5* Changes in socio-economic status (SES), 2000–2013

| SES index | Phoenix | | | Houston | | |
|---|---|---|---|---|---|---|
| | *Urban core* | *Central city* | *Suburbs* | *Urban core* | *Central city* | *Suburbs* |
| 2000 | –1.43 | –0.14 | 0.14 | –2.06 | –0.29 | 0.24 |
| 2006 | –1.39 | –0.20 | 0.17 | –1.95 | –0.36 | 0.28 |
| 2010 | –1.35 | –0.31 | 0.22 | –1.77 | –0.32 | 0.25 |
| 2013 | –1.05 | –0.25 | 0.18 | –1.18 | –0.18 | 0.14 |

Source: US Census

Note: The SES index is calculated as the average z scores of the percentage of adults college educated and households above the median income among the PUMAs in the geography.

## Conclusion

This chapter offered an initial glimpse into whether the real estate industry cemented Millennials' residence in Phoenix and Houston's urban cores and central cities over the 2000s and early 2010s. Newer housing construction and Millennials seem to be recentralizing in Phoenix and Houston. New housing became increasingly concentrated in both regions' urban core, and Millennials disproportionately moved into this housing. These trends took place on a more muted level in the central cities.

Newer housing in the urban core also was distinct from older housing in the urban core and newer housing produced in the central city or suburbs – it was largely for rent and represented a greater diversity of housing types. Millennials were more likely to live in newer homes than people who were not Millennials, particularly in the urban core. Millennials have distinct socio-economic characteristics; they tend to be more educated and racially and ethnically diverse, but also poorer than other age groups. However, their lower socio-economic status may be in part explained by Millennials being in earlier stages of their careers. Nonetheless, Millennials' recentralization may alter regional urban social structures. Our data reveals an association among new housing construction, Millennial migration, and improving socio-economic status in the urban core.

Millennials are the largest demographic cohort and are at the beginning of long professional careers, so their housing choices will likely reverberate for decades. As such, Millennial housing preferences have broad implications for planning. Urban core and central city planners will need to consider whether existing zoning codes, transportation systems, housing affordability programmes, and social services allow developers to meet demand for new housing construction, fulfill the needs of growing Millennial families, and maintain sufficient affordable housing stock as demand for city living grows. Suburban planners will need to consider whether depopulation is likely, along with policies and programmes to attract suburban-minded Millennials to their city. Finally, regional planning coordination for housing, transportation, economic development, land use and other investments is needed to ensure shifting Millennial concentrations and new housing construction do not simply move wealth from the suburbs to the urban core, but instead provide diverse housing choices at a variety of price points for all residents across the metropolitan area.

## References

Arnett, J. J. (2004). *Emerging adulthood: The winding road from the late teens through the twenties*. Oxford: Oxford University Press.
The Associated Press. (2013, December 30). Phoenix housing market shifts to central areas. *Arizona Capitol Times*. Retrieved from http://azcapitoltimes.com/news/2013/12/30/phoenix-housing-market-shifts-to-central-areas/.
Baum-Snow, N., & Hartley, D. (forthcoming). *The Causes and Consequences of Gentrification and Neighborhood Change in US Cities, 1980–2010*. Federal Reserve Bank of Cleveland working paper.
Birch, E. L. (2005). *Who lives downtown*. Metropolitan Policy Program. Brookings Institution, Washington, DC.

Bradley-Waddell, G. (2015, March 9). Millennials buying homes value upgrades, easy maintenance. *Arizona Republic*. Retrieved from www.azcentral.com/story/entertainment/home/2015/03/09/millennials-homebuyers-value-upgrades-easy-maintenance/24654601/.

Brodle, M. (2014, May 30). Resurgence in infill development evident in Phoenix. *91.5 KJZZ*. Retrieved from http://kjzz.org/content/31814/resurgence-infill-development-evident-phoenix.

City of Houston. (2015). *Houston super neighborhoods*. Houston, TX: City of Houston. Retrieved from http://mycity.maps.arcgis.com/home/webmap/viewer.html?webmap=e87cdc21ac3a43ecb2cdf2c31d75ca8e.

City of Phoenix. (2015). *Village planners*. Phoenix, AZ: City of Phoenix. Retrieved from www.phoenix.gov/pdd/pz/village-planning-committees.

Conley, D. (2015, October 8). Millennials consider buying homes as rents get more expensive. *Cronkite News Service*. Retrieved from www.azcentral.com/story/money/real-estate/consumers/2015/10/07/millennials-consider-buying-homes-as-rents-get-more-expensive/73530686/.

Dixon, C. (2015, July 6). Will Millennials keep Houston afloat? *BisNow*. Retrieved from www.bisnow.com/houston/news/other/will-millennials-keep-houston-afloat-47656.

Doherty, P. C., & Leinberger, C. B. (2010, November/December). The next real estate boom. *Washington Monthly*. Retrieved September 21, 2015 from www.washingtonmonthly.com/features/2010/1011.doherty-leinberger.html.

Downtown Houston. (2010). Residential market overview report. *Downtownhouston.org*. Retrieved from www.downtownhouston.org/site_media/uploads/attachments/2010-10-12/1-Residential_Market_Overview_Report.pdf.

Ehrenhalt, A. (2012). *The great inversion and the future of the American city*. New York: Vintage.

Ewing, R., & Cervero, R. (2010). Travel and the built environment. *Journal of the American Planning Association, 76*(3), 265–294.

Feser, K. (2015, October 22). Millennials better off buying a home than renting. *Houston Chronicle*. Retrieved from http://blog.chron.com/primeproperty/2015/10/millennials-better-off-buying-a-home-than-renting/.

Fry, R. (2014). *A record 21.6 million in 2012: A rising share of young adults live in their parents' home*. Washington, DC: Pew Research Center.

Gilger, L., & Jeffrey, C. (2015, November 25). *Downtown on top: ABC15 tracks rapid growth, development of Downtown Phoenix*. Retrieved from www.abc15.com/news/region-phoenix-metro/central-phoenix/downtown-on-top-abc15-tracks-rapid-growth-development-of-downtown-phoenix.

Goth, B. (2015, April 17). Roosevelt row highlights downtown Phoenix's growing pains. *Arizona Republic*. Retrieved from www.azcentral.com/story/news/local/phoenix/2015/04/17/downtown-phoenix-roosevelt-row-arts-district/25886609/.

Guillen, D. (2015, October 2). More millennials are buying homes in the suburbs. *Houston Chronicle*. Retrieved from www.chron.com/neighborhood/homes/article/Millennials-are-buying-home-in-the-suburbs-6545548.php.

Haldiman, P. (2015, May 27). With all this multi-family activity, how will central Phoenix look when everybody moves in? *Rose Law Group Reporter*. Retrieved from http://roselawgroupreporter.com/2015/05/with-all-this-multi-family-activity-how-will-central-phoenix-look-when-everybody-moves-in/.

Holeywell, R. (2013, October). Houston: The surprising contender in America's urban revival. *Governing*. Retrieved from http://roselawgroupreporter.com/2015/05/with-all-this-multi-family-activity-how-will-central-phoenix-look-when-everybody-moves-in/.

Huso, D. (2015, June). Mayhem in multifamily? *Realtor Magazine*. Retrieved from http://realtormag.realtor.org/commercial/feature/article/2015/06/mayhem-in-multifamily.

Joint Center for Housing Studies. (2013). *America's rental housing*. Cambridge, MA: Joint Center for Housing Studies.

Karaim, R. (2014, February 5). Instead of Suburbia: Can Phoenix discourage sprawl now that the housing market is heating up again? *Architect Magazine*. Retrieved from www.architectmagazine.com/practice/instead-of-suburbia-can-phoenix-discourage-sprawl-now-that-the-housing-market-is-heating-up-again_o.

Kneebone, E., & Lou, C. (2014, June). Suburban and poor: The changing landscape of race and poverty in the US. *Planning*, 17–21.

Kolko, J. (2012, December 13). "Renter nation" Just a myth: 93% of millennial renters plan to buy a home someday. *Forbes*. Retrieved from www.forbes.com/sites/trulia/2012/12/13/renter-nation-just-a-myth-93-of-millennial-renters-plan-to-buy-a-home-someday/.

Kotkin, J. (2013). *Retrofitting the dream: Housing the 21st century*. Pinatubo Press.

Leinberger, C. B. (2011, November 25). Death of the fringe suburb. *New York Times*. Retrieved from www.nytimes.com/2011/11/26/opinion/the-death-of-the-fringe-suburb.html.

Levin, M. (2015, July 2016). The rent is too high for Houston Millennials. *Houston Chronicle*. Retrieved from www.chron.com/news/houston-texas/article/The-rent-is-too-high-for-Houston-millennials-6388661.php.

McGraw, D. (2013, April 4). Millennials changing attitude to home buying. *Houston Chronicle*. Retrieved from www.chron.com/homes/article/Millennials-changing-attitude-to-home-buying-4406920.php.

Montgomery, A. (No Year, December 8). *Is downtown Phoenix at a tipping point?* Retrieved May 21, 2016, from This Could Be PHX: www.thiscouldbephx.com/is-downtown-phoenix-at-a-tipping-point.

Mulvaney, E. (2014, June 11). How multifamily boom in Houston, nation affects affordability. *Houston Chronicle*. Retrieved from http://blog.chron.com/primeproperty/2014/06/how-multifamily-boom-in-houston-nation-affects-affordability/.

Mulvaney, E. (2015, March 10). The mind of a Millennial: Why are young people hesitant to buy homes? *Houston Chronicle*. Retrieved from http://blog.chron.com/primeproperty/2015/03/the-mind-of-a-millennial-why-are-young-people-hesitant-to-buy-homes/.

Nelson, A. C. (2013). *Reshaping metropolitan America: Development trends and opportunities to 2030*. Washington, DC: Island Press.

Ramsey, K. (2012). *Residential construction trends in America's metropolitan regions* (2012 ed.). Washington, DC: Office of Sustainable Communities, US Environmental Protection Agency.

Reagor, C. (2015, April 25). Condo sales lead housing trend to Valley's city cores. *Arizona Republic*. Retrieved from www.azcentral.com/story/money/real-estate/catherine-reagor/2015/04/25/phoenix-area-condo-sales-up-city-core/26385639/.

Reagor, C., & Morrison, K. (2013, December 27). Housing market shifts back to metro Phoenix's core. *Arizona Republic*. Retrieved from http://archive.azcentral.com/business/realestate/articles/20131120phoenix-housing-market-core.html.

Sarnoff, N. (2005, June 26). Twentysomethings flocking to real estate sooner. *Houston Chronicle*. Retrieved from www.chron.com/business/article/Twentysomethings-flocking-to-real-estate-sooner-1952193.php.

Sarnoff, N. (2011, November 21). Apartments to multipy inside the loop. *Houston Chronicle*. Retrieved from www.chron.com/business/article/Apartments-to-multiply-inside-the-loop-2277479.php.

Sarnoff, N. (2014, July 1). Demand for urban high-density living ramps up. *Houston Chronicle*. Retrieved from www.houstonchronicle.com/business/real-estate/article/Demand-for-high-density-living-ramps-up-in-5593995.php.

Sunnucks, M. (2014a, August 6). Millennials driving Phoenix multifamily market, but a drag on housing. *Phoenix Business Journal*. Retrieved from www.bizjournals.com/phoenix/blog/business/2014/08/millennials-driving-phioenix-multifamily-market.html.

Sunnucks, M. (2014b, December 1). New infill housing slated for north-central Phoenix. *Phoenix Business Journal*. Retrieved from www.bizjournals.com/phoenix/news/2014/12/01/new-infill-housing-slated-for-north-central.html.

Suro, R., Wilson, J. H., & Singer, A. (2011). *Immigration and poverty in America's suburbs*. Washington, DC: Brookings Institution.

Takahashi, P. (2015, June 12). The kids are all right: Houston housing market braces for influx of Millennials. *Houston Business Journal*. Retrieved from www.bizjournals.com/houston/print-edition/2015/06/12/the-kids-are-all-right-houston-housing-market.html.

Tempest, R. (No Year, February 4). *My Phoenix life: Ryan tempest*. Retrieved September 18, 2015, from This Could Be PHX: www.thiscouldbephx.com/my-phoenix-life-ryan-tempest.

Waddell, G. (2015, October 21). Metro Phoenix's new-home market getting back on track. *Arizona Republic*. Retrieved from www.azcentral.com/story/money/real-estate/2015/10/21/metro-phoenix-new-home-market/74292200/.

Warren, A., Kramer, A., Biank, S., & Shari, M. (2013). *Emerging trends in real estate 2014*. Washington, DC: Urban Land Institute.

# 11 Boomers and their boomerang kids

## Comparing housing opportunities for Baby Boomers and Millennials in the United States

### Sarah L. Mawhorter

As the first wave of Millennials became adults, they encountered a starkly different landscape of housing opportunities than their parents did when they were young. Housing prices and rents increased much faster than their earnings, and many adult Millennials moved back in with their parents or stayed home rather than striking out on their own (Fry, 2015). Some have slightly pessimistically argued that Millennials have fared far worse than previous generations in the housing market (Myers, Painter, Lee, & Park, 2016; Goodman, Pendall, & Zhu, 2015). Others point optimistically toward the growing stock of smaller condominium units and Millennials' increasing educational attainment, which could bode well for their future housing consumption (see Chapters 2, 4, and 5). This chapter provides an empirical analysis to help inform this debate.

I compare US housing market conditions in 1980, when the first wave of Baby Boomers were in their late 20s and early 30s, to housing market conditions in 2015, when the first wave of Millennials were in their late 20s and early 30s. I analyse the demand growth and supply constraints that led to the severe housing shortfall and affordability crisis for Millennials. I then assess the housing market outcomes for Millennials compared with Boomers in terms of household formation, homeownership, and housing affordability. The findings point to the urgency of addressing the housing affordability crisis for the sake of an entire generation of young people, who are decidedly worse off than their parents' generation in terms of housing market opportunities and outcomes. Their glass is indeed half empty!

Millennials entered adulthood facing an array of challenges, including unprecedented levels of student debt (Fry, 2014; Stiglitz, 2013), stunted careers and stagnant wages in the wake of the Great Recession (Elsby, Shin, & Solon, 2013), and rising income inequality (Pew Research Center, 2014; Chapters 3 and 4). As this analysis shows, Millennials grew up during an ongoing housing shortfall, when housing construction had not kept pace with population growth for decades (also see Chapter 12). I find that insufficient new construction was compounded by reduced turnover of existing housing units, and the overall amount of housing available on the market shrank relative to the total housing stock and relative to sharply increasing demand as more and more Millennials sought housing (Myers, 2016).

Compared to Boomers when they were young, Millennials faced steeper competition for housing from older people, and were in a weaker economic position relative to older groups. Millennials, like other groups, encountered stiff barriers in access to mortgage credit after the housing bust, which made it more difficult for Millennials to purchase homes and also increased competition for rental housing (Xu, Johnson, Bartholomae, O'Neill, & Gutter, 2015). I find that in this context, Millennials formed households and bought homes at much lower rates than Boomers did at the same ages. Millennial renters also dealt with worse affordability problems than Boomers. Millennials without college degrees bore the brunt of the housing shortfall, which widened the existing disparities between young people with and without college educations.

## Analytical strategy

To compare housing market conditions for Baby Boomers and Millennials, I examine the periods when the first wave of each generation were in their late 20s and early 30s. The first Boomers were born in 1946, and turned 34 in 1980. The first Millennials were born in the early 1980s, and turned 33, 34, or 35 in 2015. I focus on housing market conditions and outcomes for people aged 25–34 in 1980 (Boomers) and people aged 25–34 in 2015 (Millennials) since the late 20s and early 30s are crucial periods for household formation and home buying. I compare housing market shifts during the ten-year periods leading up to 1980 and 2015: 1970 to 1980 and 2005 to 2015.

I first analyse changes in the population, housing stock, and housing prices from 1970 to 1980 and 2005 to 2015. I then turn attention to the specific situation of Boomers and Millennials in the housing market and their housing outcomes in terms of homeownership and housing affordability. I combine a wide range of data sources for this analysis, many of which are collected by the US Census Bureau and made available through the Federal Reserve Economic Data website from the Federal Reserve Bank of St. Louis, Social Explorer, or directly from the US Census Bureau website. I also use Integrated Public Use Microdata for a more detailed analysis of housing affordability by age and education (Ruggles, Genadek, Goeken, Grover, & Sobek, 2015).

In many cases, I use cohort analysis to understand changes over time as people move through their lives. Cohort analysis refers to using multiple data points to follow the trajectory of a group of people as they age over time. For example, by comparing the characteristics of 15-to-24-year-olds in 1970 and 25-to-34-year-olds in 1980, ten years later and ten years older, researchers can observe the progression of the cohort, composed of roughly the same people, from their late teens and early 20s into their late 20s and early 30s. A cohort's progress over time can then be compared with the progress of a similarly-aged cohort in a different period of time (see Myers, 1992). Cohort analysis is particularly useful to compare the housing careers of Boomers and Millennials, who grew up in such different eras that their *levels* of headship and homeownership might be quite different due to a host of social and cultural reasons, as well as economic and demographic factors, but whose *progress* towards household formation and homebuying can more reasonably be compared.

## Millennials fare worse on most accounts

### *Boomers led population growth in the 1970s and 2000s*

The first wave of Boomers came of age during the 1970s, spurring growth in demand for housing. From 1970 to 1980, the adult population grew by 21% (US Census Bureau, 1970, 1980). When the first Millennials came of age from 2005 to 2015, adult population growth was comparatively low at 11% (US Census Bureau Population Estimates Program, 2012, 2016).

Figure 11.1 shows the growth in different age groups during the 1970s and from 2005 to 2015. During the 1970s, growth was led by Boomers, who increased the ranks of young adults aged 25–34 by 49%. From 2005 to 2015, growth was strongest in the older age groups: 65-to-74-year-olds increased by 46%, and 55-to-64-year-olds increased by 33%, while 25-to-34-year-olds increased by only 12%. Boomers were the dominant force in the housing market in the 1970s, and they remained the dominant force in the housing market 35 years later.

## Millennial college graduates

Millennials' incomes were also lower relative to older age groups. In the late 1970s, Boomers aged 25–34 earned only 9% less than people aged 35–54 in the prime of their careers, and earned 18% more than people aged 55–64. In the early 2010s, Millennials aged 25–34 earned 19% less than adults aged 35–54, and

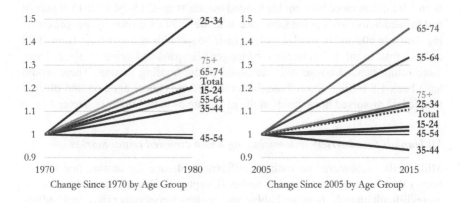

*Figure 11.1* US population growth by age, 1970 to 1980 and 2005 to 2015

Sources: US Census Bureau: Census 1970, Census 1980, retrieved from SE; US Census Bureau Population Estimates Program: Intercensal Estimates 2000–2010, Annual Population Estimates 2010–2016, retrieved from FRED

earned 10% less than people aged 55–64. Millennials' lower employment and earnings relative to older adults meant that they were in a comparatively weaker position to compete in the housing market. As Fry shows in Chapter 4, Millennial incomes also varied more widely by education level: Millennial college graduates made more than Boomer college graduates, while Millennials without a college degree made less than Boomers without a college degree. This increased divergence of incomes by education meant that not only were all Millennials more disadvantaged than Boomers relative to older groups, Millennials with lower educational attainment were at an even greater disadvantage.

### Millennials form fewer households

Over the 1970s, as the first wave of Boomers aged from their late teens and early 20s to their late 20s and early 30s, they formed new households at a rate of 42%, achieving 49% headship by 1980. Faced with rising housing prices, higher levels of unemployment, and stagnant or falling incomes, Millennials increasingly doubled up with others or did not move out of their parents' homes in the first place, and were much less likely to marry (Fry, 2015; Goodman et al., 2015). Millennials formed new households at a rate of 33% from 2005 to 2015, much lower than Boomers' household formation during the 1970s, and reached only 43% headship by 2015.

In 2015, there were 19.1 million households headed by Millennials aged 25–34. If Millennials had formed households at the same rate as Boomers did, there would have been 3.3 million (17%) more Millennial households in 2015. In fact, people at all ages under 65 formed households at lower rates in 2005–2015 than people at the same ages in the 1970s. People entering their late 40s through early 60s had net positive household formation in the 1970s, but dissolved slightly more households than they formed from 2005 to 2015. All told, there would have been 8.9 million more households headed by adults aged 15–64 in 2015 if people had formed households at the same rate as in the 1970s. Conversely, people entering their late 60s and beyond dissolved their households at lower rates from 2005 to 2015 than in the 1970s because of longer life spans and aging in place. Thus, more senior-headed households remained in the housing market. There would have been 1.8 million fewer households headed by seniors aged 65 and older if people had dissolved households at the same rate as in the 1970s (Chapter 12).

### Millennials face barriers to homebuying and a crowded rental market

Millennials encountered an entirely different landscape for renting and buying homes than Boomers had 35 years earlier (Chapter 2). Buying homes became far more difficult after the housing bubble and ensuing foreclosure crisis, and Millennials faced fiercer competition in the rental market. By contrast, Boomers had relatively easy access to credit, but their interest rates were high and went up sharply shortly after 1980. Millennials had a different problem: low mortgage interest rates, and yet their access to credit was severely restricted by tightened mortgage underwriting standards (Courchane, Kiefer, & Zorn, 2015). Many would-be homeowners and people who had undergone foreclosure were also diverted into the rental market (Myers et al., 2016; Xu et al., 2015)

Traditionally, young people entering the housing market become renters, later leaving the rental market to purchase homes. In turn, older people typically sell their homes or leave their rental units at the end of their lives, releasing housing back onto the market (Clark, Deurloo, & Dieleman, 2003; Myers, 1983). But these longstanding patterns were disrupted by economic and demographic shifts taking hold at least since the 1970s.

Figure 11.2 shows the rates at which people of different ages became new renters or owners or left the rental or for sale markets during the 1970s compared with 2005–2015, using a method adapted from Poethig et al. (2017). Rental uptake rates measure the percentage of cohort households who were not renters at the beginning of the decade and became renters by the end of the decade, and rental release rates measure the percentage of cohort renters at the beginning of the decade who no longer occupied rental units by the end of the decade. Similarly, home purchase rates measure the transition to homeownership for households in a cohort who did not own homes at the beginning of the decade. Home selloff rates measure the percentage of cohort homeowners at the beginning of the decade who were no longer homeowners by the end of the decade.

Rental uptake and home purchasing rates were similar in the 1970s and in 2005–2015 for the youngest people entering the housing market. From 1970 to 1980, 77% of the 15–24 age cohort who formed households became renters, and 23% became owners. From 2005 to 2015, 78% of the 15–24 age cohort who formed households became renters, and 22% became owners. Sharp differences between the 1970s and 2005–2015 renting and homebuying dynamics become apparent in all the older age cohorts.

People entering their late 20s through early 40s bought homes at much lower rates during the 2005–2015 period than they had during the 1970s. During the 1970s, nearly half of all renters or newly-formed households in the 25–34 and 35–44 cohorts bought homes, whereas from 2005 to 2015, only a third of renters or new households in the 25–34 cohort bought homes, and barely a quarter of renters or new households in the 35–44 cohort bought homes. Instead, the 25–34 cohort had higher rental uptake rates, and the 35–44 cohort remained in rental housing at higher rates in 2005–2015 than in the 1970s.

For middle-aged adults in cohorts between 45 and 64, the difference between the 1970s and 2005–2015 is even more dramatic. About a third of renters or new households in these cohorts bought homes during the 1970s, and a fifth to a quarter of renters left the rental market. In 2005–2015, middle-aged homebuying plummeted to just below zero and, as a corollary, middle-aged people remained in rental housing instead of leaving. With longer life expectancies and an increased tendency to age in place, senior attrition from both rental and owner-occupied housing slowed as well.

The reductions in homebuying rates, paired with longer occupancy of rental housing, led to substantial impacts on the housing market for Millennials. There

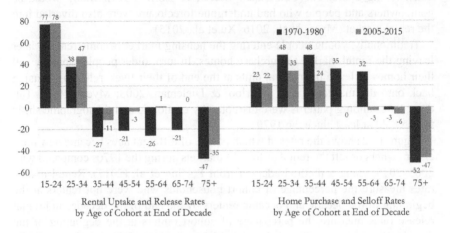

*Figure 11.2* Rental uptake and release and home purchase and selloff rates by age of cohort, 1970 to 1980 and 2005 to 2015

Sources: US Census Bureau: Census 1970, Census 1980, retrieved from SE; US Census Bureau Population Estimates Program: Intercensal Estimates 2000–2010, Annual Population Estimates 2010–2016; Current Population Survey/Housing Vacancy Survey Annual Housing Inventory Estimates

were 9.3 million (12.5%) fewer homeowners in 2015 than there would have been if people had bought homes at the same rate as they did during the 1970s. And there were 6.9 million (16.2%) more renters in 2015 than there would have been if people had left the rental market as they did in the 1970s. This shift toward renting is consistent with the argument of Myers et al. (2016), and put tremendous pressure on the rental housing market (also see Chapter 12).

### Reduced Millennial headship, homeownership, and housing affordability

Table 11.1 compares headship, homeownership, and housing affordability for Boomers aged 25–34 in 1980 and Millennials aged 25–34 in 2015. Millennials' headship rate was 43%, as compared with Boomers' rate of 49%. Millennials' homeownership rate was even lower, at a mere 39% compared with Boomers' 52%. Rental affordability was also much lower for Millennials than it had been for Boomers. A large majority of Boomer renters paid under 30% of their income for rent, and only 14% paid more than half of their income for rent. Many more Millennials had affordability problems. A full 21% paid over half of their income for rent.

The one metric that did not decline for Millennials as compared with Boomers was owner affordability, which remained strikingly similar across 35 years. The apparent sustained affordability of homeownership is related to stringent requirements Millennials faced to qualify for a mortgage; Millennials who would have had difficulty affording to buy a home were simply excluded from buying homes altogether.

*Table 11.1* Headship, homeownership, and housing affordability for people aged 25–34 by educational attainment in 1980 and 2015

| | All Aged 25–34 | | College Grads Aged 25–34 | | Non Grads Aged 25–34 | |
|---|---|---|---|---|---|---|
| | **1980** | **2015** | **1980** | **2015** | **1980** | **2015** |
| **Headship rate** | 49 | 43 | 58 | 48 | 47 | 37 |
| **Homeownership rate** | 52 | 39 | 55 | 44 | 51 | 33 |
| **Percent of renters aged 25–34 paying** | | | | | | |
| <30% Rent/Income | 68 | 55 | 75 | 69 | 65 | 47 |
| 30–50% Rent/Income | 18 | 24 | 15 | 19 | 19 | 27 |
| 50%+ Rent/Income | 14 | 21 | 9 | 12 | 16 | 26 |
| **Percent of owners aged 25–34 paying** | | | | | | |
| <30% Owner costs/income | 78 | 79 | 79 | 85 | 77 | 74 |
| 30–50% Owner costs/income | 16 | 14 | 17 | 12 | 16 | 17 |
| 50%+ Owner costs/income | 6 | 7 | 5 | 4 | 7 | 9 |

Sources: US Census Bureau: Census 1980, American Community Survey 2015, analysis of IPUMS-USA microdata.

### College makes a difference amongst Millennials

In the grim housing market of 2015, existing economic disparities between more and less advantaged young people were heightened. As shown in Table 11.1, there were bigger gaps in homeownership and housing affordability between Millennials with and without college degrees than there had been for Boomers. The gap between college graduates' and non-graduates' homeownership rates was only 4 percentage points for Boomers, but the gap widened considerably to 11 percentage points for Millennials. Rental affordability problems also disproportionately affected Millennials without college degrees. Over a quarter of Millennials without college degrees paid over half of their income for rent, while only 12% of Millennial college graduates paid over half of their income for rent. Millennials had a 14 percentage point gap between the proportion of college grads and non-graduates who paid over half their income for rent, while Boomers had only a 7 percent gap. Even Millennial college graduates had a harder time than Boomer college graduates in paying rent and buying homes, and Millennials without college degrees were in an even more difficult position. This disadvantage starts with their lower labour market prospects and earnings, as noted above and in Chapter 4, but is amplified by the difficulties they face in the housing market.

### The tough road ahead

The comparison with Boomers throws Millennials' housing market conditions into sharp relief. Millennials were worse off than Boomers on most of the dimensions examined in this chapter. To start with, Millennials had lower employment and earnings than Boomers, and Millennials also had weaker earnings relative to older age groups. Dramatically lower construction and turnover meant that a smaller pool of housing units were available on the market for Millennials compared with Boomers. Millennials faced higher barriers to home buying, and steeper competition for rental housing. In this environment, Millennial household formation, home buying, and rental affordability suffered.

Though Millennials all came of age in the same overheated housing market, some fared much better than others. Millennial college graduates got off relatively lightly, with somewhat lower household formation, home buying, and rental affordability than Boomer college graduates. But Millennials without college degrees fell far behind Millennial college graduates and Boomers without college degrees alike. In a competitive, expensive housing market, differences in employment and income can translate into much wider differences in housing outcomes.

This research highlights the importance of addressing the housing shortfall for an entire generation of Millennials, who are stuck with reduced housing opportunities compared with their parents' generation, especially those starting out with fewer resources. As Clapham (2005) emphasizes, housing careers are path-dependent, and often intertwined with decisions about marriage, fertility, and employment (Clark et al., 2003; Malmberg, 2012). That is to say that an adverse start in the housing market can have long-lasting consequences in many areas of a person's life. This research also brings up questions of equity in the provision of housing – both between generations and between more and less advantaged groups. Housing has often been the vehicle through which both wealth and disadvantage have been handed down from generation to generation (Briggs, 2005; Sharkey, 2013). And it seems that the housing market offers Millennials less wealth, more disadvantage, and more inequality than previous generations.

How can Millennials reclaim housing for a more affordable and equitable future? This chapter points to the need to build more housing. Development decisions are generally made at the local level, fought out project by project. Millennials can start there to encourage sorely-needed housing construction. Young people, especially renters, currently have relatively little influence in local political processes. Young people are less likely to vote, tend to make frequent moves, are often at a demanding and time-intensive phase of their lives when it is difficult to participate in local meetings, and may have fewer financial resources to support local politicians or advocacy groups. Boomers have more political clout, yet on the face of it, their immediate interests of older generations are often aligned with the status quo rather than growth and change. In order to build enough housing to accommodate young people, it will be crucial to build alliances between Boomers and Millennials. Millennials have a strong claim for Boomers' support: Boomers enjoyed the benefits of ample housing opportunities when they were young, and now it is Millennials' turn.

# References

Briggs, X. de S. (2005). *The geography of opportunity: Race and housing choice in metro-politan America.* Washington, DC: Brookings Institution Press.

Clapham, D. (2005). *The meaning of housing: A pathways approach.* Bristol: The Policy Press.

Clark, W. A. V., Deurloo, M. C., & Dieleman, F. M. (2003). Housing careers in the United States, 1968–93: Modelling the sequencing of housing states. *Urban Studies, 40*(1), 143–160. Retrieved from https://doi.org/10.1080/00420980220080211.

Courchane, M. J., Kiefer, L. C., & Zorn, P. M. (2015). A tale of two tensions: Balancing access to credit and credit risk in mortgage underwriting. *Real Estate Economics, 43*(4), 993–1034. Retrieved from https://doi.org/10.1111/1540-6229.12105.

Elsby, M. W., Shin, D., & Solon, G. (2013). *Wage adjustment in the great recession.* Working Paper No. 19478, National Bureau of Economic Research, Cambridge, MA. Retrieved from www.nber.org/papers/w19478.

Fry, R. (2014). *Young adults, student debt and economic well-being.* Social and Demographic Trends Project, Pew Research Center, Washington, DC. Retrieved from www.pewso cialtrends.org/2014/05/14/section-1-student-debt-and-overall-economic-well-being/.

Fry, R. (2015). *More Millennials living with family despite improved job market.* Washington, DC: Pew Research Center. Retrieved from www.pewsocialtrends.org/2015/07/29/more-millennials-living-with-family-despite-improved-job-market/.

Goodman, L., Pendall, R., & Zhu, J. (2015). *Headship and homeownership: What does the future hold?* Washington, DC: The Urban Institute. Retrieved from www.urban.org/research/publication/headship-and-homeownership-what-does-future-hold.

Malmberg, B. (2012). Fertility cycles, age structure and housing demand. *Scottish Journal of Political Economy, 59*(5), 467–482. Retrieved from https://doi.org/10.1111/j.1467-9485.2012.00590.x.

Myers, D. (1983). Upward mobility and the filtering process. *Journal of Planning Education and Research, 2*(2), 101–112. Retrieved from https://doi.org/10.1177/0739456X8300200206.

Myers, D. (1992). *Analysis with local census data: Portraits of change.* San Diego, CA: Academic Press.

Myers, D. (2016). Peak Millennials: Three reinforcing cycles that amplify the rise and fall of urban concentration by Millennials. *Housing Policy Debate, 26*(6), 928–947. Retrieved from https://doi.org/10.1080/10511482.2016.1165722

Myers, D., Painter, G., Lee, H., & Park, J. (2016). *Diverted homeowners, the rental crisis and foregone household formation.* Washington, DC: Research Institute for Housing America.

Pew Research Center. (2014). *The rising cost of not going to college.* Social and Demographic Trends Project, Washington, DC. Retrieved from www.pewsocialtrends.org/2014/02/11/the-rising-cost-of-not-going-to-college/.

Poethig, E.C., Schilling, J., Goodman, L., Bai, B., Gastner, J., Pendall, R., & Fazili, S. (2017). *The Detroit housing market: Challenges and innovations for a path forward.* Washington, DC: Urban Institute.

Ruggles, S., Genadek, K., Goeken, R., Grover, J., & Sobek, M. (2015). *Integrated public use microdata series: Version 6.0 [Machine-readable database].* Minneapolis, MN: University of Minnesota. Retrieved from https://usa.ipums.org/usa.

Sharkey, P. (2013). *Stuck in place: Urban neighborhoods and the end of progress toward racial equality.* Chicago: University of Chicago Press.

Stiglitz, J. E. (2013, May 12). *Student debt and the crushing of the American dream.* Retrieved from http://opinionator.blogs.nytimes.com/2013/05/12/student-debt-and-the-crushing-of-the-american-dream/.

U.S. Bureau of Labor Statistics. (2016a). *Consumer price index for all urban consumers: Rent of primary residence [Data file].* Retrieved from FRED, Federal Reserve Bank of St. Louis; https://fred.stlouisfed.org/series/CUUR0000SEHA.

U.S. Bureau of Labor Statistics. (2016b). *Current population survey annual social and economic supplement [Data file].* Retrieved from www.bls.gov/cps/earnings.htm#education.

US Census Bureau. (1970). *Decennial census of population and housing 1970 [Data file].* Retrieved from www.socialexplorer.com.

US Census Bureau. (1980). *Decennial census of population and housing 1980 [Data file].* Retrieved from www.socialexplorer.com.

U.S. Census Bureau. (2015). *American community survey 2015 [Data file].* Retrieved from www.socialexplorer.com.

U.S. Census Bureau. (2016a). *Current population survey, series H-111/Housing vacancy survey [Data file].* Retrieved from www.census.gov/housing/hvs/index.html.

U.S. Census Bureau. (2016b). *Housing inventory estimate: Total housing units for the United States [Data file].* Retrieved from FRED, Federal Reserve Bank of St. Louis; https://fred.stlouisfed.org/series/ETOTALUSQ176N.

U.S. Census Bureau. (2016c). *Survey of new residential construction [Data file].* Retrieved from FRED, Federal Reserve Bank of St. Louis; https://fred.stlouisfed.org/series/HOUST.

U.S. Census Bureau Population Estimates Program. (2004). *National intercensal datasets: 1990–2000 [Data file].* Retrieved from www.census.gov/data/datasets/time-series/demo/popest/intercensal-1990-2000-national.html.

US Census Bureau Population Estimates Program. (2012). *National intercensal datasets: 2000–2010 [Data file].* Retrieved from www.census.gov/data/datasets/time-series/demo/popest/intercensal-2000-2010-national.html.

US Census Bureau Population Estimates Program. (2016). *National population by characteristics datasets: 2010–2016 [Data file].* Retrieved from www.census.gov/programs-surveys/popest/data/data-sets.All.html.

Xu, Y., Johnson, C., Bartholomae, S., O'Neill, B., & Gutter, M. S. (2015). Homeownership among Millennials: The deferred American dream? *Family and Consumer Sciences Research Journal, 44*(2), 201–212. Retrieved from https://doi.org/10.1111/fcsr.12136.

# 12 Beyond "peak Millennial"

## Developing an index of generational congestion for local government use in the United States and Canada

*Jeff Henry and Markus Moos*

The Millennial generation has been reshaping cities in the USA and Canada much as their Boomer parents did before them. As Millennials enter and progress through young adulthood, their challenges in securing housing in an increasingly expensive housing market have drawn the attention of researchers and the concern of planners and policymakers. If previous generations are a guide, how these challenges manifest themselves, and are resolved, will impact Millennials' future life course and the shape of our cities for generations to come (Chapter 2).

Previous studies has explored the relationship among the changing size of the young adult cohort, household formation and progression into home ownership at a national level in the USA (Myers, 2015, 2016). This prior research suggests that, since the size of the Millennial cohort has reached its apex or "peak millennial" (Myers, 2015), the pressure on the housing market from young adults will begin to ease.

Unfortunately, national-level data can obscure local variations. While broad economic factors and national government policy have important effects on housing markets, local variations in policy and context can have differential effects (see for example Moore & Skaburskis, 2004; Quigley & Rosenthal, 2005). Moreover, Canada did not experience the same housing market collapse in 2008 as the USA, and the rise and fall of its Millennial population is different due to different demographic trajectories.

This chapter builds on the valuable work conducted by Myers. It begins by comparing the shape of the rise and fall in births between Canada and the USA and the changes in the number of young adults at the door of the housing market. Then, it makes a novel contribution to the literature in housing studies, demography, and planning through the creation of a new quantitative tool to express as an indicator what Myers (2015, 2016) recently has coined the "generational congestion" of young adults. We call this the index of generational congestion.

The objective of providing this new index of generational congestion is both academic and pragmatic. We aim to support local planners and policymakers in assessing the flows of cohorts in and out of their housing markets, and identify resulting housing policy responses. Although more sophisticated tools and datasets are available, our methodological choices keep these specific end users in mind. We use publicly available data and aim to minimize complexity in calculations to create an index that could easily be replicated by local governments and planning departments with limited resources and time.

We begin the chapter with a contextual overview, focusing on the cohort approach to population projections, as well as the affordability constraints faced by young adults. We then replicate the methods carried out by Myers for Canada (and the USA as originally done by Myers), and illustrate necessary modifications to derive the index of generational congestion at a local level.

## Forecasting demographic change using cohorts

Arguing that demography can explain "two-thirds of everything" past and future, Foot and Stoffman (1998, p. 8) discussed the cohort size effect on housing demand in their seminal work, *Boom, Bust & Echo*. The ability to forecast using demographic information is based on the premise that once each person is born, they will be a year older next year than they are this year. Subject to fertility, mortality, and migration in and out of specific geographic areas, the size of the cohort and its movement through a series of ages from birth to death is eminently predictable, and has had substantial impacts on cities.

The large size of the post-war Baby Boom cohort and the much smaller size of the Baby Bust cohort (who followed the Boomers) has an impact on the growth and subsequent crash of the downtown apartment market in the 1970s and 1990s (Myers & Pitkin, 2009). It also shaped the size, and ease, of entry into the labour force, as well as the rise and collapse of the commercial real estate market, over roughly the same time period (Foot & Stoffman, 1998; Myers & Pitkin, 2009).

Millennials are the largest cohort since the Baby Boomers. As this current generation of young adults enters the housing market, demographic forecasters expect Boomers to begin to exit homeownership, predicting a generational housing bubble from these state-by-state changes in aggregate demand (Myers & Ryu, 2008). However, since the Great Recession of 2008, Millennials trying to enter the US housing market are finding themselves stalled, with lower rates of household formation and persistently higher rates of rental tenure (Myers, 2016).

There are cautionary tales for researchers forecasting the current and future experience of Millennials as they enter and move through the housing market. Mankiw and Weil (1989) famously forecasted that the smaller Baby Bust generation arriving at their housebuying years in the 1990s would generate substantially lower housing demand, depressing real housing prices through to 2007. This did not happen. Missing from the straight line cohort projection-based forecast were changes in real income, relative prices to rent and own, and real interest rates that led to sustained and increasing housing prices (Swan, 1995). Analysts and forecasters, then, must understand cohort size effects but should not be limited by them.

## Housing, affordability, and young adults

Researchers of young adults typically focus on those aged between 25 and 34, as this is a time when transitions occur in housing, employment, and family careers

(Moos, 2015). Over the last 30 years in the USA, home ownership rates have risen sharply until age 34, at which point increases begin to level off (Myers & Lee, 2016). Age 25 is a convenient point to measure educational attainment, as a standard path through a bachelor's programme would be complete. This age cohort is also highly mobile in the early 20s. As migration levels decrease sharply with age, the location decisions of young adults can have lasting impacts on cities (Cortright, 2014; Chapter 2).

Housing affordability is viewed as the outcome of class and neoliberal housing policies, which reduce the degree of government intervention in the market (Filion & Kramer, 2011). Affordability is correlated with household income, household size, the presence of children, the type of tenure, the price of housing (and rent), and city size (Moore & Skaburskis, 2004). Researchers who have studied equitable access to housing in gentrifying cities – where working class neighbourhoods are being upgraded, leading to displacement of long-term residents – have recently also added a generational lens, observing an increasing concentration of young adults renting for longer periods in downtowns (Moos, 2015).

The traditional housing lifecycle assumes young adults move from the parental home into rental accommodations and then home ownership, predictably building their nuclear family along the way (Mumford, 1949). However, in the post-modern city, the normative expectations of the housing life cycle have fractured into multiple housing pathways with young adults living with parents, sharing accommodations, and living in couples without children for extended periods of time (Clapham, Mackie, Orford, Thomas, & Buckley, 2014; Lee & Painter, 2013; Moos, 2014).

Housing pathways at the later stages of life also affect housing markets, particularly as households transition from owning to renting or as they dissolve due to mortality or moving into care facilities. Myers and Ryu (2008) forecasted a generational housing bubble when the number of Boomers selling their homes outnumber new buyers, which would have implications for planners, policymakers, and young homebuyers. US data on household sales by age cohort suggested sales picked up for those aged 70 and older, and particularly for those aged 80 and older. In forecasting net household formation, Canada Mortgage and Housing Corporation (CMHC) assumes net household formation is driven primarily by young adults (aged 20 to 29) less those aged 75 and above dissolving households (Gabay, 2013). These older adult housing pathways highlight the role of outflow, in addition to inflow for assessing housing markets and understanding the range of potential effects on young adults attempting to enter the housing market.

## Peak millennial in Canada and the United States

Myers' (2016) exploration of the rise and fall of registered births over time in the USA identified 1990 as the highest point since the Baby Boom, which he labelled as "peak millennial". While there are various definitions for Millennials, Myers (2016) defines Millennials in the USA as those born between 1980 and 1999, which is statistically convenient given the timing of the decennial census. Canadian Millennials are defined here as those born between 1981 and 2000, which is statistically convenient given the timing of the Canadian census.

Registered births in the USA had sharply declined from a high of 4,268,326 in 1961 to a low of 3,136,965 in 1973. Births then rose steadily, hitting a peak of 4,158,212 in 1990 (to peak millennial). Births edged off slightly over the next decade, hitting a low of 3,880,894 in 1997 before bouncing back ("National Vital Statistics Reports, Volume 64, Number 1", 2015).

The Canadian picture, however, looks substantially different. Births remained relatively steady throughout the 1970s and 1980s, ranging between a low of about 342,000 in 1973–1974 to high of about 376,000 in 1984–1985. Similar to the USA, births peaked in 1990 at a level sharply higher than in the 1970s and 1980s at around 403,000. What is drastically different in Canada, however, is what happens after *peak millennial*. Unlike in the USA, registered births in Canada fell sharply through the 1990s to a low of about 327,000 in 2000–2001 (Statistics Canada, 2015b).

Births affect housing markets by eventually changing the aggregate demand of new entrants. As noted previously, researchers have focused on the 25–34 age bracket as the time when people enter the housing market (Myers & Lee, 2016). The change in the size of this age bracket is an echo of the rise and fall of births 25 years prior, less mortality and augmented by net immigration.

Generational congestion is a description of these changes in demand in the housing market from the changing size of cohorts. When measured at the entry point for young adults, it can be described as either rising or falling based on changes in the size of the 25–34 cohort. Precisely how this congestion should be defined, measured, and interpreted is described later in this chapter.

Based on negligible mortality rates and relatively flat to declining immigration in the USA, Myers (2016) projected the annual change of the 25–29 age cohort over time excluding mortality and net migration. In the 1990s, the size of this critical age group declined by approximately 2% per year, while the entry of the Millennial generation into this age group resulted in increases peaking at just under 3% in 2015, 25 years after registered births also peaked in 1990 (calculated from "National Vital Statistics Reports, Volume 64, Number 1", 2015). While the 25–29 age cohort will continue to grow through 2018, the rate of growth will slow, which Myers has suggested could end (or at least slow) the urban renaissance in downtown cores (Delgadillo, 2016).

The Canadian data shows some drastic differences as compared to the USA. Excluding mortality and net immigration effects, as Myers did, the period from the late 1990s through the mid-2000s saw modest growth, rather than decline, in the 25–29 age cohort at around 0.7%. While this growth rate approximately doubles with the arrival of the Millennials, it then falls sharply as the generation ages. The turning point is 2015 when the annual change becomes negative and ultimately declines at a rate of about 2.5% per year between 2017–2018 and 2021–2022 (calculated from Statistics Canada, 2015b).

While both countries reach "peak millennial" in 2015, the overall shape of annual registered births are almost a mirror image – a valley followed by a peak and a relative plateau in the USA, and a relative plateau followed by a peak and a deep valley in Canada. The Canadian urban renaissance, if it indeed is attributable to sustained growth in the size of the young adult cohort, may then come to a stunning halt. But is it reasonable to ignore the effects of net immigration on the overall trend in Canada as Myers did in the USA? We explore this question next.

## Assessing net immigration in Canada

Based on Canadian census data, the size of the non-immigrant population aged 25–29 drops (−334,812) between 2001 and 2006, in sharp contrast with the modest growth based on a straight projection of registered births described above. In this same period, there was an increase in the size of the immigrant population aged 25–29 (+199,115), which moderated the decline in the non-immigrant population of the same age (calculated from Statistics Canada, 1991, 1996, 2001, 2006, 2011). In the other periods, the change in immigrants and non-immigrants was in the same direction. This suggests emigration, not just immigration, is an important factor when assessing changes in the number of young adults in Canada and declaring "peak millennial".

The general trend line of immigrants as a percentage of those aged 25–29 has also been increasing in Canada, from 13.9% in 1991 to 18.5% in 2011. However, the importance of this varies across the country. For example, by 2011 the proportion of immigrants among those aged 25–29 reached 42% in Toronto and 34% in Vancouver while the share in Quebec City was only approaching 6% (calculated from Statistics Canada, 1991, 1996, 2001, 2006, 2011).

Data on immigration and the variation in non-immigrant populations from registered birth projections suggest the limitation of such projections in assessing trends at the national and local level in Canada. This may also be true at the local level in the USA since growth patterns vary dramatically by community across the nation there too. This is important because if there are indeed generational congestion effects, they would play out at the local level where housing market dynamics are most evidently affected by in and out migration, and births and deaths. It is critical, then, to quantify changes in the numbers of young adults in a way that accounts for net migration in local contexts.

## An "index of generational congestion"

An old adage in real estate is that three things matter most: location, location, location. In assessing generational congestion in housing markets, the national picture obscures important variations across the country. Migration matters both in terms of the settlement patterns of immigrants and in terms of domestic movers, whose numbers include a substantial share of young adults (Cortright, 2014). The proportion of different age cohorts in a local population also varies, particularly in areas of declining populations or retirement communities.

Local politicians, planners and policymakers require a measurement of generational congestion to assess the impacts of generational change at the level of the local housing market. Statistically, the regional level for housing markets best fit the delineations of metropolitan areas: metropolitan statistical areas (MSAs) in the USA, and census metropolitan areas (CMAs) in Canada.

CMAs and MSAs are defined as areas with a high degree of social and economic integration based on commuting patterns of municipalities and counties respectively (Statistics Canada, 2015a; US Census Bureau, n.d.). Using MSAs and CMAs also ensures socio-economic data is available over the same time period and geography, allowing researchers and local planners to conduct multivariate analysis to test the relative impact of generational congestion against other socio-economic factors that impact housing markets.

## *Methodology*

In his analysis, Myers (2016) focused on the rise and fall in the number of young adults entering the housing market at the national level. Locally, this rise and fall is shaped not just by young adults, but also by migration and mortality. Any added pressure from net new young adults in a market may be eased by older adults exiting the market.

Our generational congestion index models these factors through both an "entry" and "exit" component at the local housing market level. These factors are normalized to changes in people and households in the whole market to ensure the model isolates these generational entry and exit effects. Calculations for both the entry and exit components are provided in Table 12.1.

On the entry side (entry congestion), the change in young adults (aged 25–34) is normalized against the total population potentially seeking housing. This allows planners and policymakers to understand entry congestion in the broader context of all population changes in their metropolitan area. Data collected would include the effects of net migration and so do not need to be separately obtained or calculated. Population rather than the number of households is used to avoid undercounting young adults living at home or with roommates due to affordability concerns.

*Table 12.1* Generational congestion calculations

|  | *Population description* | *Generational congestion index calculation* |
|---|---|---|
| **Housing market entry** | Relative change of young adults (25–34) to all adults (15+) | $\left(\dfrac{P_t^{25-34}}{P_t^{15+}}\right) \Big/ \left(\dfrac{P_{t-5}^{25-34}}{P_{t-5}^{15+}}\right)$ |
| **Housing market exit** | Relative change in older adult-led household dissolutions (75+) to all households (15+) | $\left(\dfrac{HH_{t-5}^{70+} - HH_t^{75+}}{HH_t^{15+}}\right) \Big/ \left(\dfrac{HH_{t-10}^{70+} - HH_{t-5}^{75+}}{HH_{t-5}^{15+}}\right)$ |
| **Combined** | Relative change in housing demand from ratio of entry (25–34) and exits (75+) | Housing market entry/Housing market exit |

*where*
$P_t^c$ *is the population of age group c at time* t
$HH_t^c$ *is the number households headed by age group c at time* t

The construction of the measurement as a ratio quickly shows a rate of change, allowing comparisons of different sized regions as well as changes in rates of change in other factors as part of a multivariate analysis.

When the entry component of generational congestion in a metropolitan area is greater than one, the size of the young adult population has increased compared with five years prior relative to the size of the adult population as a whole. This indicates an increase in the entry component of generational congestion. When the result is less than one, the size has decreased, indicating a decrease in congestion.

On the exit side (exit decongestion), the model focuses on household dissolutions, which consist primarily of mortality or moving into care facilities. Calculating household dissolutions is similar to using the cohort-component projection method (Klosterman, 1990), except the data points are in the present and the past rather than the future. Dissolutions are measured for those aged 75 and older (Gabay, 2013).

The number of household dissolutions among older adults over a five-year period is calculated as the difference between the number of households aged 70 and older five years ago and the number of households aged 75 and older today. This is not precisely a household dissolution measurement as it includes net migration, likely from retirement moves, and the net effects of couples forming and dissolving, which fully captures the effect on housing demand. This dissolution rate is normalized against all households.

Myers' (2016) research on generational congestion used a five year rolling average of change in the size of the young adult population. A ratio where the numerator and denominator are five years apart is used in the index for both the entry and exit sides. This interval should provide enough time to recognize changes over a short enough period to be useful for local policymakers and planners.

Interpreting the exit component is the inverse of the entry component in terms of its effect on congestion. A rate greater than one represents an increased rate of household dissolutions, which creates decongestion in the metropolitan area. A rate less than one indicates a slowing rate of household dissolutions, which increases generational congestion.

Relating the two components to create a complete index of generational congestion can be obtained by dividing the entry component by the exit component. Similar to the entry component, a result greater than one means an increase in generational congestion while a result less than one means a decrease in generational congestion.

The most important data availability differences between Canada and the USA relate to the timeframe. The census in Canada occurs every five years, while the US census occurs every ten years. The US Census Bureau also releases annual data from the American Community Survey (ACS) beginning in 2005 for areas larger than 65,000 people and for all areas since 2010. This data is continuously collected, rather than at one particular time in the year, and provides a statistically valid sample for regions larger than 65,000 annually and for all areas every five years. As the sample is rolling, updated oen-year estimates for regions larger than

65,000 and five-year estimates for all regions are released every year (US Census Bureau, 2008).

Data is publicly available for all components, except that households aged 70+ is not publicly available in the USA from the ACS. Available data sources for Canada include the Census for population and household heads (including 15+, 65+, 70+, and 75+) by CMA every five years, except 2011 where this data is available instead from the National Household Survey. Unfortunately the 2011 NHS, as a voluntary survey, may be biased against lower income and minority respondents (see for example Green & Milligan, 2010). Available data sources for the USA include the decennial Census for population and household heads (including 15+, 65+, and 75+) by MSA, as well as the ACS (1-Yr) since 2005.

We now proceed to review how these components and the complete index change over time and how this varies across metropolitan areas.

## What does the index show?

The data presented here on generational congestion is focused on metropolitan areas that were larger than 500,000 people in 2015. Smaller communities are excluded since housing market and demographic trends play out quite differently in smaller metropolitan areas. Selecting metropolitan areas of at least 500,000 people provides broader insights into mid-sized areas that are not as well covered in urban and planning research (Filion, Bunting, Frenette, Curry, & Mattice, 2000). This allows those local planners and policymakers to study and observe how generational congestion applies in their context rather than debate whether research from the largest, more frequently studied, metropolitan areas apply to them.

Generational congestion ratios for these Canadian census metropolitan areas (CMAs), including both entry and exit components, are provided in Table 12.2. Given the large number of metropolitan statistical areas (MSAs) reviewed in the USA, the distribution of entry congestion is instead provided as a box plot in Figure 12.1.

Nationally, entry or young adult generational congestion, relative to the changes in the adult population, has only just appeared in 2011 in Canada. In the USA, it was present only in 2012, 2013, and 2015. However, entry congestion has generally been increasing over the selected periods in both Canada and the USA, crossing recently from easing (below 1.0) to growing (above 1.0). These results have interesting distributions at the metropolitan level, which we will explore next.

### Generational congestion in Canada

Looking at the Canadian variations in aggregate, the median entry congestion for CMAs and the national average are similar, rising substantially between 2001 and 2011. However, changes in decongestion, which increased sharply overall, is widely varied among these CMAs and differs from the national average. The range of combined generational congestion indices among these CMAs is modestly tighter than their decongestion indices, with the median continuing to vary from the national average.

Overall, the combined generational congestion index shows that the increasing rate of dissolutions is outpacing the increasing relative rate of young adults nationally and in most metropolitan areas. Entry congestion was highest in 2011 in Edmonton (1.095), followed by Quebec City (1.031), Vancouver (1.030), Calgary (1.028), and Winnipeg (1.025), with other areas experiencing an easing in entry congestion (between 0.973–0.996).

Exit congestion from dissolutions jumped between 2006 and 2011 in most metropolitan areas, with the sharpest acceleration in Edmonton (0.743 to 2.220) and Calgary (0.590 to 1.987). Kitchener-Cambridge-Waterloo, however, saw declining dissolutions (0.868 and 0.873). Variations from the trend in the combined generational congestion index occurred in Edmonton and Calgary in 2006, which returned to the general trend of accelerating rates of dissolutions in 2011. Kitchener-Cambridge-Waterloo stands alone, with entry congestion outpacing exit decongestion in 2006 and 2011.

Explaining these variations, young labour was likely attracted to Edmonton and Calgary's booming oil-and-gas economy, while Vancouver's rising concentration of young adults has been connected with rising costs, changing economies, and preferences for urban living (Moos, 2015). Winnipeg's young adult population was still 16% below 1991 levels, so its recovery has only been partial and recent.

The combination of older adult migration into jurisdictions and an aging demographic can quickly grow the number of individuals who reach a point in their

*Table 12.2* Generational congestion in medium and large canadian CMAs

| | Entry congestion | | | Exit decongestion | | | Combined index | | |
|---|---|---|---|---|---|---|---|---|---|
| | 2001 | 2006 | 2011 | 2001 | 2006 | 2011 | 2001 | 2006 | 2011 |
| **Canada** | 0.839 | 0.935 | 1.010 | 0.909 | 1.065 | 1.204 | 0.923 | 0.878 | 0.839 |
| **Median – Below CMAs** | 0.850 | 0.937 | 0.996 | 0.954 | 0.948 | 1.260 | 0.896 | 1.004 | 0.817 |
| Quebec | 0.825 | 1.025 | 1.031 | 0.949 | 0.880 | 1.626 | 0.869 | 1.166 | 0.634 |
| Montreal | 0.849 | 0.981 | 0.993 | 0.954 | 0.948 | 1.490 | 0.890 | 1.035 | 0.667 |
| Ottawa-Gatineau | 0.843 | 0.897 | 0.996 | 0.842 | 1.042 | 1.056 | 1.002 | 0.860 | 0.943 |
| Toronto | 0.864 | 0.907 | 0.973 | 1.108 | 0.994 | 1.108 | 0.975 | 0.912 | 0.878 |
| Hamilton | 0.850 | 0.897 | 0.985 | 1.161 | 0.894 | 1.198 | 0.732 | 1.004 | 0.822 |
| Kitchener-Cambridge-Waterloo | 0.864 | 0.932 | 0.979 | 1.969 | 0.868 | 0.873 | 0.439 | 1.074 | 1.122 |
| London | 0.820 | 0.937 | 0.991 | 0.916 | 1.171 | 1.559 | 0.896 | 0.800 | 0.635 |
| Winnipeg | 0.850 | 0.948 | 1.025 | 0.787 | 1.127 | 1.177 | 1.079 | 0.841 | 0.871 |
| Calgary | 0.896 | 0.966 | 1.028 | 1.092 | 0.590 | 1.987 | 0.820 | 1.637 | 0.518 |
| Edmonton | 0.860 | 0.986 | 1.095 | 0.956 | 0.743 | 2.220 | 0.900 | 1.327 | 0.493 |
| Vancouver | 0.851 | 0.900 | 1.030 | 0.880 | 0.965 | 1.260 | 0.968 | 0.933 | 0.817 |

Notes: CMAs have experienced modest geographic changes over time, usually small additions in area. Historical adjustments have not been made in this table. Geographic names have also changed, with the 2011 name used in this table.

lives where they dissolve their households. CMHC had also forecast a decline in Boomer home ownership between 2011 and 2016 after peaking at some point between 2006 to 2011 (Gabay, 2013). Edmonton and Calgary, both in Alberta, are lower tax jurisdictions, which may have attracted retirees. One theory to explain the variation in Kitchener-Cambridge-Waterloo could be due to the approximately 90,000 students in its three post-secondary institutions, a much higher relative population than other mid- and large-sized Canadian metropolitan areas. Many of these young adults are counted elsewhere at their parental home by the census (Parkin & Martin, 2012) but who may subsequently settle and be counted in the area after graduation, increasing the relative number of young adults.

### Generational congestion in the United States

Looking at these variations in the USA in aggregate, the average and median congestion for the 107 MSAs and the national average are similar, rising almost continuously between 2010 and 2015. The variation between most MSAs has been consistently small, as the size of the interquartile range has remained stable, between 0.08 and 0.10. There is still a wide variation among a small number of

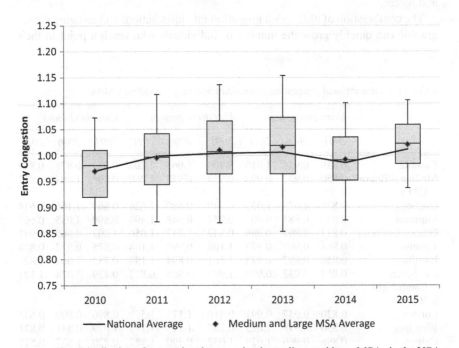

*Figure 12.1* Distribution of generational congestion in medium and large MSAs in the USA

Notes: Calculations use ACS 1-Yr data from 2005 to 2015 as described in Table 12.1. Geographic boundaries and names of MSAs have changed several times over the period presented above, including absorbing and discarding counties and micropolitan statistical areas. Adjustments have been made to reflect 2015 boundaries to the extent possible, as ACS 1-Yr data is valid only where a county is larger than 65,000 persons.

MSAs, though this has tightened, as the 5th to 95th percentile range grew to 0.30 by 2013 before falling again to 0.17 in 2015.

Entry congestion was most pronounced in New Orleans (peaking at 1.306 in 2012) followed by Milwaukee (peaking at 1.185 in 2013), San Francisco (peaking at 1.184 in 2013), and Pittsburgh (peaking at 1.166 in 2013). New Orleans, in recovering most of its 2005 pre-Hurricane Katrina population by 2014, attracted young adults faster than the whole adult-aged population. A growing technology industry in San Francisco relies on young adult workers, but understanding Milwaukee and Pittsburgh's relative gains requires further analysis to offer a reasonable hypothesis.

The largest easing of entry congestion was in North Port-Sarasota-Brandenton, Florida (dropping to 0.794 in 2011 and 0.796 in 2013) followed by Cape Coral-Fort Myers, Florida (dropping to 0.806 in 2013), Deltona-Daytona Beach-Ormond Beach, Florida (dropping to 0.830 in 2011), and Riverside-San Bernardino-Ontario, California (dropping to 0.830 in 2011). Most of the communities with the greatest easing in entry congestion are in Florida and have seen both a growing older adult population and a stagnant young adult population, signalling that they are growing retirement communities. However, all four had seen reversals by 2015 and are now experiencing entry congestion. These are caused by a recent spike in young adults, suggesting these communities are attracting more than just retirees.

The wide variation among exit congestion indices in the Canadian context suggest that the decongestion data in the USA, as it becomes available in future years through the ACS, would add further local insights and potentially affect policy recommendations. The national level is insufficient for housing market analysis since these markets are so locally specific.

## Population growth versus generational congestion

While there has been a rise and a peaking in the size of the young adult cohorts seeking to enter the housing market, some metropolitan areas' adult populations and households have continued to grow. This suggests population growth rather than generational congestion is the more important factor for local planners and policymakers to consider.

The increasing rate of household dissolutions amongst older adults in Canada is already outpacing the arrival of young adults. Should the number of young adults arriving begin to decline rapidly without offsetting net migration (net immigration nationally and a combination of net immigration and net domestic migration locally), congestion may ease substantially in the years ahead.

Attracting migrants, if it is not already a policy objective and economic development strategy of metropolitan areas, could become critical in avoiding what Myers and Pitkin (2009) termed the generational housing bubble. The early moves of President Trump in the USA suggest fewer available immigrants, leaving metropolitan areas pursuing migrant attraction to increasingly compete domestically rather than internationally for internal migrants. Subject to consistent policies and economic conditions, forecasting future changes in congestion is possible using standard cohort component projection methods, as discussed in Klosterman (1990). Limiting projections to the near term would minimize the underlying volatility in forecasting net migration, allowing the relatively uncomplicated linear extrapolation approach, which projects the past trend into the future on a straight line, to be reasonably accurate (see description in Klosterman, 1990, pp. 9–16).

## The geographer's mantra: local context matters

Peak millennial arrived in both Canada and the USA around 2015; however, the way the number of young adults grow and decline over time in the two countries are mirror opposites of each other. That shape of the Millennial cohort is also complicated in Canada by substantial variations in immigration and emigration over time and space.

In both countries, location and local context matter. The index of generational congestion shows wide variation across mid-sized and large metropolitan areas in the USA and Canada. Inclusive of net migration, the index shows that outflows may already be outpacing inflows in some Canadian housing markets, indicating young adults could be the wrong generational change on which to focus. The increasing outflow of older adults may finally bring the challenges to local housing markets that have been prematurely forecast for years.

Planners and policymakers in metropolitan areas can use this new index to assess whether their housing policies are generationally aligned with their present and future demands. Notwithstanding an easing of congestion in a local housing market, variations in the tenure, type, size, and condition of housing demanded can result in a housing market where demand and supply are mismatched. In local housing markets where the generational congestion index shows an easing but housing affordability remains a challenge, further research is required and this research would benefit from the local insights of planners and policy makers.

## References

Clapham, D., Mackie, P., Orford, S., Thomas, I., & Buckley, K. (2014). The housing pathways of young people in the UK. *Environment and Planning A, 46*(8), 2016–2031. Retrieved from https://doi.org/10.1068/a46273.

Cortright, J. (2014). *The young and restless and the nation's cities.* City report, City Observatory. Retrieved from http://cityobservatory.org/wp-content/uploads/2014/10/YNR-Report-Final.pdf.

Delgadillo, N. (2016, March 16). Could US cities soon reach "Peak Millennial," and start to experience population decline? – CityLab. *CityLab.* Retrieved from www.citylab.com/housing/2016/03/keeping-millennials-in-cities-dowell-myers/473061/.

Filion, P., Bunting, T., Frenette, S., Curry, D., & Mattice, R. (2000). Housing strategies for downtown revitalization in mid-size cities: A city of Kitchener feasibility study. *Canadian Journal of Urban Research, 9*(2), 145.

Filion, P., & Kramer, A. (2011). Metropolitan-scale planning in neo-liberal times: Financial and political obstacles to urban form transition. *Space and Polity, 15*(3), 197–212. Retrieved from https://doi.org/10.1080/13562576.2011.692567.

Foot, D. K., & Stoffman, D. (1998). *Boom, Bust & Echo 2000: Profiting from the demographic shift in the new millennium* (Rev. ed.). Macfarlane Walter & Ross.

Gabay, R. (2013). *Long-term household growth projections – 2013 update*. Research Highlight No. 13–006. Canada Mortgage and Housing Corporation, Ottawa.

Green, D. A., & Milligan, K. (2010). The importance of the long form census to Canada. *Canadian Public Policy, 36*(3), 383–388. Retrieved from https://doi.org/10.1353/cpp.2010.0001.

Klosterman, R. E. (1990). *Community analysis and planning techniques*. Savage, MD: Rowman & Littlefield Publishers, Inc.

Lee, K. O., & Painter, G. (2013). What happens to household formation in a recession? *Journal of Urban Economics, 76*(Complete), 93–109. Retrieved from https://doi.org/10.1016/j.jue.2013.03.004.

Mankiw, N. G., & Weil, D. N. (1989). The baby boom, the baby bust, and the housing market. *Regional Science and Urban Economics, 19*(2), 235–258. Retrieved from https://doi.org/10.1016/0166-0462(89)90005-7.

Moore, E., & Skaburskis, A. (2004). Canada's increasing housing affordability burdens. *Housing Studies, 19*(3), 395–413. Retrieved from https://doi.org/10.1080/0267303042000204296.

Moos, M. (2014). "Generationed" space: Societal restructuring and young adults' changing residential location patterns. *The Canadian Geographer, 58*(1), 11–33.

Moos, M. (2015). From gentrification to youthification? The increasing importance of young age in delineating high-density living. *Urban Studies*, Early view. Retrieved from http://usj.sagepub.com/content/early/2015/09/15/0042098015603292.abstract. https://doi.org/10.1177/0042098015603292.

Mumford, L. (1949). Planning for the phases of life. *The Town Planning Review, 20*(1), 5–16.

Myers, D. (2015). *Three reinforcing cycles supporting the rise and fall of the new urban Millennials*. Presented at the Annual Conference of the Association of Collegiate Schools of Planning (ACSP), Houston, TX.

Myers, D. (2016). Peak Millennials: Three reinforcing cycles that amplify the rise and fall of urban concentration by Millennials. *Housing Policy Debate*, 1–20. Retrieved from https://doi.org/10.1080/10511482.2016.1165722

Myers, D., & Lee, H. (2016). Demographic change and future urban development. In G. W. McCarthy, G. K. Ingram, & S. A. Moody (Eds.), *Land and the city* (pp. 11–58). Bolton, MA: Lincoln Institute of Land Policy.

Myers, D., & Pitkin, J. (2009). Demographic forces and turning points in the American City, 1950–2040. *The Annals of the American Academy of Political and Social Science, 626,* 91–111.

Myers, D., & Ryu, S. (2008). Aging baby boomers and the generational housing bubble. *Journal of the American Planning Association, 74*(1), 17–33.

*National vital statistics reports*, Volume 64, Number 1. (2015, January 15). National Centre for Health Statistics. Retrieved from www.cdc.gov/nchs/data/nvsr/nvsr64/nvsr64_01.pdf.

Parkin, M., & Martin, V. (2012, May). *Planning information bulletin: 2011 Year-end population and household estimates*. Region of Waterloo. Retrieved from www.regionofwaterloo.ca/en/discoveringTheRegion/resources/BulletinYearEnd2011PopandHouseholdEstimates.pdf.

Quigley, J. M., & Rosenthal, L. A. (2005). The effects of land use regulation on the price of housing: What do we know? What can we learn? *Cityscape, 8*(1), 69–137.

Statistics Canada. (1991). *Census of Canada 1991: Individual public use microdata*. Retrieved from the Canadian Census Analyzer at CHASS at the University of Toronto http://sda.chass.utoronto.ca/sdaweb/html/canpumf.htm.

Statistics Canada. (1996). *Census of Canada 1996: Individual public use microdata*. Retrieved from the Canadian Census Analyzer at CHASS at the University of Toronto http://sda.chass.utoronto.ca/sdaweb/html/canpumf.htm.

Statistics Canada. (2001). *Census of Canada 2001: Individual public use microdata*. Retrieved from the Canadian Census Analyzer at CHASS at the University of Toronto http://sda.chass.utoronto.ca/sdaweb/html/canpumf.htm.

Statistics Canada. (2006). *Census of Canada 2006: Individual public use microdata*. Retrieved from the Canadian Census Analyzer at CHASS at the University of Toronto http://sda.chass.utoronto.ca/sdaweb/html/canpumf.htm.

Statistics Canada. (2011). *National household survey, 2011: Individuals file*. Retrieved from the Canadian Census Analyzer at CHASS at the University of Toronto http://sda.chass.utoronto.ca/sdaweb/html/canpumf.htm.

Statistics Canada. (2015a). Census metropolitan area and census agglomeration definitions: 2016 census metropolitan area and census agglomeration strategic review consultation guide. No. 93-600-X. Retrieved from www.statcan.gc.ca/pub/93-600-x/2010000/definitions-eng.htm.

Statistics Canada. (2015b, September 29). Table 051-0013 – Estimates of births, by sex, Canada, provinces and territories, annual (persons), CANSIM (database). Retrieved from http://www5.statcan.gc.ca/cansim/a26?lang=eng&id=510013.

Swan, C. (1995). Demography and the demand for housing: A reinterpretation of the Mankiw-Weil demand variable. *Regional Science and Urban Economics*, *25*(1), 41–58.

US Census Bureau. (2008). *A compass for understanding and using American community survey data: What general data users need to know*. Washington, DC: US Government Printing Office. Retrieved from www.census.gov/content/dam/Census/library/publications/2008/acs/ACSGeneralHandbook.pdf.

US Census Bureau. (n.d.). *Metropolitan and micropolitan statistical areas*. Retrieved from www.census.gov/population/metro/.

# 13 Urban vacation rentals and the housing market

## Boon or bane in the Millennial city

*Junfeng Jiao and Jake Wegmann*

One of the most important insights from the *youthification* hypothesis proposed by one of this volume's authors (Moos, 2015) is that an influx of young adults into a centrally located urban neighbourhood can coexist with and reinforce gentrification processes even when many of the young adult newcomers lack high incomes. One could say that many of the gentrifiers are not actually members of the "gentry", at least when measured in terms of how much they earn or their personal wealth. But how do young adults, even when equipped with high levels of education, a parental financial safety net, and (very often) white skin, manage to translate these non-monetary privileges into a place for themselves in popular big city neighbourhoods amidst escalating rents?

At least one of the ways in which they do so is by enduring living in cramped, communal, and therefore comparatively cheap, living quarters (ibid.). While, to be sure, poor city dwellers have long used such tactics to navigate unfriendly housing markets, they are more easily and flexibly used by people who have no caregiving responsibilities to children or other relatives, who have a wide variety of options in the labour market (even if comparatively modestly paid), good health, and myriad other advantages. This chapter is an empirical exploration of a new strategy that some young adults are deploying to mitigate high housing costs in their favoured big city neighbourhoods, one that has risen to prominence within a span of only about half a decade: *urban vacation rentals*.

Urban vacation rentals, or the practice of renting out part or all of a big city housing unit to visitors, have, of course, long existed. However, there has been a recent explosion in the growth of urban vacation rentals in tandem with the broader rise of the *sharing economy* – a term that is in some ways inapposite for the case of urban vacation rentals, as we will discuss later. Nonetheless, many of the same forces that have powered the rise of other sharing economy practices in domains such as urban transportation and short-term casual labour have been central to the widespread presence of the urban vacation rental market leader, Airbnb, along with competitors such as Homeaway. These companies have used websites, smartphone apps, and other contemporary information technologies to make a far more thickly-traded market of "hosts" and "guests" (to use the emerging industry's terminology) than would otherwise be possible. But for the theme of this book, it would be scarcely necessary to mention that the direct link between IT and urban vacation rentals ensures that the customers for this new industry are disproportionately, though not exclusively, members of the Millennial generation.

For a subset of participants in the urban vacation rental economy, many of them Millennials, renting out all or part of their dwellings with the help of Airbnb or other platforms can be thought as a *boon*. Though not all Millennials – not even a majority of them – live in expensive, "youthified" city neighbourhoods, it seems plausible that some of them are using urban vacation rentals as a tactic for preserving their footholds in such locations by using the income they earn to offset the high rents they must pay. Not surprisingly, rhetoric from Airbnb has called attention to exactly this practice since the company's founding (Samaan, 2015).

But what the sharing economy giveth, it can also taketh away. In other work we have previously demonstrated that a nontrivial subset of Airbnb listings are for what could be described as *permanent urban vacation rentals*, i.e. dwellings that are occupied so frequently by out-of-town paying guests that it is unlikely that they could be plausibly serving as permanent housing (Wegmann, Jiao & Dharmadhikari, unpublished). In high-rent areas, it is conceivable that what would otherwise be housing inhabited by neighbourhood residents is being taken off the housing market to accommodate tourists, thus further aggravating housing unaffordability for residents. While this dynamic – what we think of as a *bane* of the urban vacation rental economy from the standpoint of housing markets – appeared in our previous research to be mostly minor at the citywide level (ibid.), it could well be significant within particular city neighbourhoods. It would seem likely that these neighbourhoods overlap with the youthified districts described by Moos (2015).

In this chapter we explore three empirical questions. First, what is the geographic reach of the urban vacation rental economy, particularly with respect to youthification? More specifically, in what kinds of city neighbourhoods can the urban vacation rental economy be found, and where is it largely absent? The answer allows us to determine whether the geographic patterns map onto the typical characteristics of youthified neighbourhoods as described by Moos. Second, what is the magnitude of the boon effect of urban vacation rentals? In other words, to what extent are they helping non-professional hosts offset the costs of their houses or apartments? Third, what are the city neighbourhoods, if any, in which the bane effect is large enough that it could conceivably be contributing to a tightening of the local housing market?

We present results from five large cities in the USA spanning its length and breadth: Austin, Boston, Chicago, San Francisco, and Washington, DC. Note that our analysis does not take into account other possible banes and boons of the urban vacation rental economy. For instance, we ignore the question of whether the services offered by Airbnb and its competitors leave tourists better or worse off compared to when they use traditional lodgings such as hotels. Such a question is undoubtedly important, but we are focused on the point of view of people inhabiting the neighbourhoods where urban vacation rentals take place.

We also do not focus on neighbourhood quality of life issues, such as the impacts that might accrue when a formerly all-residential block begins to take on a more transient character owing to urban vacation rentals. Anecdotal and media reports suggest that these issues are very real. Nonetheless, of all city residents, young adults – the least likely to own property or to have children – are as a group the least affected by late-night noise, security in apartment buildings compromised by a steady stream of paying guests, tighter parking, or other quality-of-life impacts compared to their Generation X, Baby Boomer, or older neighbours. We keep our analysis squarely focused on the housing market, and the extent to which urban vacation rentals are a boon or a bane – or possibly both – in relation to this source of anxiety for many Millennials.

The rest of this chapter proceeds as follows. We provide a brief overview of research on urban vacation rentals and their impacts on local housing markets. We then outline our three methods – an exploration of the geography of urban vacation rentals, and quantifications of both the boon and the bane effects as described – before summarizing their results. In brief, we find that urban vacation rentals are largely co-extensive with youthified neighbourhoods, and that they appear to be *both* a housing market boon and bane for Millennials. We conclude by exploring some of the implications of our findings.

## Emerging research on urban vacation rentals and housing markets

Contemporary urban vacation rentals, powered by platforms such as Airbnb, are not the exclusive domain of Millennials, but they have characteristics that align with young adults' generational characteristics. From the demand side, booking a place to stay via Airbnb rather than in a traditional hotel fits well with Millennials' willingness to seek alternatives while traveling (also see Chapter 8). And, although their consumption patterns are "transactional in that they make a clear exchange in terms of time and money, [nonetheless] the need for experiences and entertainment is omnipresent" (Leask, Fyall, & Barron, 2014, p. 467). Airbnb and its competitors allow guests to choose from among thousands of listings in large cities, and the unique experience of interacting with a local host is often far more important to young adults than brand loyalty for a particular hotel chain. Of course, in many cases guests and the hosts arrange for the exchange of keys without ever meeting in person. Even in these cases, urban vacation rentals offer a more customized experience, particularly given that most are located within residential neighbourhoods rather than along the well-trafficked streets or bustling districts where motels and hotels tend to be found.

While urban vacation rentals are an emerging phenomenon, a trickle of studies examining their effects on housing markets and urban neighbourhoods has begun to accumulate. Many of them are sharply critical. Perhaps the gold standard to date is the New York State Office of the Attorney General (2014), which obtained data from Airbnb via court subpoena. It found that most short-term rentals in New York City violate various local ordinances, and that listings are concentrated in the neighbourhoods experiencing the tightest housing markets, such as the Lower East Side and Chelsea in Manhattan and Williamsburg and Greenpoint in Brooklyn. Other studies relying on data scraped from the web in San Francisco and Los Angeles have found similar findings in those cities (San Francisco Budget and

Legislative Analyst, 2015; Samaan, 2015). Anticipating such critiques, an earlier Airbnb-commissioned study found housing market effects to be minimal at the citywide scale in New York, but did not address variations between neighbourhoods (Rosen, Sakamoto, & Bank, 2013).

A separate line of critique holds that the online sharing economy, and urban vacation rentals in particular, reproduces racially disparate outcomes that exist in the wider world. For instance, one study found that all else equal, Airbnb listings held by identifiably black hosts received 12% less revenue and were 16% less likely to be rented than those held by whites (Edelman & Luca, 2015). Legal scholars have begun to examine the applicability of public accommodations laws prohibiting discrimination on the basis of race and other characteristics in lodgings, and found it inadequate to emerging forms of discrimination in the sharing economy, including urban vacation rentals (Leong & Belzer, 2017).[1]

How are we to make sense of the clashing claims of harm and benefit from urban vacation rentals? Martin (2016), in reviewing hundreds of recent articles on the sharing economy, distills their perspectives to six broad framings. Three of them are positive, while the three others are critiques. The first critique is that the sharing economy creates unregulated marketplaces, one emphasized in several of the studies referenced earlier. The second and most far-reaching is that the sharing economy reinforces the neoliberal economic paradigm, in sharp contrast to true sharing, which involves the non-monetary exchange of goods and services rather than the monetization of assets with slack capacity. The last is that the sharing economy is an "incoherent field of innovation [that] . . . has little to do with sharing; is framed very differently by different actors; creates a mix of positive and negative impacts; and, is discussed using confusing and interrelated terminology" (ibid., p. 155).

One of the first peer-reviewed studies in the planning field on Airbnb is a good example of this last framing (Gurran & Phibbs, 2017. This study of Sydney finds that Airbnb is indeed making it possible for some households, though likely not low-income ones, to earn a nontrivial amount of supplementary income by renting out extra living space to guests. At the same time, certain sections of the city, though not all, are affected by a loss of housing stock and the erosion of residential character. As Martin (2016, p. 158) writes: "It is not enough to present visions of decentralized, digital utopias. Advocates need to be more specific about which socio-digital innovations could address which social and environment challenges, and more convincingly articulate how these innovations could scale up." We might add that researchers as well as advocates need to do the same. Gurran and Phibbs' (2017) study is an even-handed move in the direction called for by Martin. The study summarized in this chapter seeks to do so as well, with particular attention to the effects of the urban vacation rental economy on the housing market as experienced by Millennials.

## Methods

In addition to providing a comprehensive frame for different views of the sharing economy, Martin (2016) ends his article with a spirited call for "empirical research which critically analyzes the nature and impacts of the sharing and collaborative economies in their many and varied forms" (p. 159). But doing so is not easy. Sharing economy phenomena such as urban vacation rentals are not easily captured in official data sources such as the census, many of them exist in a legal grey area, and much of the best data is proprietary. Nonetheless, researchers, ourselves included, must find a way to gain what insights they can with the data they can obtain.

We follow the lead of others in relying on information webscraped from Airbnb's website for our analyses in this chapter. The details of how this data was obtained are described elsewhere (Wegmann, Jiao & Dharmadhikari, unpublished). While this data set is undoubtedly incomplete and not always perfectly accurate – each data point, after all, represents an *advertisement* for an Airbnb listing posted by a prospective host rather than an actual completed transaction that could affect the housing market – it nonetheless is rich and comprehensive. We use these data for the three analyses described later in this section.

## Case selection

For our five cities, we selected US cities that are well-known tourist destinations spanning the various regions of the country. They rank from moderate to high among North America's 57 metropolitan areas of 1 million people or more in terms of young adult percentage, ranging from Austin (#2) to Boston (#37) with Washington (#6), San Francisco (#13), and Chicago (#23) occupying intermediate positions (Moos, 2015; Chapter 1). All but Chicago are similar in population. Four of the five cities ranked among the world's 25 foremost Global Cities according to one well-known ranking schema, and the fifth, Austin, is lauded as a hub of culture and technology despite its relatively small size and isolation from the East and West Coasts (AT Kearney, 2016; Tretter, 2015).

## Geographic analysis

Our first task is to look at what factors within the five cities correlate with a heavy concentration of Airbnb listings. From this we can then identify types of neighbourhoods in which both the boon and bane effects we measure are likely to have their greatest effects on the local housing market.

To do so, we measured the concentration of Airbnb listings (of all kinds) within every census tract within the five cities as a ratio of the number of housing units within the tract. We then ran an Ordinary Least Squares (OLS) regression model for each city in an attempt to predict, using a number of factors, the concentration of Airbnb listings in each census tract within that city. These factors included variables related to youthification. The percentage of the tract population in 2010 that were young adults, defined as ages 25–34, following Moos (2015); and the percentage increase in young adults in the tract from 2000 to 2010.

We included a number of other tract-level factors in the model, related to the state of the local housing market; neighbourhood character; and sociodemographic makeup of the area. We used a stepwise OLS procedure that automatically selected independent variables for inclusion in each city's model (see online appendix for details; Jiao & Wegmann, 2017). Our models were generally reasonably success-ful in predicting the variation in the concentration of Airbnb listings at the tract level, with the exception of San Francisco, where we predicted only 25%. In the other four cities, our models predicted from 42% (Boston) to 71% (Austin) of the variation. A list of the variables, the model results, and other details can be found in the online appendix (Jiao & Wegmann, 2017).

### Boon analysis

Our "boon analysis" is an attempt to approximate what proportion of a given housing unit's likely monthly cost is being covered by revenue realized via urban vacation rentals facilitated by Airbnb. To do so, we began by excluding certain types of listings so as to limit our analysis to hosts who could be plausibly using their dwellings listed on Airbnb to defray their housing costs. We excluded:

- *Permanent urban vacation rentals*. We defined these as units rented out in their entirety (as opposed to only bedrooms) whose occupancy we estimated as 25% or greater. Our reasoning was that it is unlikely for most people to be able to permanently inhabit a housing unit if they must be absent from it for one out of four days of the year or more.
- *B&B listings*. We inferred that units self-described as "bed & breakfast" rent-als are likely operated as a business activity by their owners, and likely via other means aside from Airbnb.
- *Listings with multi-unit hosts*. Because hosts are assigned unique identifiers, it is possible to tell which listings are held by the same host. We viewed multi-unit hosts as actors using Airbnb as more of a business activity than as a housing affordability strategy, and therefore excluded their listings from the boon analysis (also see Wegmann, Jiao & Dharmadhikari, unpublished).

For each listing included in the boon analysis, we calculated the proportion of the unit's estimated monthly rent covered by revenue from Airbnb. (Refer to the tech-nical appendix for details of how we calculated revenue.) To calculate each unit's estimated rent, we began with the census' estimate of the median rent for hous-ing units in the same census tract occupied by tenants who had moved in since 2010. We believed that doing so best mimicked the housing market perspective of young adults, most of whom are likely to have moved into rental units relatively recently. We then adjusted this rent on the basis of 1) the average household size in the census tract and 2) the estimated number of bedrooms.

These calculations make a number of assumptions, many of which do not line up with reality. For instance, we assumed that the hosts using Airbnb are renters who have moved in since 2010. In fact, likely many are homeowners, and many have lived in their dwellings since prior to 2010. But our boon analysis allows at least a rough estimate of the extent to which recently arrived young adult renters in big city neighbourhoods *could* be defraying their housing costs, and thus the extent to which listing one's unit via Airbnb is a promising strategy to do so.

### Bane analysis

The objective of our "bane analysis" is to identify the census tracts in the five cities in which permanent urban vacation rentals may be plausibly impacting the local housing market to the detriment of renters. We therefore mapped the subset of Airbnb listings that we classified as permanent urban vacation rentals (as defined above) and tallied their number within each census tract. We compared the number of permanent urban vacation rentals to the number of units in the tract reported by the census to be vacant and available for rent. If this ratio exceeded 10% in a given tract, we deemed the tract to be one whose housing market is potentially substantially impacted by permanent urban vacation rentals. We were then able to compare the young adult population share in impacted tracts to the citywide tract average.

If a tract had no rental units available, but at least one permanent urban vacation rental, we included it in the tally of impacted tracts. We did not, however, include tracts reported by the census as unpopulated – even those with permanent urban vacation rentals (perhaps representing, for example, industrial spaces being rented out to tourists via Airbnb). Our reasoning for the latter is that unpopulated tracts do not officially have a housing market that could be impacted by Airbnb, although admittedly it could plausibly impact nearby tracts.

## Results

### Geographic analysis

The full results of the OLS regression on tract-level Airbnb listing density done for each of the five cities are summarized in the technical appendix. Here we comment on the most consistent and striking results and their connection to youthification.

While the two variables related to market conditions show no clear trend, three of the variables we included to model neighbourhood character do. Two of these – percentage of workers driving to work alone, and presence of storefront retail – exhibit a relationship consistent with our typical picture of youthified neighbourhoods. Tracts with more non-driving commuters are more likely to have high listing densities in Austin, Boston, and Washington, DC. (There is no relationship in the other two cities.) And neighbourhoods with more storefront retail – and therefore, presumably more restaurants, bars, and shops – are associated with higher listing densities in all but San Francisco (where there is no relationship). Storefront retail is an especially powerful predictor in Austin, which stands out among the five cities because of its lack of walkable neighbourhoods; presumably the few that it does have are sought after by tourists using Airbnb.

Three of the sociodemographic variables have consistent results that are also suggestive of a link between youthified neighbourhoods and high concentrations of Airbnb ads. In all cities but Boston, census tracts with a higher share of college-educated adults, more "non-family" households, and lower median family incomes generally have a greater density of listings. A "non-family" household, per the US Census, is a group of people living together within a single housing unit none of whom are related by blood, adoption, or marriage. All three of these characteristics, taken together, are suggestive of youthified neighbourhoods.

In a finding that is perhaps surprising at first blush, the two variables that *directly* model youthification do not show a consistent pattern. Only in Chicago are Airbnb listings clearly associated with tract-level young adult percentage, and only in San Francisco are the tracts that have gained more young adults from 2000 to 2010 likely to have a high density of listings. However, these results are not as surprising as they may initially seem, since all of the other patterns reported above are associated with youthified neighbourhoods. In other words, it is not surprising that the effect of youthification disappears after we control for most of the variables associated with this process (Moos, 2015). Overall, our results strongly suggest that Airbnb listing density is in fact strongly connected to youthification.

### Boon analysis

Our boon analysis, whose results are shown in Table 13.1, allows for the comparison of three different strategies whereby tenants list their units. *Whole-unit listings* are the most common, accounting in all five cities for 59% or more of the listings that were eligible for analysis (as described earlier) and that had produced at least some revenue. This strategy entails vacating one's apartment or house during the time when one or more guests stay in it, thus providing the latter an experience most akin to staying in a hotel among the three options available via Airbnb. In all cities save for Washington, it is the most economically potent choice for the hosts in the median case, covering 7% to 18% of rent. (For reasons we do not

*Table 13.1* Summary of boon analysis (data from May to July 2015)

|  | Austin | Boston | Chicago | San Francisco | Washington |
|---|---|---|---|---|---|
| Number of qualified listings | 3,445 | 1,053 | 2,230 | 3,195 | 1,727 |
| Number of listings with nonzero revenue | 1,943 | 810 | 1,633 | 2,384 | 1,374 |
| **Whole unit listings** | | | | | |
| Share of revenue-producing listings | 74.4% | 58.5% | 59.6% | 65.1% | 68.4% |
| Median share of imputed monthly rent covered by Airbnb revenue | 17.5% | 17.9% | 14.8% | 7.1% | 0.4% |
| 90% percentile | 71.1% | 68.0% | 54.5% | 27.0% | 6.4% |
| **Private room listings** | | | | | |
| Share of revenue-producing listings | 23.8% | 39.0% | 37.9% | 33.8% | 29.4% |
| Median share of imputed monthly rent covered by Airbnb revenue | 5.8% | 8.9% | 10.3% | 4.4% | 0.9% |
| 90% percentile | 31.4% | 58.1% | 42.6% | 32.7% | 20.4% |
| **Shared room listings** | | | | | |
| Share of revenue-producing listings | 1.7% | 2.5% | 2.5% | 1.1% | 2.3% |
| Median share of imputed monthly rent covered by Airbnb revenue | 7.2% | 14.0% | 4.5% | 1.5% | 0.5% |
| 90% percentile | 37.3% | 45.7% | 19.4% | 8.4% | 17.9% |

yet understand, revenue is exceptionally weak in the median case in Washington for all three strategies.) The most aggressive hosts – those in the 90th percentile in terms of share of rent covered – are able to cover as much as 71% of their rent costs in Austin and almost as much in Boston. Clearly hosts who have the ability to make their units available to out-of-town guests frequently can, in some cases, cover a hefty portion of their housing costs, just as Airbnb promotional materials suggest they can.

Making one's dwelling available as a whole unit listing, however, requires a willingness to live a somewhat peripatetic life. It may work for those who travel frequently for work, for example, or who have romantic partners with their own housing units, or relatives living nearby. For others it would be inconvenient or impossible. For some, the strategy of *private rooms* might be preferable, in which the host is present while one or more guests stay in a private bedroom while sharing the common areas and possibly a lavatory. Private room listings compromise privacy to some degree but allow the host to remain in their home. In general this appears to be a less potent strategy in terms of offsetting rent. It is also less common, accounting for only 24% to 39% of revenue-producing listings in the five cities.

Most intrusive, least common, and – for the most part least economically potent among the three strategies – are *shared room* listings. In these situations, guests share bedrooms and lavatories with strangers, as in a hostel. This strategy does appear, however, to have at least some potency, particularly in Boston, perhaps because of the high student population there – but also perhaps of less help to Millennials who are out of college or who never attended.

### Bane analysis

The results of the bane analysis are summarized in Table 13.2. Our most notable finding is that in four of the five cities – all but San Francisco – the neighbourhoods in which permanent Airbnb rentals exceed 10% of available rental units occupy a small footprint. In these four cities, the tracts where Airbnb permanent rentals could possibly be displacing a substantial quantity of housing for permanent residents account for between 3% (Chicago) and 16% (Washington, DC) of the citywide population.

With respect to the bane analysis, San Francisco is in a category unto itself. This is the only city of the five in which there is some evidence that permanent Airbnb rentals could be having an impact at something approaching the citywide scale. We can imagine a number of factors that are contributing to these conditions: first, San Francisco's booming economy at the time of writing and its status as an iconic tourist destination. Another is the city's location within the world's leading metropolitan IT industry cluster (not to mention its status as Airbnb's headquarters), which may contribute to early and enthusiastic adoption of online urban vacation rentals. Another factor may be unusual regulations, such as San Francisco's 1981 ordinance restricting the conversion of single-room occupancy apartment buildings into tourist accommodations, which restricts the supply of hotels. Finally, San Francisco has by far the strongest rent control protections

Table 13.2 Summary of bane analysis (data from May to July 2015)

| | Number of tracts citywide* | Number of tracts where permanent Airbnb rentals exceed 10% of available rental units* | Total population residing in tracts where permanent Airbnb rentals exceed 10% of available rental units** | Average real median rent increase in tracts where permanent Airbnb rentals exceed 10% of available rental units, 2006–2010 to 2011–2015*** | Average citywide median real rent increase, 2006–2010 to 2011–2015**** | Average young adult share percentile in tracts where permanent Airbnb rentals exceed 10% of available rental units** |
|---|---|---|---|---|---|---|
| Austin | 220 | 21 | 6.7% | 20.1% | 9.2% | 59.2% |
| Boston | 181 | 22 | 9.5% | 7.8% | 1.3% | 67.6% |
| Chicago | 803 | 26 | 2.8% | 5.7% | 0.2% | 85.9% |
| SF | 197 | 93 | 48.6% | 6.8% | 7.9% | 53.5% |
| Wash. DC | 179 | 29 | 16.4% | 25.4% | 14.8% | 65.1% |

* Includes tracts that lie partially but not wholly within city limits.
** Includes tracts that contain no available rental units. Young adults defined as ages 25–34.
*** Social Explorer, Table T104 ("Median Gross Rent"), 2010 dollars. Average excludes missing rent data for four tracts in Boston, one in SF, and one in Washington.
**** Social Explorer, Table T104 ("Median Gross Rent"), 2010 dollars.

among the five cities, which may increase the incentives for landlords to maximize revenue by turning to urban vacation rentals instead of renting to permanent tenants, despite the risk of enforcement (San Francisco Budget and Legislative Analyst, 2015).

In the other four cities, the tracts whose housing markets are most impacted by permanent Airbnb listings have some similarities. They tend to be at least somewhat youthified – they are, on average, in the top tercile (Boston, Chicago, Washington) or just shy of the second quintile (Austin) of tracts citywide in terms of young adult population share. In every case but San Francisco, rents have grown faster over the past half decade than they have citywide (Table 13.2). And for the most part the impacted tracts are relatively central – usually near but not quite in the Central Business District (Figure 13.1). Those familiar with these cities will likely be unsurprised by the neighbourhoods that are included: close-in East and South Austin; the North End and Allston-Brighton in Boston; the near North and Northwest Sides in Chicago; and Capitol Hill and Adams-Morgan in Washington, DC. These are quintessentially "youthified" neighbourhoods in Moos' (2015) characterization. Some of the tracts are somewhat further away and less well-known, but the general pattern is apparent.

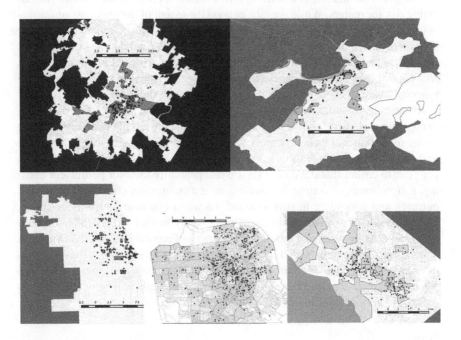

*Figure 13.1* Geographies of Airbnb

Caption: Whole-unit, high-occupancy Airbnb listings (black dots) and impacted census tracts (shaded in grey) in the five cities. Clockwise from top left: Austin; Boston; Washington, DC; San Francisco; and Chicago.

## Conclusion

Where do these various findings leave us? Overall, they suggest an urban vacation rental economy that is substantial and growing, but "spiky" in its footprint.
The places within the five cities that we studied in which this economy touches
down, so to speak, tend to coincide with neighbourhoods whose housing markets
are being most affected by youthification. In these places – a notable but relatively
small share of the territory of each city save San Francisco – there is a palpable
"bane" effect, or at least indirect evidence that urban vacation rentals are taking
what would otherwise be permanent housing units off the market. Because of
where they are, young adults are presumably disproportionately affected, although
the effects extend to everyone actively involved in buying or renting housing in
those areas, regardless of age.

At the same time, the "boon" effect is considerable. Listing one's rented or
owned dwelling unit as an urban vacation rental is not likely to be a strategy
that can reasonably be expected to help all residents cope with their housing cost
burdens within the affected areas, as the previously cited research on racial discrimination against Airbnb hosts demonstrates. But it is likely that among those it
*does* help, young adults are likely to figure prominently. They may be in the best
position to add urban vacation rentals as an additional arrow to their quiver of
other homesharing strategies that Moos (2015) identified as already in wide use.

Using Airbnb hosting as a housing affordability strategy is not without considerable risk for renters. While we lack data on the extent of the phenomenon, anecdotal and media accounts suggest that renters are sometimes evicted from their
apartments when they host urban vacation rental guests without the knowledge or
consent of their landlords. In a tight housing market – particularly in a city with
rent control, such as San Francisco or Los Angeles, where such apartments are
especially hard to come by – this can be a devastating outcome.

As cities begin to take steps to regulate urban vacation rentals, they may want
to consider ways to minimize the bane effects while preserving and extending the
boons of Airbnb and its competitors for their users. We suggest restricting urban
vacation rentals in whole units to no more than a certain number of total nights
per year, to ensure that the dwellings are still primarily used for permanent housing. Enforcement, of course, is essential, as is a companion effort towards public
outreach and education, in part to avoid the worst outcomes, such as evictions.
Success would necessitate dedicated staffing, likely funded by fees collected from
a robust licensing scheme, to avoid enforcement of regulations on urban vacation
rentals from being superseded by other priorities. At any rate, as cities begin to
ramp up their regulatory regimes, we believe that considerably more empirical
research on the boons and banes of urban vacation rentals will be needed, as this
topic of inquiry is still only in its infancy.

## Note

1  At the time of writing, in direct response to the Edelman and Luca study (2015), Airbnb
   has embarked on efforts to reduce racial discrimination. These include de-emphasizing
   (though not eliminating) the use of personal photos of guests and compelling hosts to

agree to a nondiscrimination policy (Benner, 2016). At present it is too early to tell if these measures will be effective. Meanwhile, competitors have recently sprung up that cater to would-be urban vacation rental guests who have experienced discrimination (Finley, 2016).

# References

AT Kearney. (2016). *Global cities 2016*. Retrieved from www.atkearney.com/documents/10192/8178456/Global+Cities+2016.pdf/8139cd44-c760-4a93-ad7d-11c5d 347451a.

Benner, K. (2016, September 9). Airbnb adopts rules to fight discrimination by its hosts. *New York Times*.

Edelman, B., & Luca, M. (2015). *Digital discrimination: The case of Airbnb.com*. Harvard Business School Working Paper Series, Harvard University, Cambridge, MA.

Finley, T. (2016, June 21). These Airbnb alternatives want to make travel more welcoming for black people. *Huffington Post*. Retrieved from www.huffingtonpost.com/entry/innclusive-noirbnb-airbnb-alternatives_us_5768462ae4b0853f8bf1c675.

Gurran, N., & Phibbs, P. (2017). When tourists move in: How should urban planners respond to Airbnb? *Journal of the American Planning Association, 83*(1), 80–92.

Jiao, J., & Wegmann, J. (2017). *Technical appendix*. Retrieved from http://soa.utexas.edu/work/urban-vacation-rentals-and-housing-market-boon-or-bane-or-both-millennial-city.

Leask, A., Fyall, A., & Barron, P. (2014). Generation Y: An agenda for future visitor attraction research. *International Journal of Tourism Research, 16*, 462–471.

Leong, N., & Belzer, A. (Forthcoming). The new public accommodations. *Georgetown Law Journal*. Cited with permission of the authors.

Martin, C. J. (2016). The sharing economy: A pathway to sustainability or a nightmarish form of neoliberal capitalism? *Ecological Economics, 121*, 149–159.

Moos, M. (2015). From gentrification to youthification? The increasing importance of young age in delineating high-density living. *Urban Studies, 53*(14), 2903–2920.

New York State Office of the Attorney General. (2014, October). *Airbnb in the city*. Albany, NY: Research Department and Internet Bureau. Retrieved from www.ag.ny.gov/pdfs/Airbnb report.pdf

Rosen, K. T., Sakamoto, R., & Bank, D. (2013, October). *Short-term rentals and the impact on the apartment market*. Rosen Consulting Group. Retrieved from http://publicpolicy.airbnb.com/wp-content/uploads/2014/04/Short-TermRentalsandImpactonApartment MarketNY1.pdf.

Samaan, R. (2015, April). *Airbnb, rising rent, and the housing crisis in Los Angeles*. Los Angeles Alliance For a New Economy (LAANE). Retrieved from www.laane.org/airbnb-report.

San Francisco Budget and Legislative Analyst. (2015, May 13). *Policy analysis report re: Analysis of the impact of short-term rentals on housing*. Memo to San Francisco Supervisor David Campos. Retrieved from www.sfbos.org/Modules/ShowDocument. aspx?documentid=52601.

Tretter, E. (2015). Live music, intercity competition, and reputational rents: Austin, Texas the "live music capital of the world". *Human Geography, 8*(3), 49–65.

Wegmann, J., Jiao, J., & Dharmadhikari, A. (Unpublished). *Taming Airbnb: Toward guiding principles for local regulation of urban vacation rentals based on empirical results from five US cities* (available upon request from authors).

# Part IV
# Millennial mobilities

# 14 Will Millennials remain in the city?

Residential mobility in
post-industrial, post-modern,
post-suburban America

*Markus Moos and Nick Revington*

In the face of environmental, health and fiscal concerns associated with suburban sprawl, a growing number of planning officials across the USA and Canada have incorporated Smart Growth policies into their planning frameworks to promote higher density, infill and more compact developments (Filion & Kramer, 2011). Growing concentrations of young adults (Millennials or Generation Y) in the inner cities of major metropolitan areas have thus been met with great optimism in planning circles, as this trend can be viewed as a measure of success of Smart Growth planning policies. However, it remains to be seen whether this trend will continue.

In this chapter, we examine whether we can expect young adults who are currently residing in urban neighbourhoods to remain in these neighbourhoods as they grow older. In other words, can we expect a surge in demand for suburban single-family homes once Millennials age and more of them start to have children, or will demand for higher density living continue?

Moos (2016) advances a youthification hypothesis to explain the increasing share of young adults in higher density, amenity-rich neighbourhoods. The hypothesis does not suggest that all young adults are residing in downtowns or high-density areas, but only that these are the locations where concentrations of young adults are highest and have been increasing. One interpretation of the youthification hypothesis is that young adults do not actually contribute to more compact urban forms. Youthification is conceptualized as an outcome of decreases in household size, delays in childbearing, rising housing costs and declines in economic fortunes of today's young adult generation due to structural changes in the labour market. Therefore, if the reasons for moving into higher density neighbourhoods are an outcome of changing demographic and socio-economic conditions, as opposed to changing preferences or planning policies, young adults may move away from higher density areas if their household size increases, or if their status in the labour market improves (also see Skaburskis, 2006).

Forecasts using expected earnings, changing economic conditions, demographic projections and/or historic residential mobility trajectories to make predictions are a common approach to understanding future housing preferences and trends (Sanchez & Dawkins, 2001; Myers & Ryu, 2008; Patterson, Saddier, Rezaei, & Manaugh, 2014). All forecasting methods are limited by their assumptions, but the reliance on previous generations' mobility trajectory has arguably become particularly problematic. Due to cultural, demographic and economic shifts in a post-modern, post-industrial, and post-suburban era (Phelps & Wood, 2011; Keil, 2013), we cannot assume *a priori* that young adults today will follow their parents' residential mobility trajectory from more urban to increasingly lower density suburban locations as they age. Thus, we attempt to shed further light on the question of future location patterns by drawing on contemporary sociological theories and empirical evidence that demonstrate that long-term housing and location preferences of individuals are shaped by a) the residential environment where they grew up; and b) their current residential environment.

The analysis draws on results from an online survey of almost 600 people aged 18 to 40 in the USA and Canada. Using statistical tools, the analysis adds much needed empirical evidence to the debate on whether Millennials will remain in urban areas, and denser neighbourhoods, over time. The results have important implications for understanding changing urban geographies, as well as housing demand forecasts and planning policies that shape housing supply. In considering the residential trajectories of particular age groups, we contribute to understanding how urban space is increasingly "generationed", in addition to being shaped by the trifecta of race, gender and class that have historically received more attention (Moos, 2016).

We begin with an overview of the research examining residential location decisions and the changing societal context within which these take place. Next, we explain our methods in further detail, followed by the empirical analysis. Based on our findings, we conclude by offering alternative parameters that policymakers need to consider in their forecasts of Millennials' future residential locations.

## Post-industrial, post-modern, post-suburban America

From the late 1970s onward, post-industrialism has been commonly used to describe both the changing character of urban economies, moving from manufacturing to more service-oriented activities, as well as resulting societal structures that elevate those with access to information, technology and education, but further marginalize those that do not (Bell, 1973/2008; Ley, 1986). Cities historically reliant on manufacturing started to see decline as jobs began to be outsourced to places with lower labour costs, predominantly in the developing world.

Some North American cities, such as New York or Toronto, managed this transition by becoming financial centres and attractive places to highly educated service workers, often dubbed the "creative class" (Florida, 2002) or "new middle class" (Ley, 1996). Post-industrialism thus set the stage for gentrification in these cities as the share of jobs in industrial sectors decreased and lower income and blue collar workers were displaced, often by those in the quaternary sector of the economy, comprising those working in social sciences, education, government, arts and recreation related occupations (Zukin, 1982; Hutton, 2008; Ley, 1996). Gentrification allowed for youthification, where the inner city became home to young adults from across the income spectrum (Moos, 2016). However, other North American cities, such as Detroit or Cleveland, saw more persistent decline in their economies.

The concurrent shift to a post-modern society is also important in the gentrification process. Among other things, post-modernity has engendered an increasing diversity of lifestyles, life courses and household types as societal norms have evolved (Bell, 1973/2008; Ley, 2003). Gender equality efforts, changing laws and norms regarding sexual identities, increasing educational attainment, lower birth rates, higher divorce rates, blended families, and increasing life expectancy have all contributed to greater variation in household types and sizes, as well as an overall decline in average household size (Rose & Villeneuve, 2006; Townshend & Walker, 2010). Smaller households increased demand for smaller housing spaces that could most aptly be served in the inner city and urban areas in general (Van Diepen & Musterd, 2009). Arguably, post-modernism also aligns with the current neoliberal context, in making generational and age dimensions of cities more pronounced through an emphasis on individuality and place making, as in the case of targeted marketing of specific urban spaces and housing developments to particular age and lifestyle groups.

Due to its prevalence in North America, some have called gentrification "the fifth migration" due to population growth in central cities (Fishman, 2005). However, in terms of overall growth patterns, suburban development remains most common in both the USA and Canada (Gordon & Janzen, 2013). And while suburbs have historically been conceptualized as predominantly new, low-density residential developments at the outskirts of metropolitan areas, suburbs have long been more differentiated than is often acknowledged (Forsyth, 2012). More recent understanding of suburbs as variegated spaces (Phelps & Wood, 2011), "edge cities" (Garreau, 1991/2011), and higher density suburban nodes served by transit (Moos & Mendez, 2014) has led some writers to adopt the term "post-suburban" to label an emerging period of metropolitan transformation without clear distinction between suburbs and city as mutually exclusive categories (Keil, 2013; Walks, 2013a).

Post-suburban landscapes may offer more housing and lifestyle options than the stereotypes, such as the auto-oriented cul-de-sac and the "spacious" single-family home, would have us believe. Indeed, some young adults may find more housing options matching their household profiles in suburban areas, and amenity provisions in some suburban areas may be quite similar to those found in urban areas. The dominance of suburbs, as far as total population counts and growth rates are concerned, also means that most young adults today actually grew up in a suburban setting. It is reasonable to speculate that they may return to the suburbs, if and when they decide to raise their own children or their space needs increase for other reasons. While the blurring between the suburban and the urban may suggest inherent difficulties in measuring urban/suburban location decisions, it is important to note that low-density, automobile-oriented characteristics, as well as distinct voting patterns, still by-and-large delineate suburbs as those areas outside a central city (Walks, 2007; Gordon & Janzen, 2013; Harris, 2014).

The emergence of a post-industrial, post-modern, post-suburban era is likely to affect residential mobility patterns in such a way that renders them more difficult to predict on the basis of previous generations. Residential mobility decisions often involve economic, social, cultural and other external factors, which have been restructured in the development of a post-industrial, post-modern, post-sub-urban society, as well as internal and even emotional factors. It is important, then, to understand what these factors are and how they may have changed.

## Residential mobility

As of 2011, approximately 12% of both the US and Canadian population had moved in the previous year (US Census Bureau, 2016; Statistics Canada, 2013). However, residential mobility is closely linked to life-course events and thus indirectly with age (Clark, Deurloo, & Dieleman, 2003). In particular, young adulthood is often a period of higher residential mobility associated with events such as moving out of the parental home, potentially going away to university, coupling or marriage, finding a new job, and having children (Beer, Clower, Faulkner, & Paris, 2011). Moreover, as people age, they are more likely to give up their aspirations to move (Coulter, 2013). As the set of factors influencing the residential location decision process is inherently complex, it is very difficult to isolate the range and relative importance of each one. Given the breadth and depth of literature on residential mobility and location decisions, we will only describe the broad trends and focus on the most commonly discussed factors (see Coulter, van Ham, & Findlay, 2015, for a more comprehensive review).

Building on Alonso (1964), many have conceptualized location decisions as a sorting process whereby households end up in neighbourhoods containing housing stock that matches their household size and composition, making trade-offs between housing and commuting costs, and expressing preferences for different kinds of amenities. The traditional housing trajectory portrayed households moving from smaller – perhaps shared – apartments during young adulthood, to larger single-family homes after marriage or before childbearing, and eventually downsizing with age. The diversity in household types, lifestyles and life course trajectories that has accompanied the shift to a post-modern era, along with the extension of the young adult life cycle stage and growing post-secondary school enrollment, means that – in many cases – the traditional housing trajectory no longer holds and location decisions have become more difficult to predict (Beer Clower, Faulkner, & Paris, 2011; Coulter et al., 2015). That said, childbearing remains one of the most important factors initiating a move away from urban areas (Feijten, Hooimeijer, & Mulder, 2008; Moos, 2014a; see also Boterman, 2012).

The rapidly growing condominium market, often targeting young adults (see Chapter 10), has redefined the inner city by enabling a high concentration of young adults in the centre of many North American metropolitan areas (Lehrer & Wieditz, 2009; Rosen & Walks, 2015; Moos, 2016). Obviously not *all* young adults live or want to live in cities, as is sometimes sensationalized in the popular press (e.g. Frizell, 2014), but the highest and increasing concentrations are found in urban areas. For some young adults, condominiums might represent "the new starter home". However, while young adults across the income spectrum are found in central areas, access to housing markets is highly differentiated: Some own, but many others rent condominiums from investors, or live in basement apartments, with or without roommates.

Indeed, age but also class, gender and ethnicity are structuring elements of young adults' residential decisions (Walks, 2013b, 2014; Rosen & Walks, 2013). Furthermore, sociological models point to people's tendencies to move into neighbourhoods with socio-economic and demographic profiles matching their own (Bourne & Ley, 1993). Studies in geography also point to social class, racial composition, mortgage lending practices, tenure differences, ethnic clustering and affective values attached to place as structuring factors (Lees, Slater, & Wyly, 2010; Townshend & Walker, 2010).

In the USA, young adults with post-secondary education (indirectly related to social class) are most likely to reside in urban neighbourhoods, whereas young adults without at least one degree are decreasing in urban neighbourhoods (Weissman, 2015). However, as noted above, the socio-economic divide between urban and suburban locations appears to be dissipating in many large, growing metropolitan areas, with both the inner city and suburban areas seeing increasing divisions and diversity of neighbourhoods, segregated by class, race, and/or ethnicity (Moos & Mendez, 2015; Keil, 2013). For instance, the immigrant landscape in both Canada and the USA has become more suburban, partly due to gentrification and partly due to growth in suburban job opportunities, but also due to the higher socio-economic standing of some more recent immigrants who are able to enter homeownership markets (Singer, Hardwick, & Brettell, 2008; Ley, 2010).

There is an emerging literature that considers current residential location decisions as a function of past residential environments, particularly the residential environment experienced during childhood (Blaauboer, 2011). Feijten et al. (2008) show that "the place of birth turns out to play a decisive part in shaping residential environment choices later in life" (p. 156). Their results point to social networks and family connections as important factors drawing people to return to the urban or suburban areas where they lived as children (see also Hedman, 2013). Feijten et al.'s (2008) study also shows that those who grew up in a rural setting were also more likely to move back to any rural area later in life, not just the specific place where they grew up. A recent USDA study that included 300 interviews with rural residents, also found that childrearing and family connections were important reasons for returning to a rural area (Cromartie, von Reichert, & Arthun, 2015).

Finally, there is a large body of literature that examines the complex linkages between preferences for specific built environments and transportation patterns. For instance, Handy, Cao, and Mokhtarian (2006) find that the built environment characteristics influence walking after taking personal preferences and attitudes into account. The work suggests that while a self-selection effect may be at play in terms of why people choose to live in particular locations (i.e. in that these locations enable particular kinds of transport mode decisions), the built form also influences subsequent travel behaviour regardless of self-selection.

Whether these findings can be extrapolated to all kinds of residential environments in a North American context is largely unknown due to a dearth of research in this field. As Coulter et al. (2015) show, even the topic of "residential mobility" in general has received much less attention than global or regional migration patterns and immigration within specific national contexts. They argue for more empirical research and also "re-thinking residential mobility and immobility as relational practices that link lives together and connect people to structural conditions through time and space" (p. 16).

## Connecting past, current and future locations

Inspired by Coulter et al. (2015), we aim to link young adults' urban versus suburban location decisions as an outcome of specific "structural conditions through time and space" (p. 16) as we hope to observe how people's past and current residential environments influence future aspirations. We use multivariate regression to test whether past residential environments influence current decisions, and whether past and current location characteristics influence stated future preferences. The data are drawn from the survey of young adults residing in cities across the USA and Canada (see Chapter 9).

Survey respondents were asked to identify their residential environment as urban, suburban, small town or rural. In this chapter, we focus solely on the urban versus suburban residential environments. Table 14.1 shows the residential environment of survey respondents over time (past, current and future) by housing types for urban and suburban neighbourhoods. Almost 70% of survey respondents grew up in suburban neighbourhoods in single-family homes. This pattern corresponds with the general population in the USA and Canada, where most of the population lives in low-density suburban neighbourhoods, earning them the appropriate title of "suburban nations" (Gordon & Janzen, 2013; Keil, 2013). After reaching the age of majority, respondents' primarily moved into urban areas, with about 80% currently residing in urban neighbourhoods, although it should be noted that about a quarter resides in single-family homes in urban neighbourhoods. Respondents expect this drastic change in residential locations to persist in the future. Expected residential environments ten years from now change only modestly from the current situation, with a slight increase in expected urban living. More respondents expect to reside in urban neighbourhoods in either low density or medium density environments (i.e. single-family homes or attached/mid-rise buildings).

Additional variables are included to understand the relative importance of various factors influencing residential mobility. The variables are the presence of children, household size, gender, visible/racial minority status, housing tenure,

*Table 14.1* Residential location and housing type of respondents (%)

| Residential environment | When growing up | Current | 10 yrs from now |
|---|---|---|---|
| Urban, single-family home | 19.8 | 27.0 | 32.4 |
| Urban, attached or mid-rise | 4.7 | 38.8 | 40.1 |
| Urban, high-rise | 2.6 | 13.6 | 12.2 |
| Suburban, single-family home | 68.7 | 13.9 | 13.1 |
| Suburban, attached or mid-rise | 4.0 | 6.2 | 1.6 |
| Suburban, high-rise | 0.2 | 0.7 | 0.6 |
| *N cases* | *470* | *583* | *549* |

*Table 14.2* Select characteristics of survey respondents

| Variable | Percent |
|---|---|
| Currently resides in urban neighbourhood | 77 |
| Grew up in urban neighbourhood | 21 |
| Preference to reside in urban neighbourhood ten years in future | 78 |
| Currently resides in Canada (versus USA) | 54 |
| Visible/racial minority | 16 |
| Female | 54 |
| Renters | 69 |
| Children present | 13 |
| Quaternary sector occupations | 38 |

N=516

quaternary sector occupations, and country of residence (USA versus Canada) (Table 14.2).[1] Three different models are used to test the linkages among past, current and future residential locations. This is done to avoid issues of multicollinearity; using separate models also aids with ease of interpretation of the results.

We need to approach these numbers with caution. As we do not necessarily have a representative sample of young adults (see Chapter 9), we cannot generalize the distribution of actual location decisions from the survey. Since the survey sample is much more highly educated than the general population, and educational attainment is an important variable in residential mobility patterns, we can likely, at best, generalize findings to highly educated young adults only.

## Path dependency in residential location decisions

Table 14.3 shows the results from the logistic regression analysis. Coefficients are presented as odds ratios, so that a value greater than 1 indicates a higher likelihood of a given outcome. All three models are statistically significant as shown by the Wald chi-squared tests. The first model includes stated preference for future urban living as the dependent variable. The results show that having grown up in an urban neighbourhood has the largest effect on whether respondents see themselves living in urban neighbourhoods in the future. Those who grew up in urban neighbourhoods have twice the odds of expressing preferences for future urban living than those who grew up in other kinds of residential environments. Being female and the presence of children in the household reduces the likelihood of expressing a preference for urban living. Those employed in quaternary sector occupations have an increased tendency to express urban preferences as well.

The second regression model shown in Table 14.3 includes preferences for future urban living as the dependent and introduces the current residential location as an independent variable. As in the first model, being female and the presence of children decrease the likelihood of expressing future urban location preferences whereas quaternary sector employment increases the likelihood. Those currently residing in urban neighbourhoods have seven times higher odds of expressing

*Table 14.3* Logistic regression of respondents' residential locations in different time periods (odds ratios shown)

**Stated preference to live in urban neighbourhood ten years from now**

| | |
|---|---|
| Grew up in urban neighbourhood | 2.3** |
| Visible/racial minority | 0.9 |
| Female | 0.4*** |
| Canada | 1.2 |
| Renter | 1.0 |
| Household size | 1.0 |
| Presence of children | 0.4** |
| Quaternary sector occupation | 1.8* |
| Constant | 4.5*** |
| Wald chi2(8) | 37.3*** |
| Log pseudolikelihood | −292.0 |
| N cases | 582 |

**Stated preference to live in urban neighbourhood ten years from now**

| | |
|---|---|
| Currently lives in urban neighbourhood | 7.4*** |
| Visible/racial minority | 1.3 |
| Female | 0.5** |
| Canada | 1.2 |
| Renter | 0.7 |
| Household size | 1.1 |
| Presence of children | 0.5* |
| Quaternary sector occupation | 1.6* |
| Constant | 1.1 |
| Wald chi2(8) | 89.7*** |
| Log pseudolikelihood | −260.0 |
| N cases | 582 |

**Currently lives in an urban neighbourhood**

| | |
|---|---|
| Grew up in urban neighbourhood | 1.5 |
| Visible/racial minority | 0.5* |
| Female | 0.7 |
| Canada | 1.0 |
| Renter | 2.2** |
| Household size | 0.8* |
| Presence of children | 0.5* |
| Quaternary sector occupation | 1.4 |
| Constant | 4.7*** |
| Wald chi2(8) | 48.1*** |
| Log pseudolikelihood | −285.2 |
| N cases | 582 |

Notes: ***p<0.001; **p<0.01; *p<0.05 Robust standard errors are used.

future preferences for urban living than those currently residing in suburban, small town or rural areas.

The third and final model in Table 14.3 includes the current residential location as the dependent variable and includes the past residential location of respondents as an independent variable. Perhaps not surprisingly, being a renter is associated with a higher likelihood of urban living. The presence of children, visible minority status and larger household size decrease the likelihood of residing in an urban neighbourhood. Importantly, the variable capturing residential environment when growing up was not statistically significant; where one grows up does not have an effect on current residential location.

Further insight into these results is provided from value statements about why respondents chose their current residential location. Table 14.4 shows the importance respondents place on different characteristics by urban versus suburban residential locations. Perhaps not too surprisingly, those currently residing in urban locations are more likely to point to the importance of proximity to restaurants/cafes/bars, public transit, friends, cultural amenities and walkable neighbourhoods and cycling infrastructure. Those currently residing in suburban locations are more likely to point to the importance of proximity to freeways, good schools, family members and being in safe and quiet neighbourhoods.

While a larger share of suburbanites indicated that a large yard was "very important" in their location decision, it is interesting that more than 50% of respondents in the urban *and* suburban locations ranked this characteristic as "not very important" or "not important at all". There are no statistically significant differences between how urbanites versus suburbanites answered the question regarding the importance of proximity to work/college/university. It would appear that preferences for different kinds of amenities and built environments are important factors in distinguishing urban from suburban residents in the sample.

*Table 14.4* How important was each of the following in selecting your current residential location?

| Percentage of respondents | Current location | | p-level |
|---|---|---|---|
| **Proximity to restaurants/cafes/bars** | **Suburban** | **Urban** | |
| Not important at all | 15 | 3 | |
| Not very important | 19 | 4 | |
| Neutral | 15 | 7 | |
| Somewhat important | 41 | 47 | |
| Very important | 9 | 39 | *** |
| **Proximity to public transit** | | | |
| Not important at all | 30 | 5 | |
| Not very important | 17 | 4 | |
| Neutral | 9 | 6 | |
| Somewhat important | 20 | 25 | |
| Very important | 24 | 60 | *** |
| **Walkability of neighbourhood and cycling infrastructure** | | | |
| Not important at all | 10 | 2 | |
| Not very important | 21 | 2 | |
| Neutral | 20 | 6 | |
| Somewhat important | 32 | 29 | |
| Very important | 18 | 61 | *** |
| **Location with good highway/freeway access** | | | |
| Not important at all | 13 | 38 | |
| Not very important | 14 | 23 | |

(*Continued*)

*Table 14.4* (Continued)

| Percentage of respondents | Current location | | p-level |
|---|---|---|---|
| Neutral | 18 | 15 | |
| Somewhat important | 40 | 19 | |
| Very important | 14 | 5 | *** |
| **Proximity to where my friends live** | | | |
| Not important at all | 18 | 6 | |
| Not very important | 15 | 11 | |
| Neutral | 26 | 24 | |
| Somewhat important | 32 | 43 | |
| Very important | 8 | 16 | *** |
| **Proximity to work/college/university** | | | |
| Not important at all | 10 | 5 | |
| Not very important | 6 | 6 | |
| Neutral | 16 | 14 | |
| Somewhat important | 30 | 35 | |
| Very important | 38 | 41 | |
| **Proximity to good schools** | | | |
| Not important at all | 31 | 46 | |
| Not very important | 12 | 17 | |
| Neutral | 20 | 21 | |
| Somewhat important | 20 | 11 | |
| Very important | 17 | 6 | *** |
| **Lowest cost option you could find** | | | |
| Not important at all | 10 | 10 | |
| Not very important | 10 | 20 | |
| Neutral | 23 | 21 | |
| Somewhat important | 35 | 32 | |
| Very important | 21 | 17 | |
| **Quiet neighbourhood** | | | |
| Not important at all | 7 | 12 | |
| Not very important | 7 | 24 | |
| Neutral | 21 | 24 | |
| Somewhat important | 47 | 33 | |
| Very important | 18 | 7 | *** |
| **Safety of neighbourhood** | | | |
| Not important at all | 4 | 2 | |
| Not very important | 4 | 7 | |
| Neutral | 10 | 17 | |
| Somewhat important | 41 | 50 | |
| Very important | 41 | 24 | * |
| **Proximity to a waterfront or other natural amenity** | | | |
| Not important at all | 25 | 17 | |
| Not very important | 19 | 21 | |
| Neutral | 31 | 28 | |
| Somewhat important | 16 | 27 | |
| Very important | 9 | 8 | * |
| **Large yard** | | | |
| Not important at all | 32 | 47 | |
| Not very important | 16 | 23 | |

*Table 14.4* (Continued)

| Percentage of respondents | Current location | | p-level |
|---|---|---|---|
| Neutral | 18 | 18 | |
| Somewhat important | 22 | 10 | |
| Very important | 11 | 3 | *** |
| **Cultural amenities nearby and/ or music/arts scene** | | | |
| Not important at all | 26 | 8 | |
| Not very important | 28 | 12 | |
| Neutral | 21 | 23 | |
| Somewhat important | 21 | 45 | |
| Very important | 4 | 12 | *** |
| **Proximity to family members** | | | |
| Not important at all | 31 | 39 | |
| Not very important | 15 | 19 | |
| Neutral | 20 | 23 | |
| Somewhat important | 23 | 16 | |
| Very important | 11 | 4 | ** |
| **Ethnic/racial composition of area** | | | |
| Not important at all | 47 | 35 | |
| Not very important | 18 | 16 | |
| Neutral | 21 | 33 | |
| Somewhat important | 12 | 13 | |
| Very important | 1 | 2 | * |

Notes: ***p<0.001; **p<0.01; *p<0.05

## Will young adults remain in urban areas?

In this chapter, we aimed to provide new evidence to help answer the question as to whether young adults currently living in urban neighbourhoods would remain there as they age. We approached the question in a non-traditional way by linking past, current and future residential locations. This approach is inspired in part by Coulter et al. (2015) who have argued that we should conceptualize residential mobility as a process that needs to be understood by looking at individuals, not aggregate patterns, over time. The findings in this paper do indeed provide some, although tentative, evidence that we can expect higher demand for urban neighbourhoods as young adults age than would have been the case in previous generations. This is because more Millennials are spending their young adult years living in urban settings than was the case for Baby Boomers. We found that living in an urban neighbourhood is a strong predictor of expecting to live in an urban neighbourhood in the future. Previous literature in other contexts arrived at similar conclusions in terms of current location preferences shaping future desires.

Some observers have argued that more young adults living in urban areas than in the past, and remaining there longer due to an expanding young adult life cycle stage, does not ultimately reduce demand for suburban living (e.g. Kotkin, 2013). This perspective assumes an inherent preference for suburban lifestyles, and delayed but ultimately linear and traditional life course trajectory. However, not only are life course trajectories no longer linear in post-modern society, but as the analysis shows, our current residential experiences influence our future decisions, perhaps as we grow accustomed to a particular lifestyle or residential experience.

The increase in urban living during young adulthood could indeed have lasting effects, as there is an association between currently living in an urban area and expressing a desire to continue living in an urban area in the future. Again, the linkages between current and future residential location decisions was confirmed in prior studies as well. This trend may contribute to attaining Smart Growth planning goals toward higher-density and more compact living generally associated with urban neighbourhoods. Further research is required to see whether the findings from this survey can be generalized to all young adults, beyond our highly educated sample. Additionally, there are many variables that can influence residential location decisions not considered in the model here such as a partner's job location and employment characteristics, or neighbourhood quality (Rabe & Taylor, 2010).

There are several other factors that require consideration if we are to tentatively extrapolate from the findings to the impacts on aggregate residential location patterns. First, there is the issue of using stated preferences to forecast future demand. Preferences do not necessarily translate into actual outcomes. There is friction to residential mobility. People may not be able to truly anticipate their future housing needs as circumstances change; they may not be able to afford their stated preference; or they may simply change their mind.

Second, there is the housing market and metropolitan context. As discussed earlier, post-suburban landscapes include an increasingly higher diversity of housing types and neighbourhoods, in some cases even transit coverage approaching levels traditionally only seen in urban neighbourhoods. The changes mean that distinctions between what constitutes an urban versus a suburban neighbourhood will become more blurred (Walks, 2013a; Moos & Mendez, 2015), but also that people with urban preferences might become more willing to move into suburbs that have specific urban characteristics such as enhanced transit or walkability.

Third, racial/visible minorities are less likely to reside in urban neighbourhoods, likely in part due to the development of ethnoburbs and suburban enclaves (Teixeira & Wei, 2015). Immigration rates by country of origin and domestic visible minority birth rates should therefore be used in predictions of future housing demand for urban versus suburban living. However, it must also be remembered that in the Canadian context many immigrant families reside in suburban neighbourhoods at much higher densities than the rest of the population and use transit more frequently (Teixeira & Wei, 2015). This finding points to an inherent shortcoming of relying solely on urban versus suburban categories to study and potentially forecast Smart Growth planning goals that have become prominent in planning documents in Canada and the United States.

Fourth, the women in the survey expressed decreased likelihood of current and stated future preferences for urban living. Despite shifts towards a more postmodern society, this might reflect the persistence of traditional gender roles, and the residential decisions that follow from these roles, which have historically subordinated women to suburban domestic spaces (Rose & Villeneuve, 2006; Kern, 2010). Sense of security and safety were also important factors for those currently living in the suburbs (Table 14.4) and is often connected to childrearing. On the other hand, there has been an increase in the number of women living alone, thus increasing demand for smaller apartments in urban neighbourhoods; and an increasing share of inner city condominium apartments are explicitly marketed to women, in part by creating a heightened sense of security (Kern, 2010). Further research is required to understand and verify the gender dimensions found in the survey data.

Fifth, the analysis shows that those with children have a lower likelihood of currently living or wanting to move to an urban neighbourhood in the future. A very pertinent question for residential forecasts and urban geographies is therefore at what rate Millennials' will have children. Those in the suburbs currently are also more likely to rank proximity to good schools as important (Table 14.4). Whether or not Millennials with children will remain in urban locations over time thus depends not only on whether they have children but also on how they perceive the quality of the schools, among other qualitative factors such as safety, proximity to family and quiet settings.

In some ways it is too early to predict fertility rates for the Millennial generation. However, post-modern society also presents an inherent conundrum for population and demographic forecasting in that the decline in traditional values makes it much more difficult to anticipate household formation and fertility behaviour. Nonetheless, an overall delay in family formation and childbearing has been occurring for some time, which contributes to a decrease in fertility rates, and thus – based on the survey – an increase in demand for urban living.

It should be noted that the presence of children decreases the likelihood of an urban location in addition to the effect of household size. Young adults with children are therefore living in suburban locations not only because of the larger housing available there. There are other factors at play. One plausible interpretation is that young adults with children are moving closer to family, which for most would mean a suburban location based on where they grew up. The findings about the values motivating current suburban residential locations certainly validate this point (Table 14.4).

Another possible factor is a continuing cultural norm that has long associated childrearing with suburban living in North America. However, preliminary analysis of a survey question that asked respondents whether they see suburbs as better places to raise children does not suggest an inherent "suburban bias" in childrearing in the survey. Yet home ownership, access to a yard and access to a car are seen as important for childrearing among a large share of respondents. Recall too that the largest share of respondents expressed future preference for low- or medium-density living. These preferences are inherently difficult to realize in urban neighbourhoods, especially as development trends in the largest metropolitan areas have largely produced high-rise apartment towers in urban areas (Lehrer & Wieditz, 2009).

Given rising housing costs in most major metropolitan areas, and the relatively lower earnings of young adults compared to previous generations (Moos, 2014b), most will not be able to afford their aspirations of having a car, a yard and home-ownership in an urban setting. This trend could have several implications, some of which are opposing in terms of their effect on urban versus suburban location decisions. Inability to acquire housing seen as suitable for childrearing can contribute to a delay in childbearing, which ultimately reduces fertility rates (Lauster, 2010). Some might adapt and have children in higher density neighbourhoods than what they would actually prefer. Some may abandon (or trade off) their urban preferences and move to suburbs where they can attain their homeownership aspirations. Yet others may move to smaller cities and towns where single-family homes are more affordable in urban locations. It is difficult to predict the relative importance of each of these trends due to a lack of research, and therefore the net effect remains largely unknown.

It is important to note that the desire for urban living among Millennials clearly exists in the survey findings; this has also been confirmed in prior studies of revealed preferences (Moos, 2016). The bulk of young adults in post-modern North American society grow up in an increasingly post-suburban landscape. After the age of majority, many more than was the case in previous generations move into urban neighbourhoods, the post-industrial playground for young adults (Chatterton & Hollands, 2002). This experience appears to be influencing their future residential preferences, a finding that requires more explicit consideration in housing demand forecasts, transportation planning and study of residential geographies. Given the importance of post-secondary education for the quaternary sector employment that is so central to the post-industrial urban economy, universities and colleges may also play a role in shaping young adults' residential geographies. However, the relative importance of the location of jobs and post-secondary institutions requires further analysis in this context.

The preference for urban living that can achieve the outcomes desired using Smart Growth planning principles appear to exist in post-industrial, post-modern, post-suburban North America. Two questions remain. First, will planners and other policymakers be able to lay the groundwork to accommodate continued urban living as young adults age? Second, will young adults perceive the inner city, and the condominium apartment, to be a safe and suitable environment for childrearing, as more of them have children? Further research is also required to assess whether the trends observed here apply across a broader range of education levels (since our survey has an overrepresentation of those with post-secondary education), and to provide a more nuanced view of the role of intra-generational differences in gender, ethnicity (and other categories of social difference) in shaping residential preferences.

## Note

1 Home-ownership aspirations and income were initially included but results for these variables were not robust in that their coefficients changed signs depending on model specifications. Also, the sample of non-educated, non-student young adults was not large enough to draw conclusions about differences in education level. Coefficients for other variables did not change significance or sign regardless of whether these class-based variables were included. Analysis of housing types (single-family home, attached, mid-rise or high-rise apartment) in conjunction with the type of residential environments also did not yield robust results.

Will Millennials remain in the city? 197

# References

Alonso, W. (1964). *Location and land use: Toward a general theory of land rent*. Cambridge, MA: Harvard University Press.

Beer, A., Clower, T., Faulkner, D., & Paris, C. (2011). *Consuming housing? Transitions through the housing market in the 21st century*. Portland: The Policy Press.

Bell, D. (1973/2008). *The coming of post-industrial society: A venture in social forecasting*. New York: Basic Books.

Blaauboer, M. (2011). The impact of childhood experiences and family members outside the household on residential environment choices. *Urban Studies, 48*(8), 1635–1650.

Boterman, W. R. (2012). *Residential practices of middle classes in the field of parenthood*. PhD Thesis, Amsterdam Institute for Social Science Research (AISSR). Retrieved from http://dare.uva.nl/document/2/146064.

Bourne, L. S., & Ley, D. (1993). *The changing social geography of Canadian cities*. Montreal: McGill-Queen's University Press.

Chatterton, P., & Hollands, R. (2002). Theorizing urban playscapes: Producing, regulating and consuming youthful nightlife city spaces. *Urban Studies, 39*(1), 95–116.

Clark, W. A. V., Deurloo, M. C., & Dieleman, F. M. (2003). Housing careers in the United States, 1968–93: Modelling the sequencing of housing states. *Urban Studies, 40*(1), 143–160.

Coulter, R. (2013). Wishful thinking and the abandonment of moving desires over the life course. *Environment and Planning A, 45*, 1944–1962.

Coulter, R., van Ham, M., & Findlay, A. M. (2015). Re-thinking residential mobility: Linking lives through time and space. *Progress in Human Geography, 40*(3), 352–374. doi:10.1177/0309132515575417.

Cromartie, J., von Reichert, C., and Arthun, R. (2015). *Factors affecting former residents' returning to rural communities*. United States Department of Agriculture (USDA). Economic Research Service (ERR-185). Retrieved from www.ers.usda.gov/media/1844084/err185.pdf.

Feijten, P., Hooimeijer, P., & Mulder, C. (2008). Residential experience and residential environment choice over the life-course. *Urban Studies, 45*(1), 141–162.

Filion, P., & Kramer, A. (2011). Metropolitan-scale planning in neo-liberal times: Financial and political obstacles to urban form transition. *Space and Polity, 15*(3), 197–212.

Fishman, R. (2005). The longer view: The fifth migration. *Journal of the American Planning Association, 71*(4), 357–366.

Florida, R. (2002). *The rise of the creative class: And how it's transforming work, leisure, community and everyday life*. New York City: Basic Books.

Forsyth, A. (2012). Defining suburbs. *Journal of Planning Literature, 27*(3), 270–281.

Frizell, S. (2014, April 25). The new American dream is living in a city, not owning a house in the suburbs. *Time*. Retrieved from http://time.com/72281/american-housing/.

Garreau, J. (1991/2011). *Edge city: Life on the new frontier*. New York: Anchor Books.

Gordon, D., & Janzen, M. (2013). Suburban nation? Estimating the size of Canada's suburban population. *Journal of Architectural and Planning Research, 30*(3), 197–220.

Handy, S., Cao, X., & Mokhtarian, P. (2006). Self-selection in the relationship between the built environment and walking: Empirical evidence from Northern California. *Journal of the American Planning Association, 72*(1), 55–74.

Harris, R. (2014). Using Toronto to explore three suburban stereotypes. *Environment and Planning A*, advance online publication. doi:10.1068/a46298. Retrieved from www.envplan.com/abstract.cgi?id=a46298.

Hedman, L. (2013). Moving near family? The influence of extended family on neighbourhood choice in an intra-urban context. *Population, Space and Place, 19*(1), 32–45.

Hutton, T. (2008). *The new economy of the inner city*. New York: Routledge.

Keil, R. (Ed.). (2013). *Suburban constellations: Governance, land and infrastructure in the 21st century*. Berlin: Jovis Verlag.

Kern, L. (2010). Gendering re-urbanisation: Women and new-build gentrification in Toronto. *Population, Space and Place, 16*(5), 363–379.

Kotkin, J. (2013). *Retrofitting the dream: Housing the 21st century*. Pinatubo Press.

Lauster, N. (2010). A room to grow: The residential density-dependence of childbearing in Europe and the United States. *Canadian Studies in Population, 37*(3–4), 475–496.

Lees, L., Slater, T., & Wyly, E. K. (2010). *The gentrification reader*. London: Routledge.

Lehrer, U., & Wieditz, T. (2009). Condominium development and gentrification: The relationship between policies, building activities and socio-economic development in Toronto. *Canadian Journal of Urban Research, 18*(1), 140–161.

Ley, D. (1986). Alternative explanations for inner-city gentrification: A Canadian assessment. *Annals of the Association of American Geographers, 76*(4), 521–535.

Ley, D. (1996). *The new middle class and the remaking of the central city*. Oxford: Oxford University Press.

Ley, D. (2003). Artists, aestheticisation and the field of gentrification. *Urban Studies, 40*(12), 2527–2544.

Ley, D. (2010). *Millionaire migrants: Trans-pacific life lines*. Chichester: Blackwell-Wiley.

Moos, M. (2014a). "Generationed" space: Societal restructuring and young adults' changing residential locations patterns. *Canadian Geographer, 58*(1), 11–33.

Moos, M. (2014b). Generational dimensions of neoliberal and post-fordist restructuring: The changing characteristics of young adults and growing income inequality in Montreal and Vancouver. *International Journal of Urban and Regional Research, 38*(6), 2078–2102.

Moos, M. (2016). From gentrification to youthification? The increasing importance of young age in delineating high-density living. *Urban Studies, 53*(14), 2903–2920. doi:10.1177/0042098015603292.

Moos, M., & Mendez, P. (2015). Suburban ways of living and the geography of income: How homeownership, single-family dwellings and automobile use define the metropolitan social space. *Urban Studies, 52*(10), 1864–1882.

Myers, D., & Ryu, S. (2008). Aging baby boomers and the generational housing bubble: Foresight and mitigation of an epic transition. *Journal of the American Planning Association, 74*(1), 17–33.

Patterson, Z., Saddier, S., Rezaei, A., & Manaugh, K. (2014). Use of the urban core index to analyze residential mobility: The case of seniors in Canadian metropolitan regions. *Journal of Transport Geography, 41*, 116–125.

Phelps, N., & Wood, A. (2011). The new post-suburban politics? *Urban Studies, 48*(12), 2591–2610.

Rabe, B., & Taylor, M. (2010). Residential mobility, quality of neighbourhood and life course events. *Journal of the Royal Statistical Society A, 173*(3), 531–555.

Rose, D., & Villeneuve, P. (2006). Life stages, living arrangements and lifestyles: A century of change. In T. Bunting & P. Filion (Eds.), *Canadian cities in transition: Local through global perspectives* (3rd ed., pp. 138–153). Don Mills: Oxford University Press.

Rosen, G., & Walks, A. (2013). Rising cities: Condominium development and the private transformation of the metropolis. *Geoforum, 49*, 160–172.

Rosen, G., & Walks, A. (2015). Castles in Toronto's sky: Condo-ism as urban transformation. *Journal of Urban Affairs, 37*, 289–310.

Sanchez, T., & Dawkins, C. (2001). Distinguishing city and suburban movers: Evidence from the American housing survey. *Housing Policy Debate, 12*(3), 607–631.

Singer, A., Hardwick, S. W., & Brettell, C. B. (Eds.). (2008). *Twenty-first century gateways: Immigrant Incorporation in Suburban America.* Washington, DC: Brookings Institution Press.

Skaburskis, A. (2006). New urbanism and sprawl: A Toronto case study. *Journal of Planning Education and Research, 25,* 233–248.

Statistics Canada (2013). *National household survey: Mobility and migration.* Statistics Canada catalogue number 99-013-X2011028. Retrieved from http://www5.statcan.gc.ca/olc-cel/olc.action?objId=99-013-X2011028&objType=46&lang=en&limit=0.

Teixeira, C., & Wei, L. (Eds.). (2015). *The housing and economic experiences of immigrants in US and Canadian cities.* Toronto: University of Toronto Press.

Townshend, I., & Walker, R. C. (2010). Life course and lifestyle changes: Urban change through the lens of demography. In T. Bunting, P. Filion, & R. Walker (Eds.), *Canadian cities in transition: New directions in the twenty-first century* (4th ed., pp. 131–149). Don Mills: Oxford University Press.

US Census Bureau (2016). *Table A-1. Annual geographical mobility rates, by type of movement: 1948–2016.* CPS Historical Migration/Geographic Mobility Tables. Retrieved from www.census.gov/data/tables/time-series/demo/geographic-mobility/historic.html.

Van Diepen, A., & Musterd, S. (2009). Lifestyles and the city: Connecting daily life to urbanity. *Journal of Housing and the Built Environment, 24,* 331–345.

Walks, A. (2007). The boundaries of suburban discontent? Urban definitions and neighbourhood political effects. *Canadian Geographer, 51*(2), 160–185.

Walks, A. (2013a). Suburbanism as a way of life, slight return. *Urban Studies, 50*(8), 1471–1388.

Walks, A. (2013b). Mapping the urban debtscape: The geography of household debt in Canadian cities. *Urban Geography, 34*(2), 153–187.

Walks, A. (2014). From financialization to socio-spatial polarization of the city? Evidence from Canada. *Economic Geography, 90*(1), 33–66.

Weissman, J. (2015, April 8). *Young adults are getting more suburban. So why does your city seem full of twentySomethings?* Retrieved from www.slate.com/blogs/moneybox/2015/04/08/young_adults_and_cities_college_graduates_are_becoming_more_urban_high_school.html.

Zukin, S. (1982). Loft living as "historic compromise" in the urban core: The New York experience. *International Journal of Urban and Regional Research, 6*(2), 256–267.

# 15 Meet the four types of US Millennial travelers

*Kelcie M. Ralph*

The travel patterns of Millennials – defined here as those ages 16 to 36 – have gar-nered substantial interest from planning practitioners and scholars alike, primarily because they appear to travel differently than previous generations of young peo-ple. Miles driven and trip-making both declined 17% among Millennials between 1995 and 2009 (−7.2 miles and −0.86 trips, respectively).[1] During this period licensing rates also declined by 6% and the share of miles by walking, biking, and riding transit increased by 4%. These figures align with an extensive literature documenting a decline in mobility, trip-making, and licensing and an increase in multimodality in the USA (McDonald, 2015; Polzin, Chu, & Godfrey, 2014) and abroad (Delbosc & Currie, 2013; Kuhnimhof, Buehler, & Dargay, 2011; Le Vine & Polak, 2014; Noble, 2005).

These changes raise a number of important questions about Millennial travel patterns, four of which are discussed in this chapter. How do Millennials in the USA travel? How have travel patterns changed over time? How do travel patterns vary across neighbourhoods? What has caused changes in travel patterns over time for this group? Answering these questions is important given the myriad ways in which transportation contributes – positively and negatively – to society. While travel by automobile is associated with a host of societal ills including congestion, collisions, and pollution, transportation is also central to accessing opportunities. Understanding how young people travel and how travel patterns vary over time and space will make for more informed policy making.

This chapter synthesizes a body of work. For more details about any of the motivations, methods, results, or conclusions presented here see: Ralph (2015, 2016) and Ralph, Voulgaris, Taylor, Blumenberg, and Brown (2016).

## How do Millennials travel?

Travel behaviour research tends to focus on just one aspect of travel at a time and little is known about how multiple aspects of travel interact for young peo-ple. Before we can confidently weigh in on the meaning of changes to Mil-lennial travel patterns, we need a rich description of those patterns, one that incorporates multiple dimensions of travel including trip-making, automobility, and day-to-day variation. To fulfill that need, I developed a multi-faceted trav-eler typology.

## Change over time

We know that annual miles driven, licensing, and car ownership are down and that transit ridership is up among young people. By analysing multi-faceted traveler types instead of individual variables, this analysis provides a richer view of how travel patterns changed over time.

## Geographical variation

How do travel patterns differ in Manhattan, New York and Manhattan, Kansas? Answering questions about geographic variation in travel is particularly important given that many transportation planners hope to use planning tools to reduce automobile use. At the same time, young people live in urban areas in greater numbers than before (Blumenberg, Brown, Ralph, Taylor, & Turley, 2015). Will those urban residents drive less than their peers in the suburbs? This question is fraught with issues of causality, but a good first step is to explore how travel patterns vary in different neighbourhoods.

## Preferences of constraints?

It is well established that travel patterns are indeed changing for Millennials, yet we know very little about why the decline in driving took place. Broadly speaking, there are two competing explanations for these changes. In the first view, young people prefer to drive less for one or more of the following reasons: 1) they have strong environmental beliefs and choose to drive less to "save the environment"; 2) they do not need to travel as much because the internet and mobile communication technologies have reduced the need for face-to-face communication; or 3) they simply find non-automobile modes more convenient in their neighbourhoods. In this view, the observed changes in travel are benign and should be encouraged by programmes, plans, and policies to help more young people realize their preferences.

The second explanation is more pernicious. In this view, young people would prefer to drive as often as before, but cannot due to economic constraints. The Pew Research Center refers to the 2000s as a "lost decade for the middle class" due to the combined forces of the Dot Com bubble, the housing crash, and the Great Recession (Pew Research Center, 2012). Combined with the well-established fact that economic resources (e.g. household income and employment status) are strong predictors of travel for adults (Pucher & Renne, 2003) and young people (McDonald, 2015; Thakuriah, Menchu, & Tang, 2010), it should come as no surprise that Millennials responded to a decade of economic struggles by driving less. With widespread un- and under-employment, many young people likely travelled less because they did not have to travel to a job or to the activities that a steady income pays for. Even for those who were employed, stagnating wages and rising costs of automobility may mean that many young people could not afford to own and operate a vehicle.

While both preferences and constraints are likely at work, it is important to distinguish the relative contribution of each. Despite the allure of the first view, the scholarly literature provides little evidence that preferences have undergone a fundamental change. For instance, young people tend to report that economic factors feature more prominently than environmental concerns in their travel behaviour decisions (Le Vine, Jones, Lee-Gosselin, & Polak, 2014; Tefft, Williams, & Grabowski, 2013). Similarly, while information communication technologies (ICTs) may reduce the need for some trips (such as visiting a bank), ICTs may also generate more travel by freeing up time and resources, making travel easier (Mokhtarian, 2002), or by expanding the size and geographic scope of social networks (Schwanen, Dijst, & Kwan, 2008). Indeed, young people tend to report that ICTs help them meet up with friends (Jorritsma & Berveling, 2014) and that, on balance, ICTs actually increase their trip-making (Delbosc & Currie, 2012). Moreover, disaggregate analyses from Blumenberg, Ralph, Smart, and Taylor (2016) and Le Vine, Latinopoulos, and Polak (2014) that control statistically for potentially confounding factors such as household income, discredit the claim that using ICTs reduces travel.

## Method

This chapter includes four analytical components. First, to describe the travel patterns of Millennials in a multi-faceted manner, I classified young people into four distinct traveler typologies. After defining the traveler typologies, I determined the prevalence of the typologies in three periods (1995, 2001, and 2009) and explored how the prevalence of each type changed over time. As expected, I find that the two auto-centric types became less prevalent over time, and the other types became more prevalent. Next, I examined how the prevalence of the typologies varies across the rural to urban continuum with and without controlling for other factors. The final analytical component indirectly investigated two competing explanations for these changes – preferences and constraints – by determining whose travel patterns have changed the most.

### Sample

This analysis drew on data from the United States National Household Travel Survey (NHTS), which periodically interviews respondents from all 50 states about their travel patterns during a 24-hour period (US Department of Transportation, 2009). The NHTS also includes questions about longer-term travel behaviours, such as transit use in the past month, which were used in this analysis to capture day-to-day variation in travel. Finally, respondents also provide personal information on household income, race, life cycle, residential location, and other characteristics.

To analyse changes in travel over time, I drew on survey data from three years (1995, 2001, and 2009). Respondents were excluded from the analysis if they were missing personal information or travel data. To focus on typical travel, I excluded respondents if they flew in an airplane or traveled over 400 miles (99.5th percentile) on the survey day. This left a useable sample of 27,384 young people in 2009, 22,339 in 2001, and 18,779 in 1995. All results presented here use the survey weights provided by the NHTS to reflect the US population.

## Traveler types

The traveler types were identified using latent profile analysis, a method used to identify relatively homogenous groups in data. To enable comparison across time periods, I began by identifying variables that were measured consistently in all three survey years. From those I selected seven variables that parsimoniously captured mobility, automobile access, use of alternative travel modes, and day-to-day variations in travel. These variables are listed in Table 15.2 below. Three variables measure travel on the survey day: miles of travel by any mode; share of miles by a non-automobile mode; and number of trips. Two variables measure durable investments: automobiles per adult in the household and licensed to drive (y/n). The final two input variables measure long-term travel patterns: annual miles driven and use of public transit in the last month (never, sometimes, once a week or more). These long-term variables were double-weighted relative to the survey day variables.

With the input variables in hand, I then used latent profile analysis to identify a series of increasingly complex classifications. First, the two-class solution divided the sample into an auto-centric group and a non-auto group. Then, a three-class solution added a group whose members used a roughly even mix of modes. The fourth and fifth classes, by contrast, added groups of auto-centric respondents who travel vast distances.

To select between these alternative classifications, I compared statistical criteria for each set of classes (Akaike Information Criterion [AIC] and the Bayesian Information Criterion [BIC], and entropy score). Based on statistical criteria alone, the five-class model was optimal. However, the fifth class represented just 1% of young adults and the low sample size of the fifth class would have hampered subsequent statistical analysis. In cases like these, analysts must use their judgment about the most appropriate set of classes based on characteristics like interpretability, sample size, and reasonableness (Lanza, Collins, Lemmon, & Schafer, 2007). After carefully considering the options, I proceeded with a four-class model.

## Change over time

To assess change over time I determined the prevalence of each traveler type nationwide in 1995, 2001, and 2009 using the appropriate survey weights for each year (see results in Table 15.2).

## Geographical variation

Next, I analysed geographic differences in the travel patterns of Millennials. To do so, I employed a multi-faceted neighbourhood typology developed by Turley Voulgaris, Blumenberg, Taylor, Brown, and Ralph (2017). Turley et al. categorized every census tract in the USA into one of seven neighbourhoods types based exclusively on built environment attributes such as population density, job access, and transit service. As Table 15.1 illustrates, moving along the rural to urban continuum is associated with gradual increases in population and employment density, access to a greater number of jobs, and improved transit service.

Table 15.1 also contains information on the prevalence of each neighbourhood. One in five Americans live in a rural neighbourhood. Nearly six in ten live in suburban neighbourhoods, with fully half of the suburbanites residing in the low-density, job poor New Developments. Finally, just over one in five Americans live in urban neighbourhoods. As we will see later in the analysis, one type of neighbourhood has dramatically different travel patterns than the others. These Old Urban neighbourhoods are characterized by densities, job access, and transit service many times higher than in all other neighbourhoods. Unsurprisingly, these Old Urban neighbourhoods are few and far between; they are home to just 4% of the US population.

The geographic analysis includes two components: a descriptive comparison and a multivariate component. The descriptive component simply compares the prevalence of the traveler types in each type of neighbourhood. Yet this simple comparison may overstate the relationship between the built environment and travel because the people who live in various neighbourhoods differ in many respects and these characteristics likely shape travel patterns. For this reason, the multivariate component of the analysis includes statistical controls for household income quintile, employment status, race/ethnicity, gender, and the attainment of adult roles (lives independently, marriage status, and has a child). The multivariate results are presented relative to travel patterns in New Developments, the low-density, far-flung, suburban neighbourhoods that are home to the largest share of young people (27%). Formally this approach to presenting the regression results is known as average marginal effects and can be thought of roughly as the effect on travel of moving from one neighbourhood (New Developments) to another, everything else equal.

*Table 15.1* Characteristics and prevalence of seven neighbourhood types in the USA (2010)

| | Homes per acre | Jobs-housing balance | Rental homes (%) | Homes > 40 years old (%) | Transit supply index | Jobs within 45-minute drive | Share of US population |
|---|---|---|---|---|---|---|---|
| **All neighbourhoods** | *3.5* | *0.4* | *34%* | *46%* | *0.5* | *118* | *99.9%* |
| Rural | 0.1 | 0.3 | 19% | 42% | 0.0 | 14 | 19.1% |
| **Suburban** | | | | | | | |
| New development | 1.4 | 0.2 | 19% | 17% | 0.0 | 68 | 26.6% |
| Patchwork | 1.7 | 0.7 | 35% | 46% | 0.1 | 94 | 17.6% |
| Established suburbs | 4.1 | 0.3 | 25% | 74% | 0.6 | 186 | 13.4% |
| **Urban** | | | | | | | |
| Urban residential | 5.9 | 0.3 | 58% | 56% | 0.8 | 147 | 13.9% |
| Old urban | 27.5 | 0.3 | 76% | 74% | 4.2 | 533 | 4.3% |
| Mixed-use | 5.2 | 0.7 | 65% | 49% | 1.1 | 181 | 5.0% |

Note: Higher scores on the transit supply index indicate more transit service. Source: *Turley Voulgaris et al. (2017)*.

*Preferences of constraints?*

Finally, to shed light on the cause of changes in travel patterns and to assess the relative contribution of preferences and constraints, I compared changes over time by economic resources. This analysis rests on the assumption that people with more resources – that is, people who are employed, have moderate or high household incomes, or who have completed college or graduate education – will be better able to act on their preferences than people who are not employed, who have low incomes, or who have limited educational attainment. If I found that young people with many resources experienced large declines in driving, I would attribute the changes to a fundamental shift in preferences. If, on the other hand, the declines in driving were restricted to those with the fewest resources, I would attribute the change to economic constraints. I employ three measures of resources: employment status (employed v. not), household income quintile (adjusted for number of people in the household), and educational attainment (only for those who may be reasonably expected to have completed their education: ages 26 to 36).

*Caveats*

Before discussing the results, it is important to note four important limitations of the study. First, and most importantly, the NHTS collects very little data on attitudes and preferences. For this reason, when it comes to analysing the cause of the decline in driving, this work relied on an indirect test of preferences. The data only reveals where people currently live and how they travel, not their preferred options. It is quite possible that there is an unmet demand for affordable urban living that is both not being met and not being captured by this analysis. Second, the data is cross-sectional and, as a result, the analysis of geographical variation can only illustrate association, not causality. That being said, a large literature on travel and the built environment supports the view that there is a causal relationship between the built environment and travel (Ewing & Cervero, 2010). Third, the 2009 version of the NHTS had a very low response rate and failed to sample households without a landline telephone. Together these sampling issues may hamper the generalizability of the findings. Fourth and finally, the 2009 data predates the introduction of Uber and other ridesharing applications, which may have increased multimodality.

# Results

In a nutshell, I find that most young people in the USA drive for nearly every trip. A small minority of young people rely on non-automobile modes, but those young people tend to make very few trips. In stark contrast to expectations, very few young people use a mix of travel modes (Multimodals). Second, when it comes to changes between 1995 and 2009, auto-centric types became less prevalent and non-auto types became more prevalent. Third, when it comes to variation in travel patterns by neighbourhood, the differences are surprisingly muted. In the sole exception – Old Urban neighbourhoods – the majority of young people primarily use non-automobile modes. Yet, those neighbourhoods are extremely rare in the USA. Fourth, young people with few economic resources experienced dramatic declines in automobile travel over time, while declines in driving were fairly muted for those with relatively more economic resources. This indicates that economic constraints may play more of a role in the decline in driving compared to shifts in preferences. The following sections provide greater detail about these findings.

### Traveler types

The latent profile analysis resulted in four distinct traveler types, which are detailed in Table 15.2:

- **Drivers** traveled almost exclusively by automobile on the survey day and throughout the year.
- **Long-Distance Trekkers** also traveled almost exclusively by automobile, but traveled many more miles each day than Drivers to complete the same number of trips.
- **Multimodals** used a mix of travel modes throughout the day or week and, because they made more trips on average, were able to engage in more activities outside the home than Drivers.
- **Car-less** rarely, if ever, used an automobile and made far fewer trips each day than Drivers.

In 2009, well over eight in ten young adults nationwide used an automobile for essentially every trip as either a Driver (79%) or a Long-distance Trekker (3%). Car-less young people were the second most prevalent traveler type, comprising 14% of the population in 2009. Multimodals made up the remaining 4% of the population.

The age range analysed here is rather wide and one may wonder how the traveler types evolve as teens mature into young adults. Unfortunately, the data used here is cross-sectional, so we cannot observe changes in the prevalence of each type over time. Nevertheless, we can see a strong positive association between age and automobile travel. While just 2/3 of 16-year-olds were Drivers, the share of Drivers increases steadily through the teen years before stabilizing at roughly 80% among those in their 20s and 30s. At the same time, travel by multiple modes or entirely without an automobile becomes less prevalent at older ages. This is broadly in line with Clifton's finding that young people tend to give up non-automobile modes rather quickly and dramatically once driving becomes an option (Clifton, 2003).

### Change over time

As Table 15.2 indicates, the share of Drivers and Trekkers declined between 1995 and 2009 while the share of Multimodals and Car-less young people increased.

*Table 15.2* Travel patterns and prevalence of the four types of Millennial travelers

|  | Drivers | Long-Distance Trekkers | Multimodals | Car-less |
|---|---|---|---|---|
| **TYPICAL TRAVEL PATTERNS** | | | | |
| **Travel on the survey day** | | | | |
| Miles traveled by any mode (median) | 24 | 50 | 12 | 2 |
| Trips per day (median) | 4 | 4 | 5 | 2 |
| Share of miles by auto (%) | 100 | 100 | 52 | 0 |
| **Durable investments** | | | | |
| Annual miles driven (median) | 9,000 | 50,000 | 300 | 0 |
| Licensed to drive (%) | 90 | 100 | 67 | 47 |
| **Long-term travel patterns** | | | | |
| Use transit (%) | | | | |
| Once a week or more | 5 | 5 | 26 | 32 |
| Never | 84 | 89 | 58 | 50 |
| Automobiles per adult (%) | | | | |
| None | 1 | 1 | 7 | 24 |
| Less than one | 23 | 13 | 34 | 43 |
| One or more | 76 | 86 | 59 | 33 |
| **SHARE BY AGE IN 2009 (%)** | | | | |
| **Ages 16–36** | 79 | 3 | 4 | 14 |
| **By age** | | | | |
| Age 16 | 66 | 0 | 11 | 23 |
| Age 17 | 71 | 0 | 8 | 21 |
| Age 18 | 78 | 1 | 4 | 17 |
| Age 19 | 76 | 2 | 4 | 18 |
| Age 20 to 25 | 82 | 3 | 3 | 13 |
| Age 26 to 36 | 81 | 5 | 3 | 11 |
| **SHARE NATIONWIDE BY YEAR (%)** | | | | |
| 2009 | 79 | 3 | 4 | 14 |
| 2001 | 80 | 5 | 3 | 12 |
| 1995 | 83 | 4 | 3 | 10 |

Source: 1995 NPTS, 2001 and 2009 NHTS, weighted values

## *Geographical variation*

The geographic analysis presented in Table 15.3 illustrates two important find-ings. First, unsurprisingly, Drivers were most prevalent in the lowest-density sub-urban neighbourhoods (New Developments) and in rural areas. Second, and more surprisingly, moving along the Rural to Urban continuum was associated with only modest declines in the share of Drivers.

Notably, these findings were consistent for both the descriptive and multivari-ate analyses. The top half of Table 15.3 displays the descriptive results – the share of Drivers (and Trekkers, and so on) in each type of neighbourhood nationwide.

The bottom half of the table displays the share of Drivers (and the other traveler types) relative to the distribution of Drivers in New Developments, where the plurality of young people live. The analysis featured in the bottom half of the table controls statistically for important differences between the populations in each type of neighbourhood (e.g. household income, racial and ethnic makeup, the attainment of adult roles, and age).

To better understand the remarkable stability in the prevalence of the traveler types across most of the rural to urban continuum, consider the case of New Developments and Established Suburbs. These neighbourhoods differ in that Established Suburbs have three times the population density and 2.7 times the jobs within a 45-minute drive than New Developments. Despite these dramatic built environment differences, Drivers constituted the vast majority of young people in both types of neighbourhoods. More specifically, living in an Established Suburb was associated with a reduction in the share of Drivers of just 9.4 percentage points, everything else equal.

While Drivers made up the majority of young people in most neighbourhoods, they were in the minority in Old Urban neighbourhoods. In those neighbourhoods, Car-less young people constituted the majority. Moreover, because Old Urban neighbourhoods are so amenable to travel by non-automobile modes, young Car-less people there made more trips and used transit far more frequently than Car-less people in other types of neighbourhoods. In those other neighbourhoods, Car-less young people faced automobile-centred developments that often offered little to no transit service. As a result, Car-less young people outside Old Urban neighbourhoods primarily relied on walking for all of their travel. As

*Table 15.3* Geographic variation in traveler type among young people (16 to 36) in 2009

|  | Drivers | Long-Distance Trekkers | Multimodals | Car-less |
|---|---|---|---|---|
| **Descriptive results: Traveler type in each neighbourhood nationwide** | | | | |
| Rural | 83 | 6 | 3 | 8 |
| New Development | 87 | 3 | 3 | 7 |
| Patchwork | 84 | 2 | 2 | 11 |
| Established Suburbs | 77 | 2 | 5 | 16 |
| Urban Residential | 72 | 2 | 5 | 20 |
| Old Urban | 39 | 1 | 4 | 56 |
| Mixed-use | 71 | 3 | 6 | 19 |
| All neighbourhoods | 79 | 3 | 4 | 14 |
| **Model results: Percentage point change in population share relative to New Developments** | | | | |
| Rural | −3.4 | +3.3 | −0.5 | +0.5 |
| Patchwork | −2.2 | −0.8 | −0.5 | +3.6 |
| Established Suburbs | −9.2 | −1.1 | +1.9 | +8.5 |
| Urban Residential | −11.9 | −0.7 | +3.0 | +9.6 |
| Old Urban | −39.6 | −1.5 | +1.4 | +39.7 |
| Mixed-use | −13.6 | 0.0 | +3.6 | +10 |

Source: 1995 NPTS, 2001 and 2009 NHTS, weighted values

I discuss below, the 7–16% of young people who are Car-less in auto-oriented, suburban neighbourhoods should raise serious equity concerns for planners and policymakers.

Unsurprisingly Long-distance Trekkers were much more common in Rural neighbourhoods where destinations tend to be highly dispersed. This finding was consistent in both the descriptive and multivariate analyses.

Many planners aim to encourage people who drive to make some of their trips by other modes, yet there is little evidence of such multimodality in the data. While few would be surprised by the paucity of Multimodals in rural or low-density suburban neighbourhoods, many may be surprised to find that there were also very few Multimodals in urban neighbourhoods during the survey period. Multimodals were most common in Mixed-use neighbourhoods and, even there, just 6% of young people were Multimodal.

## *Preferences or constraints?*

Table 15.4 presents data on the distribution of the traveler types by economic resources in 1995, 2001, and 2009. This analysis illustrates two key findings. First, the results provide further evidence for the well-established connection between economic resources and travel patterns. Second, and most importantly, the results reveal that young people with relatively few economic resources experienced more dramatic declines in driving and more dramatic increases in Car-less-ness than did young people with more economic resources. In other words, people with the greatest ability to act on their preferences (i.e. those with the most resources) were the least likely to give up driving and become Car-less. This suggests that much of the decline in driving between 1995 and 2009 was the result of economic constraints rather than preferences. The following sections provide more details about these results.

Columns a, b, and c of Table 15.4 exhibit the close association between economic resources and travel patterns. In 2009, for example, 24% of unemployed respondents were Car-less, but just 9% of employed respondents were. Similarly, the propensity to be Car-less increased steadily as household income decreased. Of those in the lowest income quintile, one in four young people were Car-less in 2009. At the highest income quintile, by contrast, fewer than one in ten were Car-less. Although it is not pictured in the table, the close association between resources and travel patterns was reinforced in the analysis of educational attainment for 26- to 36-year-olds. I find that fewer than one in ten college-educated respondents were Car-less, whereas more than one in three of those with less than a high school degree were.

Columns d and e of Table 15.4 focus on change over time. While the prevalence of Drivers fell among both those with and without jobs, the decline was two times larger for those without jobs than those with them. Similarly, gains in Car-less-ness were two times larger among unemployed respondents than among employed respondents.

We turn next to change over time by household income, where I find that changes in the prevalence of driving and Car-less-ness were restricted to those in

*Table 15.4* Distribution of the traveler types by year and resources (ages 16 to 36)

|  | *(a) 1995* | *(b) 2001* | *(c) 2009* | *(d) Percentage Point Change 95:09* | *(e) Sig* |
|---|---|---|---|---|---|
| **DRIVERS ONLY** |  |  |  |  |  |
| **Employment status** |  |  |  |  |  |
| Not employed | 75 | 72 | 71 | −4.4 | *** |
| Employed | 85 | 82 | 83 | −1.8 | ** |
| **Household income quintile** |  |  |  |  |  |
| Q1 | 75 | 69 | 68 | −6.3 | *** |
| Q2 | 84 | 82 | 79 | −5.0 | *** |
| Q3 | 86 | 85 | 85 | −0.6 | n.s. |
| Q4 | 86 | 86 | 85 | −1.8 | n.s. |
| Q5 | 86 | 83 | 85 | −0.5 | n.s. |
| **CAR-LESS ONLY** |  |  |  |  |  |
| **Employment status** |  |  |  |  |  |
| Not employed | 19 | 22 | 24 | 4.9 | *** |
| Employed | 8 | 9 | 9 | 1.6 | ** |
| **Household income quintile** |  |  |  |  |  |
| Q1 | 19 | 22 | 25 | 6.2 | *** |
| Q2 | 9 | 10 | 14 | 5.2 | *** |
| Q3 | 8 | 8 | 8 | −0.1 | n.s. |
| Q4 | 7 | 6 | 8 | 2.0 | * |
| Q5 | 7 | 9 | 9 | 1.2 | n.s. |

Note: Values are unadjusted (not the result of a statistical model) and are weighted using the provided survey weights. N=18,799 in 1995, 22,339 in 2001, and 27,384 in 2009.

Source: 1995 NPTS and 2001/2009 NHTS.

the lowest two income quintiles. At the other end of the income spectrum, young people with moderate to high incomes did not experience any meaningful (or statistically significant) change in the share of Drivers or Car-less between 1995 and 2009. Indeed, the decline in Drivers was 12 times larger for those with low household incomes (Q1) than for those with high incomes (Q5).

When it comes to educational attainment, young people (ages 26–36) with the fewest resources once again experienced more dramatic shifts in travel patterns over time than those with more resources. Namely, for those with less than a high school degree, the share of Drivers decreased by 20 percentage points and the share Car-less increased by 20 percentage points. By contrast (and contrary to expectations), the travel patterns of highly-educated young people were relatively stable over time. In other words, Bachelor's and advanced degree holders were just as likely to be Drivers (or Trekkers, or Multimodals, or Car-less) in 2009 as they were in 1995.

While Trekkers and Multimodals constituted an extremely small share of the population (3 and 4% of young adults respectively), it is nevertheless worth briefly mentioning changes over time for those groups because their pattern of change over time differed from that of the Drivers and Car-less respondents. It was young people with the most resources, not the fewest, who experienced increases in multimodality and decreases in trekking. The household income analysis provides the clearest illustration of this trend. Low-income young adults (Q1 and Q2) experienced no change in the share of Trekkers or Multimodals over time. Among high income respondents, by contrast, the share of Multimodals increased and the share of Trekkers decreased between 1995 and 2009.

## Synthesis and conclusion

What do these findings imply for policymakers and practitioners? Below I draw on the literature and the empirical findings presented here to identify two distinct challenges for planners and policymakers.

I found that the vast majority of young people rely on automobiles for nearly every trip and tend to live in areas where automobile travel is the norm. Yet in focusing on *observed* patterns of travel and residential location, we may miss an unexpressed latent demand for walking, biking, and transit and for living in areas supportive of those travel modes. In other words, some young people may desperately want to ditch their cars, but may live in places that lack the supportive infrastructure – bike lanes, high-quality transit service, and a variety of nearby destinations – to do so. During the first decade of the 21st century many young people seemed to fulfill their preferences by locating in urban neighbourhoods, which were home to 4.2 million more young people in 2010 than in 2000 (Blumenberg, Brown, Ralph, Taylor, & Turley Voulgaris, 2015).

During the same period, however, growth in suburban neighbourhoods was even more dramatic – 14.3 million more young people lived in suburban neighbourhoods in 2010 than in 2000. So while an urban renaissance was indeed underway, the era of suburbanization was far from over. Of course there are two related reasons why the pace of suburbanization outpaced that of urban growth. First, there are simply more new homes in the suburbs than in urban areas and, partially as a result, homes in neighbourhoods with supportive infrastructure command a price premium. Together these forces may keep many young people from realizing their preferences. If this view is correct, then building more housing in transit-rich, walkable neighbourhoods should increase the number of young people choosing to be Car-less or Multimodal.

Despite the alluring possibility of latent demand, a second key take-away of this work is that many young people in the USA are already Car-less and that they tend overwhelmingly to be those with the fewest economic resources. In other words, many Car-less young people in the USA are not fulfilling some preference for car-free lifestyle, but rather they simply have no other choice but to live without a car in a largely inhospitable built environment. In thinking about and planning for these individuals it is useful to employ a transportation disadvantage and social exclusion lens. Social exclusion occurs when individuals cannot fully participate in the typical activities of daily life because they lack access to safe, reliable, and convenient transportation (Kenyon, Lyons, & Rafferty, 2002; Lucas, 2012). The risk that many Car-less young people likely suffer from transportation disadvantage and social exclusion is evidenced by their staggeringly low level of mobility and trip-making. Indeed, the typical Car-less individual participated in a single activity outside the home per day. By employing a transportation disadvantage lens, we may begin to view the decline in driving by young people in the 2000s not as an unmitigated success story, but potentially as an early indication of a problem.

These results lead to two distinct challenges for planners and policymakers. On the one hand, we should develop programmes and policies to help those with an unrealized preference for urban living and multimodal or car-free travel to make their preferences a reality. At the other extreme, for those forced into Car-less-ness by economic constraints, our challenge is to provide adequate access to opportunities so that those groups are not excluded from the essential elements of society. The scale of this challenge becomes ever greater as poverty suburbanizes (Raphael & Stoll, 2010). As planners and policymakers in 21st-century cities, we must tackle both challenges if we are to meet the needs of a diverse Millennial generation.

## Note

1 Author's analysis of data from the 1995 Nationwide Personal Transportation Survey and the 2009 National Household Travel Survey.

## References

Blumenberg, E., Brown, A., Ralph, K., Taylor, B., & Turley, C. (2015). *Typecasting neighborhoods and travelers: Analyzing the geography of travel behavior among teens and young adults in the US* Retrieved from www.lewis.ucla.edu/wp-content/uploads/sites/2/2015/10/Geography-of-Youth-Travel_Final-Report.pdf?mc_cid=68d255b9a1& mc_eid=cc6e1ea4c7.

Blumenberg, E., Brown, A., Ralph, K., Taylor, B., & Turley Voulgaris, C. (2015). *Back to the city? Do Millennials favor cities and what does this mean for the future of travel?* Paper presented at the Annual Meeting of the Association of Collegiate Schools of Planning, Houston, Texas.

Blumenberg, E., Ralph, K., Smart, M., & Taylor, B. (2016, November). Who knows about kids these days? Analyzing the determinants of youth and adult person-miles of travel in 1990, 2001, and 2009? *Transportation Research Part A: Policy and Practice, 93*, 39–54.

Clifton, K. (2003). Independent mobility among teenagers: Exploration of travel to after-school activities. *Transportation Research Record, 1854*, 74–80.

Delbosc, A., & Currie, G. (2012). *Using online discussion forums to study attitudes toward cars and transit among young people in Victoria.* Paper presented at the Australian Transport Research Forum, Perth, Australia.

Delbosc, A., & Currie, G. (2013). Causes of youth licensing decline: A synthesis of evidence. *Transport Reviews, 33*(3), 271–290. doi:10.1080/01441647.2013.801929.

Ewing, R., & Cervero, R. (2010). Travel and the built environment: A meta-analysis. *Journal of the American Planning Association, 76*(3), 265–294.

Jorritsma, P., & Berveling, J. (2014). *Not carless, but car-later.* Netherlands. Retrieved from www.kimnet.nl/en/publication/not-carless-car-later.

Kenyon, S., Lyons, G., & Rafferty, J. (2002). Transport and social exclusion: Investigating the possibility of promoting inclusion through virtual mobility. *Journal of Transport Geography, 10*(3), 207–219. doi:http://dx.doi.org/10.1016/S0966-6923(02)00012-1.

Kuhnimhof, T., Buehler, R., & Dargay, J. (2011). A new generation: Travel trends for young Germans and Britons. *Transportation Research Record, 2230*, 58–67.

Lanza, S. T., Collins, L. M., Lemmon, D. R., & Schafer, J. L. (2007). PROC LCA: A SAS procedure for latent class analysis. *Structural Equation Modelling, 14*(4), 671–694.

Le Vine, S., Jones, P., Lee-Gosselin, M., & Polak, J. (2014). Is heightened environmental sensitivity responsible for drop in young adults' rates of driver's license acquisition? *Transportation Research Record: Journal of the Transportation Research Board, 2465*(1), 73–78. doi:10.3141/2465-10.

Le Vine, S., Latinopoulos, C., & Polak, J. (2014). What is the relationship between online activity and driving-licence-holding amongst young adults? *Transportation, 41*(5), 1071–1098. doi:10.1007/s11116-014-9528-3.

Le Vine, S., & Polak, J. (2014). Factors associated with young adults delaying and forgoing driving licenses: Results from Britain. *Traffic Injury Prevention, 15*(8), 794–800. doi:10.1080/15389588.2014.880838.

Lucas, K. (2012). Transport and social exclusion: Where are we now? *Transport Policy, 20*, 105–113. doi:http://dx.doi.org/10.1016/j.tranpol.2012.01.013.

McDonald, N. C. (2015). Are Millennials really the "Go-Nowhere" generation? *Journal of the American Planning Association, 81*(2), 90–103. doi:10.1080/01944363.2015.1057196.

Mokhtarian, P. L. (2002). Telecommunications and travel: The case for complementarity. *Journal of Industrial Ecology, 6*(2), 43–57. doi:10.1162/108819802763471771

Noble, B. (2005). *Why are some young people choosing not to drive?* Paper presented at the European Transport Conference, Strasbourg, France.

Pew Research Center. (2012). *The lost decade of the middle class: Fewer, poorer, gloomier.* Retrieved from www.pewsocialtrends.org/2012/08/22/the-lost-decade-of-the-middle-class/.

Polzin, S. E., Chu, X., & Godfrey, J. (2014). The impact of millennials' travel behavior on future personal vehicle travel. *Energy Strategy Reviews, 5*, 59–65.

Pucher, J., & Renne, J. (2003). Socioeconomics of urban travel: Evidence from the 2001 NHTS. *Transportation Quarterly, 57*(3), 49–77.

Ralph, K. (2015). *Stalled on the road to adulthood? Analyzing the nature of recent travel changes for young adults in America, 1995 to 2009.* Ph.D., University of California, Los Angeles, Los Angeles.

Ralph, K. (2017). Multimodal Millennials? The four traveler types of young people in the United States in 2009. *Journal of Planning Education and Research, 37*(2), 150–163. doi:10.1177/0739456X16651930.

Ralph, K., Voulgaris, C. T., Taylor, B. D., Blumenberg, E., & Brown, A. E. (2016). Millennials, built form, and travel insights from a nationwide typology of US neighborhoods. *Journal of Transport Geography, 57*, 218–226. doi:http://dx.doi.org/10.1016/j.jtrangeo.2016.10.007.

Raphael, S., & Stoll, M. A. (2010). *Job Sprawl and the suburbanization of poverty.* Retrieved from www.brookings.edu/research/job-sprawl-and-the-suburbanization-of-poverty/

Schwanen, T. I. M., Dijst, M., & Kwan, M.-P. (2008). ICTs and the decoupling of everyday activities, space and time: Introduction. *Tijdschrift voor economische en sociale geografie, 99*(5), 519–527. doi:10.1111/j.1467-9663.2008.00489.x.

Tefft, B. C., Williams, A. F., & Grabowski, J. G. (2013). *Timing of driver's license acquisition and reasons for delay among young people in the United States, 2012.* Washington, DC.

Thakuriah, P., Menchu, S., & Tang, L. (2010). Car ownership among young adults. *Transportation Research Record: Journal of the Transportation Research Board, 2156*(1), 1–8. doi:10.3141/2156-01.

Turley Voulgaris, C., Blumenberg, E., Taylor, B. D., Brown, A., & Ralph, K. (2017). Synergistic neighborhood relationships with travel behavior: An analysis of travel in 30,000 US neighborhoods. *Journal of Transport and Land Use, 10*(2), 1–25.

US Department of Transportation. (2009). *National household travel survey.* Washington, DC: Federal Highway Administration.

# 16 Emerging mobility patterns of the Millennials in Canada

*Ajay Agarwal*

As discussed in the previous chapter (Chapter 15), the Millennials – those born between 1980 and 2000 – have attracted a lot of attention in various media for their mobility patterns, which may be different from the earlier generations. Scholars claim that the Millennials have lower affinity for automobile travel as compared to their parents and therefore could truly be a sustainable generation in terms of their travel behaviour. Such arguments are not completely baseless. After decades of strong growth in aggregate travel demand, driven mainly by increasing car use, growth is slowing down in North America, Europe, and Australia (Newmann & Kenworthy, 2011; Le Vine, Jones, & Polak, 2009). Since 2007, total annual vehicle kilometres travelled (VKT) have declined remarkably in the USA and somewhat less dramatically in Canada (Giuliano & Agarwal, 2014; Natural Resources Canada, 2009). This reduction in VKT could be attributed to multiple factors including, but not limited to, the deep economic recession during the second half of the past decade, ongoing retirement of Baby Boomers from the workforce, and perhaps different mobility patterns of the Millennials (Blumenberg, Taylor, & Smart, 2012). Building on the research in Chapter 15, this chapter lays out a detailed theoretical framework to explain why we might expect the Millennials to travel differently than prior generations. The chapter then offers an analysis of Canadian data on Millennial mobility patterns.

## Why should we expect the Millennials to travel differently?

Before attempting to address whether or not the Millennials travel differently, let us review the theoretical framework of travel behaviour. In theory, travel is a derived *demand*. That is, an individual's demand for travel originates from their desire to participate in an activity located some distance away from the individual's present location. To be able to travel, individuals must bear some travel *cost*. Travel cost consists of both time and money spent in making a trip. The *benefit* that an individual derives from travelling to (and therefore participating in) a particular activity would vary by the activity itself and is likely to be different for different individuals. For example, the benefit of travelling to work would be different from travelling for a haircut. Furthermore, different individuals would likely value a haircut differently. The final decision on whether to make the trip or not will depend on an implicit cost–benefit analysis that the individual must make: whether benefits of making the trip outweigh the associated costs.

According to standard economic theory, individuals tend to maximize their *utility* while operating under a *budget constraint*. In context of travel, utility could be construed as the net of all benefits associated with all the activities participated in, minus the associated travel costs. The budget constraint comprises of both time available to spend on travel (out of 24 hours in a day) and money available to spend on travel. Clearly, higher income individuals and households are more likely to have more money available to spend on travel and are therefore likely to travel more than lower income individuals and households. Life cycle stage will have a strong influence on time available to spend on travel. For example, parents with young children will have less time to travel since they have to spend time on childcare as opposed to childless couples, all else equal. Single-worker households are likely to have different travel budgets as compared to dual-worker households, all else equal.

The theoretical framework described above suggests several reasons for expecting different travel behaviour from the Millennials.[1] First, adoption of new information and communication technologies (ICT) influences how people use their time and may affect how much they travel (Golob & Regan, 2001; Kwan, 2006). For example, use of ICT increases flexibility not only in terms of where but also when work could be performed. Teens and young adults, i.e. the majority of the Millennials, tend to be the earliest and most frequent adopters of latest technologies including the internet-based applications (Lenhart, Madden, & Hitlin, 2005; Pew Research Center, 2010; Mans, Interrante, Lem, Mueller, & Lawrence, 2012). The rapid profusion of ICT amongst the Millennials is likely to influence their activities and therefore their mobility patterns. An important piece in this regard is time spent on social media websites or mobile phone apps. Online social media is a relatively new phenomenon, dating back to 2002 or so. Again, the Millennials were the earliest adopters and most intense users of social media (Mans et al., 2012). Engaging with social media is likely to influence the Millennials' travel patterns. Time spent on virtual social networks means there is less time available for travel on physical networks. However, expansion of one's virtual social networks may also lead to higher travel demand, as one may desire physical interaction with new "friends". Smart mobile phones and mobile internet access are not cheap. Money spent on smartphones and data plans means much less money available to spend on physical travel.

Second, unemployment rates were very high during the most recent economic recession, and did not return to pre-recession levels in Canada until 2012 (Bloskie & Gellatly, 2012). High unemployment influences travel demand in two ways; first, unemployment leads to reduced demand for journey-to-work and work-related travel; and, second, unemployment limits the resources available to spend on non-work activities and associated travel. In addition to an overall reduction in travel demand, the prolonged economic downturn may also have influenced the Millennials' travel patterns indirectly. Several studies suggest that youth struggling to find work increasingly "boomerang" back home to live with parents (Kaplan, 2009; Wiemers, 2011; Galarneau, Morissette, & Usalcas, 2013), drawn by free or steeply discounted accommodation, meals, and access to their parents' vehicle(s). Where the location of the parent's house is suburban, subsidized access to a car may promote more automobile travel among the Millennials. However, living with parents also implies delayed independent household formation among the Millennials and therefore a delay in progressing to more mature life cycle stages (such as forming familial partnerships, bearing children), which have profound implications on travel patterns.

Third, transport policy during the past two decades may have influenced the mobility patterns of the Millennials. Various states and provinces in the USA and Canada have implemented graduated driver's licensing programmes and made driver's licensing in general more difficult and restrictive than previously (Marzoughi, 2011). These restrictions can be viewed as an added "cost" of travel, not experienced by earlier generations. Additionally, after a gap of several decades, some North American cities have started reinvesting in public transit (Giuliano & Agarwal, 2014). Improved public transit service coupled with restricted driving ability is likely to reduce automobile travel and increase public transit use among the Millennials, particularly teens and young adults who have yet to obtain a driving licence.

## Empirical evidence

Existing empirical evidence tends to support the theoretical arguments presented above in general. The evidence establishes that, in general, the Millennials tend to acquire driving licences later in life, have lower rates of car ownership, have lower automobility, and use public transit more as compared to their preceding generations. However, it is still unclear whether these trends will persist in the future as the Millennials progress to later life cycle stages and their economic condition improves. Table 16.1 presents findings from some of the recent studies of the Millennials' spatial trends from North America.

*Table 16.1* Recent studies of the Millennial's spatial trends from North America

| Publication | Study area (Study period) | Data source | Key findings |
| --- | --- | --- | --- |
| Klein and Smart (2017) | USA (1999–2013) | Panel Study of Income Dynamics (PSID) | In general young adults from the Millennial generation own about 13% fewer cars than previous generations at same age. However, economically independent young Millennials have higher than expected car ownership rates, surprising in view of their comparatively lower incomes. |

*(Continued)*

*Table 16.1* (Continued)

| Publication | Study area (Study period) | Data source | Key findings |
|---|---|---|---|
| Newbold and Scott (2017) | Canada (1992–2010) | StatsCan General Social Survey (GSS) | The Millennials are more likely to use public or active transport modes than previous generations. However, they are "catching up" with older generations, as evident from increases in the proportion holding a valid driver's licence and auto trips. |
| Blumenberg et al. (2016) | USA (1990–2009) | Nationwide Personal Transportation Survey (NPTS), National Household Travel Survey (NHTS) | Decline in vehicle miles travelled is significantly associated with lower employment rate among youth. No evidence that declining personal travel is associated with changes in travel preferences. |
| Brown et al. (2016) | USA (2001–2009) | NHTS, American Community Survey, Environmental Protection Agency's Smart Location Database | Young adults are indeed more likely to ride public transit than older adults. However, there is little evidence that these patterns would persist as teens and young adults age. |
| McDonald (2015) | USA (1995, 2001, 2009) | NHTS | There has been a general decline in automobility between 1995 and 2009 among both the Millennials and Generation X. The decline in automobile use among the Millennials is partially explained by delays in household formation, partnering, childbirth and low employment. |
| Garikapati, Pendalya, Morris, Mokhtarian, and McDonald (2016) | USA (2003–2013) | American Time Use Survey (ATUS) | Overall, the Millennials drive less. As the Millennials age into their 30s, they are increasingly exhibiting activity-time use patterns that resemble Gen X individuals when they were in their early 30s. |

*Table 16.1* (Continued)

| Publication | Study area (Study period) | Data source | Key findings |
|---|---|---|---|
| Atherwood (2015) | San Francisco Bay Area (2007–2011) | American Community Survey | The older Millennials (25+ years old) tend to concentrate in more urban and central residential locations. These locations are characterized by comparatively greater racial diversity, and higher population density. |
| Grimsrud and El-Geneidy (2014) | Greater Montreal (1998–2008) | O-D Surveys from Agence Metropolitaine de Transport (AMT) | Recent cohorts of youth (age 20–24 in 2008) are using transit more as compared to earlier cohorts of youth. However, their transit use decreases as they age to their 30s and 40s, |
| Moos (2014) | Montreal and Vancouver (1981–2006) | StatsCan | Young adults are increasingly concentrating in high-density neighbourhoods with high degree of walkability to recreational amenities, and within close proximity of public transit. Such neighbourhoods are mostly, but not necessarily, located in central cities. |
| Blumenberg, Taylor, and Smart (2012) | USA (1990, 2001, 2009) | NPTS and NHTS | The Millennials were travelling fewer miles per day (−18%), and making fewer trips per day (−4%) as compared to the previous generations, all else equal. The reductions were a result of lower employment and not changes in travel preferences per se. |

Blumenberg et al. (2016) investigated a series of hypotheses concerning factors associated with the observed decline in vehicle miles travelled by teens and young adults during the last decade. The factors include increasingly stringent driver licensing regulations, the economic recession during the late 2000s, early adoption of ICTs among youth, and delayed formation of familial households. Using 2001 and 2009 National Household Travel Survey (NHTS) data from the USA, the authors estimated nine multivariate regression models to test their hypotheses. The results showed that after controlling for personal, household, and location attributes, only the declining employment rate – due to economic recession – was significantly associated with declining automobile travel among teens and young adults. The authors conclude that the observed decline in automobile travel among teens and young adults in the USA is not due to changes in their travel preferences, but due to economic effects. Hence, the presently observed decline in private automobile travel may not continue in the future.

Brown et al. (2016) examined whether the decline in vehicle miles travelled by teens and young adults in the last decade was associated with an increased reliance on public transit. The authors used 2001 and 2009 NHTS data to estimate logistic regression models in their analysis. The results showed that after controlling for individual, household, and neighbourhood characteristics teens and young adults are indeed more likely to use public transit than older adults. However, the authors did not find evidence that the likelihood of greater transit use by teens and young adults will persist in the future, as they age.

McDonald (2015) examined 1995, 2001, and 2009 NHTS data from the USA to analyse mobility trends and identify factors associated with changes in automobility amongst the Millennials. The author finds that there has been a general decline in automobility, measured as daily vehicle trips and daily vehicle kilometres travelled, between 1995 and 2009. Further, the trend is not limited to the Millennials but also includes Generation X. The regression results showed that lifestyle-related demographic shifts, including decreased employment and delayed formation of households, are responsible for a 10 to 25% decline in automobility; Millennial-specific factors such as increased online interaction and changes in attitudes towards mobility account for 35–50% decline in automobility; and the remaining decrease (40%) comes from general reduction in automobility between 1995 and 2009. The author concluded that since the decline in automobile use among the Millennials is partially explained by delays in household formation, partnering, childbirth and low employment, the Millennials' automobile use could potentially increase substantially as these conditions are reversed in the future and if the Millennials rely solely on the personal automobile to meet their increased travel demand.

Blumenberg et al. (2012) examined the 1990 National Personal Travel Survey (NPTS) and the 2001 and 2009 NHTS data from the USA to compare mobility patterns of the Millennials to that of the previous generations at a similar life stage. The authors found that the Millennials were travelling fewer miles per day (−18%), and making fewer trips per day (−4%) as compared to the previous generations, all else equal. The authors found that the reductions were a result of "economic effects" (i.e. lower employment rates) rather than "generational effects" (i.e. changes in travel preferences).

Using 1981 and 2006 data from Vancouver and Montreal, Moos (2014) found that the young adult population – defined as 25–34 years old – is increasingly concentrated in high-density neighbourhoods. Using multivariate regression analysis, the author established that the trend is supported by young, childless, non-family households living consumption-oriented lifestyles, i.e. an extended young adult life cycle stage. The study found that the young adults are drawn to such neighbourhoods because of proximity to public transit and high levels of walkability to recreational amenities. In Vancouver, concentration of young adults extends outward from the downtown (decentralization) along transit lines because housing is too expensive in the centre. In Montreal, centralization is dominant.

Using 1998, 2003, and 2008 origin-destination survey data from the Greater Montreal region, Grimsrud and El-Geneidy (2014) examined public transit's travel mode share of commute trips. The authors restricted their analysis to persons aged 20–65 years. They divided their sample into five-year age groups (i.e. 20–24, 25–29, 30–34, and so on). The authors found that persons in younger age groups tended to use public transit more as compared to older age groups. The authors observed that public transit use declined for individuals between 20 and 30 years of age, but tended to stabilize as the person reached their early 30s. Further, 2008 data showed higher transit use among persons in their early 30s. Since travel pattern was observed to stabilize as a person reached 31–35 years age group, the authors surmise that "high transit use rates observed among Greater Montrealers currently in their early 30s are expected to continue as they replace older, lower-use birth cohort commuters" (p. 123). Findings of Grimsrud and El-Geneidy (2014) are consistent with those of Moos (2014); centralization of young adults should indeed lead to greater transit use. Conversely, those with higher affinity for public transit would seek central locations where public transit access is much better compared to suburban locations (Giuliano & Agarwal, 2014).

Similar changes to the mobility patterns of Millennials have also been reported outside of North America. Kuhnimhof et al. (2012) examined national travel survey data from Germany, for multiple years between 1976 and 2009, to compare changes in travel trends of young adults (defined as persons 18–29 years age) over time. The authors found that automobile travel among persons 18–29 years age consistently increased between 1976 and 1997. However, since 1997 automobile travel decreased by more than 20% in this age group. The decline in car use was similar for persons both with and without access to a car. Further, public transit use of this age group almost doubled between 1997 and 2009. The authors also observed that the share of licensed drivers in this age group has remained stagnant since 1997 and car-ownership rates have declined in this age group since early 1980s. The authors argued that both economic factors and public policy have contributed to these trends (also see Chapter 15). Both car-ownership and driving costs have increased much more than public transit cost in Germany since the early 1990s. Additionally, policies to discourage driving such as high parking cost, traffic calming etc. have all contributed to the decline in car use among youth.

## Greater Toronto area

Let us now compare mobility pattern of the Millennials to Generation X, their preceding generation, using data from the Greater Toronto Area (GTA), the largest metropolitan region in Canada in terms of both population and spatial extent. Three cross-sections – 2001, 2006, 2011 – of *Transportation Tomorrow Surveys* (TTS) are used in this analysis. TTS are household travel surveys from the GTA,

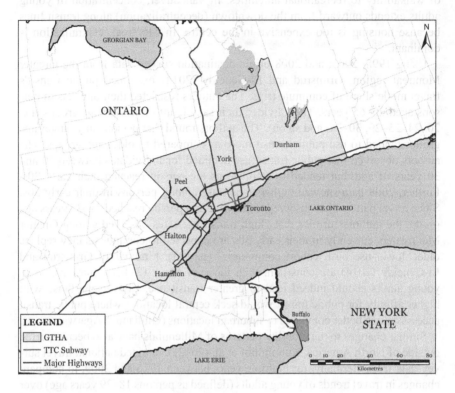

*Figure 16.1* The study area: Greater Toronto

conducted every five years since 1986. Geographical boundaries of TTS vary across different survey years. To maintain consistent study area boundaries, the study area is restricted to the City of Toronto, and municipalities of Durham, York, Peel, Halton, and Hamilton in the Province of Ontario, Canada (see Figure 16.1).

The numbers of households included in the final sample were approximately 120,000, 117,000 and 121,000 in 2001, 2006, and 2011 respectively. All persons 16–40 years of age were categorized in five groups: 16–20, 21–25, 26–30, 31–35, and 35–40. This matches the five-year intervals of the three data cross-sections, thereby allowing us to examine changes in mobility patterns within the select age-group cohorts. Three age groups will show travel trends of the Millennials: 16–20 in all the three cross-sections, 21–25 in 2006 and 2011, and 26–30 in 2011.

Table 16.2 shows changes in household attributes in the study area between 2001 and 2011. The household size steadily declined from 2.75 persons per household in 2001 to 2.59 persons in 2011. The declining household trend has persisted in Canada for quite some time. The number of full-time workers per household declined from 1.15 in 2001 to 0.96 in 2011, the effect of the deep economic recession during the past decade. The number of part-time workers per household declined marginally between 2006 and 2011. The number of licensed drivers per household declined from 1.78 in 2001 to 1.74 in 2011. Declining numbers of driving licence holders per household is indicative of lower licensure rates, which is likely due to more restrictive licensing requirements in Ontario. The average number of vehicles per household declined marginally between 2001 and 2006, and remained steady between 2006 and 2011.

Table 16.3 presents percentages of persons in different age groups who possessed a valid driving licence in years 2001, 2006, and 2011. The table shows that the Millennials are getting their driving licences later in life as compared to Generation X. The licensure rate among the 21–25 years age group in 2011 (the Millennials) is 3% lower than the same age group in 2001 (Generation X). However, licensure rate for the 26–30 years age group from both generations is not significantly different. This result is similar to those from other North American studies (Blumenberg et al., 2012; McDonald, 2015).

Table 16.4 presents average numbers of daily automobile trips by age group for the three cross-sections. The table shows that the Millennials were making significantly fewer auto trips in 2011 as compared to 2001, up to 3% less. However, the reduction in daily automobile trips was not exclusive to the Millennials during

*Table 16.2* Changes in household attributes between 2006 and 2011

| Attributes | 2001 | 2006 | 2011 |
| --- | --- | --- | --- |
| Household size | 2.75 | 2.69 | 2.59 |
| Full-time workers | 1.15 | 1.07 | 0.96 |
| Part-time workers | 0.25 | 0.25 | 0.24 |
| Driving licence holders | 1.78 | 1.75 | 1.74 |
| Vehicles | 1.46 | 1.44 | 1.44 |

*Table 16.3* Driving licensure by age group

| Age, Years | 2001 | 2006 | 2011 | 2001–2011 | t | p value | |
| --- | --- | --- | --- | --- | --- | --- | --- |
| 16–20 | **0.59** | **0.54** | **0.53** | −0.06 | 11.76 | 0.00 | *** |
| 21–25 | 0.87 | **0.85** | **0.84** | −0.03 | 6.22 | 0.00 | *** |
| 26–30 | 0.90 | 0.89 | **0.90** | 0.00 | 1.37 | 0.17 | |
| 31–35 | 0.92 | 0.91 | 0.92 | 0.00 | 0.92 | 0.52 | |
| 36–40 | 0.92 | 0.92 | 0.93 | 0.01 | 2.76 | 0.01 | *** |

Note: Bold typeface represents the Millennials

*Table 16.4* Daily automobile trips by age group

| Age Group | 2001 | 2006 | 2011 | 2011–2001 | t | p value | |
| --- | --- | --- | --- | --- | --- | --- | --- |
| 16–20 | **3.04** | **3.01** | **2.97** | −0.08 | 8.17 | 0.00 | *** |
| 21–25 | 3.01 | **2.99** | **2.94** | −0.07 | 7.82 | 0.00 | *** |
| 26–30 | 3.03 | 3.00 | **2.95** | −0.08 | 8.47 | 0.00 | *** |
| 31–35 | 3.11 | 3.04 | 3.02 | −0.10 | 10.63 | 0.00 | *** |
| 36–40 | 3.20 | 3.13 | 3.09 | −0.11 | 11.99 | 0.00 | *** |

Note: Bold typeface represents the Millennials

*Table 16.5* Percentage of all daily trips made using public transit

| Age Group | 2001 | 2006 | 2011 | 2011–2001 | t | p value | |
|-----------|------|------|------|-----------|------|---------|----|
| 16–20 | **0.25** | **0.28** | **0.32** | 0.07 | 16.68 | 0.00 | *** |
| 21–25 | 0.24 | **0.31** | **0.31** | 0.07 | 14.55 | 0.00 | *** |
| 26–30 | 0.18 | 0.20 | **0.19** | 0.00 | 0.85 | 0.39 | |
| 31–35 | 0.15 | 0.17 | 0.15 | 0.00 | 0.80 | 0.42 | |
| 36–40 | 0.13 | 0.14 | 0.14 | 0.02 | 5.76 | 0.00 | *** |

Note: Bold typeface represents the Millennials

this period. There was an overall reduction in automobility during this period. This trend is similar to that reported in the USA (McDonald, 2015).

Table 16.5 presents percentages of all daily trips made using public transit by age group. The table shows significant increases in transit use between 2001 and 2011 among the youngest two cohorts of the Millennials included in the analysis. The table also shows decline in public transit use as the 21–25 years age cohort (2006) moves to the 26–30 years age cohort (2011). This is consistent with findings from other studies that show the Millennials are "catching up" with Generation X as they age. Transit use of the Millennials in the 26–30 years age group is not significantly different from that of Generation X.

The descriptive statistics described above show that the mobility trends from the GTA are similar to those reported elsewhere in North America. We also compared mobility patterns of those living inside the central city (City of Toronto) to those located outside the central city (results not shown). The results were as expected; there was lower automobile use and higher transit use inside the central city for all age groups. The general trends in both the central city and outside the central city were consistent, i.e. the young Millennials were driving less and using public transit more as compared to the Generation X, and the older Millennials were "catching up" with Generation X.

## Discussion

Multiple studies have established the lower automobile travel demand of the Millennials during the past decade. However, as several authors have argued, there is little evidence that the lower demand reflects a fundamental shift in travel behaviour. Lowering of travel demand coincided with a deep economic recession, and scholars argue that this lower automobility was an outcome of unemployment and lower household incomes. Therefore, it may be premature to celebrate the lower rates of automobile travel by the Millennials (also see Chapter 15).

Several studies have established that the Millennials prefer central cities as compared to suburbs in terms of household locations (Chapters 2, 13, 14). If this trend continues, then it would have a profound effect on metropolitan travel trends. The built environment in central cities is quite different from suburban areas, for example high street intersection density, streets laid in grid pattern, short block lengths, and narrow rights of way (Southworth, 1995). These built environment features have been known to influence travel behaviour, particularly for walking and bicycling trips (Voorhees Yan, Clifton, & Wang, 2011; Dalton et al., 2011; Agarwal & North, 2012). Despite substantial decentralization of employment over the past few decades, urban cores continue to be the largest single concentration of employment in large metropolitan area (Giuliano, Redfearn, Agarwal, Li, & Zhuang, 2007). To the extent that those living in central cities also work in the downtown, centralization of the Millennials would lead to shorter commutes, greater use of public transit and walking/bicycling, and reduced automobile use. Furthermore, there are several nascent movements, such as "complete streets" and "sprawl repair", presently underway that advocate retrofitting suburbia to accommodate more walking, cycling and transit. If indeed suburbs in the future are built and/or rebuilt to become less autocentric, then we could expect greater use of public transit and active transport modes in suburbs as well.

Given that the young Millennials are obtaining driver's licence later in life and using public transit more, there is an excellent opportunity to inculcate the habit of public transit use in the longer term in this still young generation. Evidence from other Canadian cities has shown that a subsidized transit pass is a potent facilitator of public transit use (Agarwal & Collins, 2016; Collins & Agarwal, 2015). Discounted long-term public transit passes could be offered to teens and young adults by transit agencies, perhaps in collaboration with schools and employers, as one way of promoting public transit.

Although the existing mobility trends of the Millennials are somewhat desirable from a sustainable transportation perspective, they must be taken with caution. If the Millennials continue to rely less on the private automobile to meet their travel needs in the future, then there is both hope and opportunity for reducing automobile dominance, quite the opposite of travel trends during the past half century (Chapter 2). If these trends do not hold, then we may have to re-think policies to prevent this still young generation from becoming as automobile dependent as the previous two generations.

Public policy could play a critical role in encouraging persistent use of public transit and active transport modes by the Millennials as they age and progress to more mature life cycle stages. This would require a coordinated, multi-pronged approach. First, steps must be taken to retain the Millennials living in central cities. It means making central city living more palatable to families with young children, just as it is for the young childless Millennials now. This would involve improving central-city schools, encouraging convenient shopping and grocery stores to locate downtown, and creating amenities desired by young children such as public parks and open spaces, etc. Second, efforts to provide reliable, convenient and safe public transit service should continue. We are presently witnessing a general support for investment in public transit across North America (Giuliano & Agarwal, 2014) and transportation planners must ensure that the investments made are strategic, which may not always be glamorous (for example rail is not always the best transit option given the existing built environment). Third, as early adopters of technology, the Millennials are likely to embrace driverless or autonomous cars in the future. The future impacts of driverless cars on the built environment and urban travel patterns are still unknown, and regulations for such vehicles still being contemplated (Levinson, Cao, & Fan, 2016). If done correctly, policies concerning driverless vehicle technology could be used to encourage lower personal car ownership, and reduce parking demand in central cities. This would free up highly valuable real estate in central cities presently dedicated to parking to provide for more housing and public amenities discussed above.

Finally, there are several determinants of travel behaviour that are exogenous to transport policy, for example the prevailing health of the economy. Such factors may have stronger influence on travel behaviour as compared to land use and transportation policies, and planners should be cognizant of this.

## Note

1 For a detailed discussion on the topic see Blumenberg et al. (2012).

## References

Agarwal, A., & Collins, P. (2016). Opportunities and barriers for promoting public transit use in midsize Canadian cities: A study of Kingston, Ontario. *Canadian Journal of Urban Research, 25*(2), 1–10.

Agarwal, A., & North, A. (2012). Encouraging bicycling among university students: Lessons from Queen's University, Kingston, Ontario. *Canadian Journal of Urban Research, 21*(1), 151–168.

Atherwood, S. (2015). An ecological exploration of generation Y: Residence location choice in the San Francisco bay area. *Applied Spatial Analysis, 8*, 325–349.

Bloskie, C., & Gellatly, G. (2012). *Recent developments in the Canadian economy: Economic insights, No. 19*. Ottawa: Statistics Canada.

Blumenberg, E., Ralph, K., Smart, M., & Taylor, B. (2016). Who knows about kids these days? Analyzing the determinants of youth and adult mobility in the US between 1990 and 2009. *Transportation Research Part A, 93*, 39–54.

Blumenberg, E., Taylor, B., & Smart, M. (2012). *What's youth got to do with it? Exploring the travel behavior of teens and young adults*. Berkeley, Ca: University of California Transportation Center.

Brown, A., Blumenberg, E., Taylor, B., Ralph, K., & Voulgaris, C. (2016). A taste for transit? Analyzing public transit use trends among youth. *Journal of Public Transportation, 19*(1), 49–67.

Collins, P., & Agarwal, A. (2015). Impacts of public transit improvements on ridership, and implications for physical activity, in a low density Canadian city. *Preventive Medicine Reports, 2*, 874–879.

Dalton, M., Longcare, M., Drake, K., Gibson, L., Adachi-Mejja, A., Swain, K., Xie, H., & Owens, P. (2011). Built environment predictors of active travel to school among rural adolescents. *American Journal of Preventive Medicine, 40*(3), 312–319.

Galarneau, D., Morissette, R., & Usalcas, J. (2013). *What has changed for young people in Canada? Insights on Canadian society*. Ottawa: Statistics Canada. Retrieved from www.statcan.gc.ca/pub/75-006-x/2013001/article/11847-eng.pdf.

Garikapati, V., Pendalya, R., Morris, E., Mokhtarian, P., & McDonald, N. (2016). Activity patterns, time use, and travel of millennials: A generation in transition? *Transport Reviews, 36*(5), 558–584.

Giuliano, G., & Agarwal, A. (2014). Landuse impacts of Transportation investments: Highway and transit. In S. Hanson & G. Giuliano (Eds.), *The geography of urban transportation* (4th ed.). New York: The Guilford Press.

Giuliano, G., Redfearn, C., Agarwal, A., Li, C., & Zhuang, D. (2007). Employment concentrations in Los Angeles, 1980–2000. *Environment and Planning A, 39*, 2935–2957.

Golob, T., & Regan, A. (2001). Impacts of information technology on personal travel and commercial vehicle operations: Research challenges and opportunities. *Transportation Research*, 87–121.

Grimsrud, M., & El-Geneidy, A. (2014). Transit to eternal youth: Lifecycle and generational trends in Greater Montreal public transport mode share. *Transportation, 41*, 1–19.

Kaplan, G. (2009). *Boomerang kids: Labor market dynamics and moving back home.* Federal Reserve Bank of Minneapolis, Working Paper, 675.

Klein, N., & Smart, M. (2017). Millennials and car ownership: Less money, fewer cars. *Transport Policy, 53*, 20–29.

Kuhnimhof, T., Armoogum, J., Buehler, R., Dargay, J., Denstadli, J., & Yamamoto, T. (2012). Men shape a downward trend in car use among young adults – evidence from six industrialized countries. *Transport Reviews, 32*(6), 761–779.

Kwan, M. (2006). Transport geography in the age of mobile communication. *Journal of Transport Geography, 14*, 384–385.

Le Vine, S., Jones, P., & Polak, J. (2009). *Has the historical growth in car use come to an end in Great Britain?* Paper presented at the European Transport Conference, Leeuwenhorst, The Netherlands.

Lenhart, A., Madden, M., & Hitlin, P. (2005). *Youth are leading the transition to a fully wired and mobile nation.* Pew Internet & American Life Project. Retrieved from pewinternet.org

Levinson, D., Cao, J., & Fan, Y. (2016). *The transportation futures project: Planning for technology change.* Minnesota Department of Transportation.

Mans, J., Interrante, E., Lem, L., Mueller, J., & Lawrence, M. (2012). Next generation of travel behavior. *Journal of the Transportation Research Board, 2323*, 90–98.

Marzoughi, R. (2011). Teen travel in the greater Toronto area: A descriptive analysis of trends from 1986 to 2006 and the policy implications. *Transport Policy, 18*, 623–630.

McDonald, N. (2015). Are Millennials really the "Go-Nowhere" generation? *Journal of the American Planning Association, 82*(2), 90–103.

Moos, M. (2014). "Generationed" space: Societal restructuring and young adults' changing residential location patterns. *The Canadian Geographer, 58*(1), 11–33.

Natural Resources Canada. (2009). *Canadian vehicle survey 2009 summary report.* Retrieved from http://oee.nrcan.gc.ca/publications/statistics/cvs09/pdf/cvs09.pdf.

Newbold, K., & Scott, D. (2017). Driving over the life course: The automobility of Canada's Millennial, generation X, baby boomer and greatest generations. *Travel Behaviour and Society, 6*, 57–63.

Newmann, P., & Kenworthy, J. (2011). Peak car use: Understanding the demise of automobile dependence. *World Transport Policy and Practice, 17*, 31–42.

Pew Research Center. (2010). *Millennials: Confident. Connected. Open to change.* Retrieved from http://pewsocialtrends.org/files/2010/10/millennialsconfident-connected-open-to-change.pdf.

Southworth, M. (1995). *Walkable suburbs? An evaluation of neotraditional communities at the urban edge.* Berkeley, CA: Center for Real Estate and Urban Economics.

Statistics Canada. (2011). *Estimates of population by age and sex for Canada, provinces and territories (Table 051-0001).* Retrieved from http://www5.statcan.gc.ca/cansim/pic kchoisir?lang=eng&searchTypeByValue=1&id=0510001.

Voorhees, C., Yan, A., Clifton, K., & Wang, M. (2011). Neighborhood environment, self-efficacy, and physical activity in urban adolescents. *American Journal of Health Behavior, 35*(6), 674–688.

Wiemers, E. (2011). *The effect of unemployment on household composition and doubling up.* National Poverty Center, University of Michigan, Working Paper Series, 11–12. Retrieved from http://npc.umich.edu/publications/u/working_paper11-12.pdf.

# 17  I drive to work, sometimes

## Motility capital and mode flexibility among young adult gentrifiers

*Markus Moos, Khairunnabila Prayitno and Nick Revington*

There is growing interest in young adults' changing transportation patterns as an increasing share of young adults delay getting their driver's licence, instead relying more frequently on transit, walking and cycling as their main modes of transport (Polzin, Chu, & Godfrey, 2014). While this trend is being observed in North America and Europe, there is some disagreement as to the permanence of this shift or its underlying causes. Moreover, the majority of young adults continue to use the automobile as their main mode of transport (Chapters 15 and 16).

Recent research suggests that – at least for one segment of the young adult population (such as those associated with the youthification and gentrification of high density areas) – there is reason to believe that we are observing the beginning of a more permanent shift toward less auto-intensive lifestyles (Moos, 2016). However, there is also increasing concern that central city living, where car-free lifestyles are most plausible, has become a luxury that not everyone can afford, and that not all gentrifiers moving into these areas actually reduce their car use (Danyluk & Ley, 2007).

In this chapter, we aim to make a contribution to the research exploring young adults' changing transportation patterns in the context of gentrification and youthification. Specifically, we are interested in the phenomenon of multi-modality in commuting behaviour, or what we call mode flexibility. An example of the multi-modal commute is the same person driving to work one day, biking the next and taking the bus for the remainder of the week versus using the same mode for all commuting trips.

We conduct statistical analysis of novel primary data from an online survey of almost 700 people aged 18 to 40 in the USA and Canada. In contrast to the classic survey question that transportation researchers rely on to examine individual transportation decisions, which often asks people about their "most frequent" mode of transport (Schwanen & Mokhtarian, 2005; Pinjari, Pendyala, Bhat, & Waddell, 2007), this survey allowed respondents to choose multiple modes within a week. Although there is a segment of the transport literature that considers multi-modal behaviour, these generally have to rely on smaller, primary survey data (Molin, Mokhtarian, & Kroesen, 2016). The purpose of this research is to critically examine how differences in mode flexibility (i.e. not being constrained to any one transportation mode for a given trip) may relate to structural inequalities arising from gentrification and the workings of housing markets. These trends are increasingly seen as factors that displace lower income earners from walkable and transit-accessible locations.

We draw on the sociological concept of mobility, which holds that socio-spatial movement is intertwined with both social meanings and relations of power (Urry, 2000, 2007). This concept is useful as it allows us to consider socio-economic differences in modal split and mode flexibility as outcomes of not only individual choices but also as the result of systemic constraints (and privileges) with unequal impacts for different social classes. The ability to be mobile can therefore be considered a form of capital (much like social or cul-tural capital (Bourdieu, 1984)) in that it can be exchanged for social advantage or other types of capital (Kaufmann, Bergman, & Joye, 2004). Kaufmann (2002) calls this form of capital "motility," defined as *"the way in which an individual appropriates what is possible in the domain of mobility and puts this potential to use for his or her activities"* (p. 37, italics in original), while Urry (2007) uses the term "network capital". We position mode flexibility (e.g. having the choice in the morning to drive or bike to work depending on the weather) as a form of motility capital.

While it is likely that there have always been daily variations of some people's commute mode, we propose that there is a new segment of the population, young gentrifiers to be specific, that has the flexibility to walk, bike or take transit while also having sufficient income to own a car. In other words, there is a strong link among Millennials between motility capital, as expressed in mode flexibility, and other forms of social status. This may be a result of several factors such as socio-economic restructuring resulting in gentrification, the changing characteristics of work, prolonged young adult lifestyles and youthification, and increasing blend-ing of work and education (e.g. part-time work while going to school).

The implications of our research for planning lay in a) the limitations of current transportation surveys that only ask about dominant transport modes; and b) the ways policy ought to consider how the geographies of existing and new transpor-tation infrastructures produce different levels of motility capital.

## An interdisciplinary transport research agenda

Our analysis requires us to bring together several different literatures. Our approach is framed by studies from research on transportation, gentrification, urban geography and equity in transportation infrastructure provision. We rely on research that considers mode choice as an outcome of individual level factors to develop our statistical analysis. However, we interpret differences in mode across groups as an outcome of structural conditions, as well as individual decisions. We use the prior research on gentrification, young adult residential location in urban geography, and the concept of motility capital as a means to explore the equity implications of the way urban restructuring may be creating an urban landscape that allows some people (those with more motility capital) greater mode flexibil-ity compared to others.

### Demography and lifestyle

The determinants of travel mode are dependent on personal factors (e.g. lifestyle, preferences), demography and neighbourhood characteristics, as well as other factors such as trip purpose and weather. Younger people and children generally bike and walk more, and the use of these modes decreases with age (Transport for London, 2012). Once children enter adolescence, they typically transition to public transport as their most common mode (Nobis, 2007). Public transit is most common among adolescents out of necessity and the inability to afford a car. Post-secondary students also often rely on transit, at times supported by a university transit pass; they might also walk or cycle since post-secondary students often reside close to campus and have constrained income-earning opportunities. Once young adults enter the workforce, the car generally becomes the most common mode of transport in North America.

However, there is an increasing proportion of mature adults (above the age of 45) with high car availability who report using public transit as their most common mode of travel (Roche-Cerasi, Rundmo, Sigurdson, & Moe, 2013; Nobis, 2007). Once these adults retire and enter old age, they take fewer and shorter trips and their most prominent mode of travel returns to walking (Boschmann & Brady, 2013). There is also growing evidence that young adults are less likely to drive, and less likely to obtain a driver's licence compared to previous generations although the factors explaining this trend are not yet entirely clear (Moos et al., 2015). The factors influencing the obtainment of a driver's licence include increasing environmental concern, declining or stagnant incomes, increasing central city locations and transit system improvements.

Studies indicate that lifestyle decisions are typically better indicators of modal choice than age alone (Boschmann & Brady, 2013; Hildebrand, 2003). Young adults with busy lifestyles are most likely to need flexibility in their mode of travel, hence the choice of travelling by car, while those who reside in urban centres would likely choose cycling over driving, as cycling provides greater flexibility and autonomy (Simons et al., 2014).

### Identity (class, race, and gender)

When considering class and income factors, several studies indicate that individuals with higher income have a higher likelihood of travelling by car (Simsekoglu, Nordfjaern, & Rundmo, 2015; Susilo & Cats, 2014; Polzin et al., 2014). In addition to the perceived need for flexibility (and ability to remain flexible due to their higher incomes), individuals holding high level positions, such as managers, travel by car due to the symbolic status of power and prestige. The car not only provides reliability and punctuality for commuters, but it also offers symbolic motivations (Simsekoglu et al., 2015; Li, 2003).

Low-income households have lower access to cars than high-income households, whereby the low-income are more likely to have no access to cars as opposed to high-income earners – 21% to only 4% (Blumenberg & Pierce, 2014;

US Census Bureau, 2005–09). In addition, having access to a car generally leads to higher chances of finding employment. Since white Americans have greater access to vehicles, they are more likely to find employment than non-whites who are more likely to rely on public transit (Zax & Kain, 1996; Patacchini & Zenou, 2005; Preston, Mclafferty, & Liu, 1998).

Racial background is generally less important than gender in determining transportation modes; however, class, race, and gender intersect in complex ways to produce geographies of transportation modes (Preston et al., 1998). When considering gender, a contradiction exists between preferences and reality. Women are more likely to use public transportation than men, but have higher preferences to use a car due to safety concerns (Preston et al., 1998; Susilo & Cats, 2014). They often avoid walking or taking transit in the evening, and see driving as the safest option (Preston et al., 1998).

Perceptions of safety and security are also more general determinants of mode choice, aside from the gender dimension. Studies indicate that the car and public transit come with different perceptions of risk (Roche-Cerasi et al., 2013; Rundmo, Nordfjærn, Iversen, Oltedal, & Jørgensen, 2011). Typically, those who use the car are less worried about their safety than those using public transit. However, the perception of risk is low in determining mode choice until a certain threshold of risk is reached where people then consider alternative modes (Simons et al., 2014; Rundmo et al., 2011).

## Multi-modality

It seems intuitive that the determinants of mode choice are dynamic and that daily variations would exist. Multi-modality is defined as a traveler who makes a decision among modes based on context as opposed to a habitual traveler who uses a single mode regardless of context (Aarts, Verplanken, & Van Knippenberg, 1998). Although the phenomenon of multi-modality appears to be growing (Nobis, 2007), analysis of this aspect of transportation behaviour has proven difficult due to a shortage of available data and difficulty of developing complex models (Clifton & Muhs, 2012; Blumenberg & Pierce, 2014). Existing studies find that individuals who are mode flexible use public transit for specific reasons only, such as commuting and long-distance travel, whereby cars are used more flexibly for different purposes (Kuhnimhof, Chlond, & Von der Ruhren, 2006), and that personal characteristics have only a minor influence on multi-modality (van Nes, 2002).

A key determinant of multi-modality is the ability to access several modes. Thus, not surprisingly, individuals who live in urban centres, which have high levels of transit connections and increased walkability, are also more likely to be multi-modal (Nobis, 2007). One study found that older adults living in areas built following transit-oriented development models are less likely to be auto dependent, finding a 61% increase in use of other modes (Boschmann & Brady, 2013). But since young adults, and particularly Millennials, are more likely to reside in urban centres than older adults and previous generations (Polzin et al., 2014; Moos, 2016), there is reason to suspect that young adults have greater ability to choose multi-modality.

Neighbourhoods with characteristics supportive of multi-modality, such as having greater access to public transit and walkability, have been associated with gentrification. Jones and Ley (2016) have pointed to a number of studies indicating that homes located in close proximity to transit stations are associated with higher property values (Baum-Snow and Kahn, 2005; Debrezion, Pels, & Rietveld, 2007; Billings, 2011; Chatman, Tulach, & Kim, 2012), while Gilderbloom, Riggs and Meares (2015) point to similar results using walkability index measurements (Cortright, 2009; Diao & Ferreira, 2010; Meares, 2014; Pivo & Fisher, 2011).

There is also a higher likelihood of college graduates moving into neighbourhoods that are within one mile of a transit station (Kahn, 2007). Although there is little study to indicate the role of transit accessibility in processes of gentrification (Revington, 2015; see Grube-Cavers & Patterson, 2015, for a notable exception), several scholars associate the expansion of transit-oriented developments with gentrification (Rayle, 2015; Jones & Ley, 2016; Kahn, 2007; Quastel, Moos, & Lynch, 2012).

As these prior studies suggest, neighbourhoods that provide the greatest opportunity for mode flexibility are increasingly targeted towards the privileged, potentially increasing the differentiation of multi-modality by socio-economic status. Indeed, the groups identified as being the most multimodal include higher income Americans, adolescents, individuals who live in urban centres, and older adults with high car availability (Blumenberg & Pierce, 2014; Nobis, 2007). In addition, a study in Germany indicates that there is an increasing proportion of young adults between the age of 18–29 who have high mode flexibility, as many young adults are reducing car use and taking alternative modes such as transit and cycling while still driving at times (Kuhnimhof, Buehler, Wirtz, & Kalinowska, 2012).

### Access to multiple modes as motility capital

Greater ability to be mobile, including through the use of multiple modes, is a form of social advantage. High motility capital can also hedge against a variety of everyday risks, including car breakdowns, transit service disruptions, or unexpected travel needs (such as in an emergency) (Kaufmann, 2002). Therefore, we prefer the term mode flexibility since it is explicit about multi-modality being a form of flexibility, which is generally perceived as beneficial, and that the ability to be flexible may not be equally distributed.

Kaufmann's (2002) concept of motility describes the capacity for mobility as a form of capital, exchangeable or substitutable for other forms of social, cultural, or economic capital in producing socio-economic differentiation. Motility capital depends on three interdependent elements: access, competence, and appropriation (Kaufmann et al., 2004). Access refers to possibilities for mobility. In the case of mode flexibility, this includes both the ability to buy a car and to live in a neighbourhood that offers alternative mode options. Competence, meanwhile, accounts for the ability of individuals themselves to access these possibilities. For example, driving a car requires a set of demonstrated skills evaluated through a licensing procedure, walking and cycling rely on a certain level of physical ability, and transit use depends on knowledge of routes and timetables. Finally, appropriation refers to how individuals perceive and react to both access and competence. For instance, someone may live in a transit-accessible neighbourhood and possess the competencies to use the transit network, yet perceive transit as unsafe or overly complicated, and therefore will not use it.

Because transportation infrastructure is distributed unevenly across space, some households have more motility capital than others, even if they are similar in competence and appropriation. The ability to be mobile can take various forms from having access to highways, subways or even bicycle lanes. As we have seen, walkable central city areas and places with good access to transit, which would facilitate mode flexibility as an expression of motility capital, are increasingly gentrified. We thus propose that the ability to be mode flexible would be unevenly distributed among young adults based on their ability to afford a car and buy into central neighbourhoods that offer alternative mode options. In other words, not only does motility capital offer advantages but, aside from individual-level factors, we expect its distribution to be a product of underlying structural conditions.

## Travel data

We use responses from an online survey of young adults 18 to 40 years of age living in the USA and Canada. The survey included 80 questions on demography, employment, housing, residential and work location, transport patterns, including commute mode and distances, residential preferences and value statements regarding respondent's rationale behind their housing, location and transport decisions. The online survey was available from February 2015 to April 2016 and 1,413 people attempted the survey during this time frame. However, not everyone completed all questions, thus, our effective sample size is smaller (n=689) based on the variables included in the analysis.

Information regarding survey recruitment strategies, sample representativeness and generalizability are discussed in much more detail in Chapter 9. In general, the survey over-represents the university educated. It also has some regional under-representation in the western and southeastern USA, as well as underrepresentation in Quebec (Canada). Caution should be exercised if attempting to generalize aggregate trends to the general population. Instead the survey results are most relevant when utilized to study relationships among variables within the sample.

For our analysis, respondents were divided into three categories based on their responses on mode choice (Figure 17.1). Multi-modality can be observed for different kinds of trips but here we focus on the commute to work. Respondents were asked to indicate how many times in the past week they relied on one or more of the following transportation modes to travel to work: car, public transit, walking, cycling, other. The survey allowed respondents to choose more than one type of mode per week but only one mode per day.

After initial exploratory analysis, statistically significant differences were found almost exclusively among those traveling solely by car as compared to those with no cars. Differences also arose between exclusive drivers and those

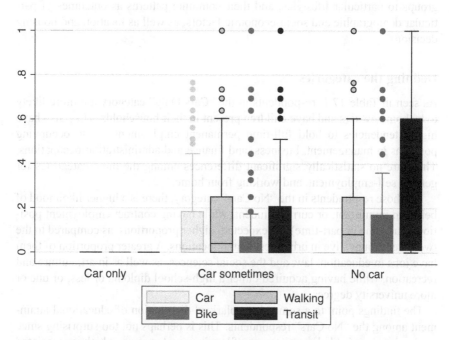

*Figure 17.1* Comparing mode share distributions among respondents commuting by "car only", "car sometimes" and "no car"

with cars that also use other modes. Thus, we create the following three categories of respondents to capture their modality: "Car Only", "Car Sometimes" and "No Car". Respondents in the "Car Only" category only use the automobile to travel to work. The "Car Sometimes" category are still car owners but they do not use a car for all their commuting trips in a given week. Those in the "No Car" category never use a car to travel to work in the reference period.

To test our hypothesis that there is a link between mode flexibility (a form of motility capital) and social status, we compare the demographic, occupational, education, residential location (urban versus suburban) characteristics of the three groups, as well as their home ownership and student status. Following prior gentrification research, we use higher income and educational attainment, occupations in the "Education, law and social sciences" and "Art, culture and recreation" categories, and urban residential locations as indicators of higher social and cultural capital, and thus likely gentrification (Ley, 1996). We also compare the work arrangement (full/part-time, permanent/temporary jobs), and amount of time spent working from home (part-time) as the permanency of job location may impact decisions about commute mode. Finally, we compare the three groups in terms of the value respondents placed on specific attributes in their residential location decisions, as well as their subjective views of the car (e.g. "a burden on the environment", "a necessity", "a hobby" etc.). This allows us to link the three groups to particular lifestyles, and their commute patterns as outcomes of particular demographic and socio-economic factors, as well as location and housing decisions.

## Defining the categories

As seen in Table 17.1, respondents in the "Cars Only" category are more likely to be home owners and have children present in their households. They also have higher tendencies to hold full-time permanent employment, while occupying positions in management, business, and finance and administration occupations. There are no statistically significant differences among the three categories for gender, self-employment, and working from home.

For those respondents in the "No Cars" category, there is a higher likelihood of being an immigrant, or current student, while having contract employment positions (full-time or part-time). As expected, higher proportions, as compared to the two other groups, live in urban residential locations. A greater proportion of them have jobs in education, law, and the social sciences, as well as in art, culture and recreation, while having acquired either a high-school diploma or less, or one or more university degrees.

The findings point to a somewhat polarized distribution of educational attainment among the "No Cars" respondents. This is perhaps not too surprising since not having a car is likely an outcome of low income for some, which is associated with lower educational attainment, but also a specific choice of some central city gentrifiers (Quastel, Moos, & Lynch, 2012).

The "Cars Sometimes" category is perhaps the most complex to make sense of. Similar to the "No Cars" category, the "Cars Sometimes" group is also more likely to be composed of immigrants, current students and those living in urban residential locations. A higher proportion of them hold occupations in the education, law, and social sciences field, and have attained one or more university

*Table 17.1* Respondent characteristics by commuter type

| Percentage (%) | Cars only | Cars sometimes | No cars | p-value |
|---|---|---|---|---|
| Car owners | 100 | 100 | 0 | |
| Average share of commute trips by car | 100 | 13 | 0 | |
| US residents (vs Canada) | 45 | 44 | 67 | <0.00 |
| Female | 48 | 52 | 42 | |
| Children present | 24 | 17 | 2 | <0.00 |
| Immigrant | 4 | 11 | 11 | <0.05 |
| Self-employed | 7 | 9 | 6 | |
| Works from home, part-time | 30 | 36 | 35 | |
| Current student | 16 | 28 | 29 | <0.01 |
| Home owners | 46 | 36 | 11 | <0.00 |
| Urban residential location (vs suburban) | 59 | 81 | 92 | <0.00 |

*Table 17.1* (Continued)

| Percentage (%) | Cars only | Cars sometimes | No cars | p-value |
|---|---|---|---|---|
| **Work arrangement** | | | | <0.01 |
| Part-time contract | 2 | 11 | 13 | |
| Part-time permanent | 6 | 5 | 5 | |
| Full-time contract | 12 | 12 | 19 | |
| Full-time permanent | 80 | 72 | 63 | |
| **Occupation** | | | | <0.01 |
| Management | 12 | 8 | 6 | |
| Business, finance and administration | 26 | 21 | 14 | |
| Natural and applied sciences and related | 10 | 9 | 10 | |
| Health | 6 | 5 | 3 | |
| Education, law and social sciences | 29 | 40 | 38 | |
| Art, culture and recreation | 5 | 7 | 18 | |
| Sales and services | 9 | 9 | 7 | |
| Trades, transport and equipment operator | 1 | 2 | 0 | |
| Natural resources, agriculture and related | 0 | 1 | 1 | |
| Manufacturing and utilities | 3 | 1 | 1 | |
| **Education** | | | | <0.01 |
| High-school diploma or less | 8 | 7 | 14 | |
| Trades certificate or college diploma | 17 | 10 | 7 | |
| One or more university degrees | 75 | 83 | 79 | |
| *Number of respondents* | *166* | *236* | *287* | |

degrees. However, similar to the "Cars Only" category, a higher proportion of respondents in the "Cars Sometimes" category are homeowners and hold full-time permanent positions.

In addition, there are evident differences in individual income among the groups (Figure 17.2). A large cluster of "No Cars" respondents fall around the lowest spectrum of the income distribution. In contrast, higher proportions of "Cars Sometimes" respondents fall within the highest spectrum, while "Cars Only" respondents are more likely to be in the middle of the income distribution.

Perhaps the most striking differences among categories appear when respondents are asked about the subjective value and purpose of cars (Table 17.2). Naturally, those who use the car at least minimally ("Cars Only" and "Cars Sometimes" respondents) value the car as a "way of life", while those who solely use the car are more likely to agree that the car is a "necessity" and, to a lesser extent, a "hobby". Moreover, those who take alternative modes of transport indicate that the car is a "burden on the environment" and a "burden on the budget".

Respondents were also asked value questions about the reasons behind their commute mode decisions, and some interesting, and statistically significant, differences appear here as well. Figure 17.3 shows respondents who "strongly agree" with the value statements provided. Respondents who only commute by car are more likely to "strongly agree" that having access to a car is important

*Table 17.2* Respondents that "strongly agree" with value statements regarding cars, by "cars only", "cars sometimes", and "no cars" commutes

| *To me a car is. . .* | | | | |
| --- | --- | --- | --- | --- |
| *Percentage (%)* | *Cars only* | *Cars sometimes* | *No cars* | *p-value* |
| Status symbol | 17 | 11 | 15 | |
| A way of life | 96 | 86 | 58 | *** |
| A hobby | 16 | 8 | 9 | * |
| A burden on the environment | 37 | 39 | 60 | *** |
| A burden on my budget | 42 | 42 | 72 | *** |
| Non essential | 2 | 19 | 71 | *** |

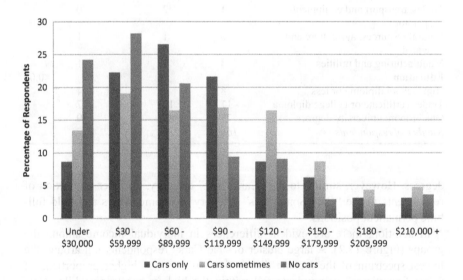

*Figure 17.2* Income distribution of respondents commuting by "car only", "car sometimes", and "no car"

when having children, and that a car is a necessity in the neighbourhood where they reside.

The respondents' answers also suggest that they would take public transit when the car is inconvenient, such as due to traffic congestion. Higher proportions of the "Cars Sometimes" and "No Cars" respondents "strongly agree" that taking alternative modes (public transit, walking or cycling) are convenient ways to get around on a daily basis. But notably, high proportions of "No Cars" respondents take public transit because it is the cheaper option, suggesting that taking transit is a decision being made due to a convenience factor as well as due to budget constraints.

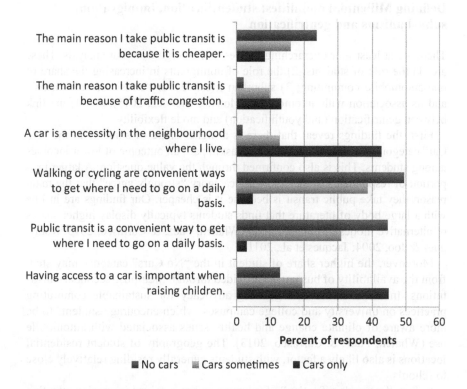

*Figure 17.3* Respondents that "strongly agree" with value statements regarding the commute, by "car only", "car sometimes", and "no car" commutes

Table 17.3 in the Appendix shows respondents' answers to "how important was each of the following in selecting your current location?" Respondents' answers illustrate how the commute mode is intertwined closely with decisions about lifestyle and residential location attributes. Those who do not use the car or use it as an alternative to other modes are more likely to value proximity to social and cultural amenities such as restaurants/cafes/bars and music/arts scenes. They also have higher likelihood to point to the walkability of the neighbourhood, cycling infrastructure and proximity to public transit as having been important in their location decisions.

Quite contrary, the respondents who solely rely on the car for their commute are more likely to point to proximity to good schools, quiet and safe neighbourhood, and proximity to family members and friends as having been important in their residential location decisions. There are no differences that are statistically significant among the three categories in terms of deciding on a residential location based on proximity to universities/colleges and waterfront/natural amenities, as well as nightlife/trendy neighbourhood, lowest cost option, and ethnic/racial composition of the area.

## Defining Millennial mobilities: studentification, immigration, suburbanisms and gentrification

There are at least four overarching themes that emerge from our analysis. These are: 1) the role of students; 2) the role of immigrants in increasing the share of non-automobile commuting; 3) suburban lifestyles, or "classic suburbanisms", and its association with automobility (Moos & Mendez, 2015); and 4) the link between gentrification (and youthification) and mode flexibility.

First, the findings reveal that there is a higher share of students in the "No Car" category. The finding is likely, at least in part, an outcome of lower incomes among students. This is also confirmed through the value question: A larger proportion of respondents in the "No Car" category "strongly agree" that the main reason they take public transit is because it is cheaper. Our findings are in line with a large body of literature that finds students typically display higher shares of alternative modes of transportation (Whalen, Paez, & Carrasco, 2013; Rodríguez & Joo, 2004; Jacques et al., 2011).

Moreover, the higher share of student in the "No Cars" category may stem from the availability of bus passes provided through their post-secondary institutions. In addition, there is growing advocacy for sustainable commuting practices on university and college campuses, which encourage students to be more aware of climate change and health issues associated with automobile use (Whalen, Paez, & Carrasco, 2013). The geography of student residential locations is also likely a factor, with students generally residing relatively close to schools.

In fact, the "studentification" of urban areas has received increasing attention as it transforms the demography and housing markets in neighbourhoods close to major post-secondary educational institutions (Smith, 2005). The growing share of post-secondary enrolment among young adults (as an outcome of growing demand for workers in the knowledge sectors of the economy) is thus indirectly contributing to growing shares of non-automobile commutes among young adults (Moos, 2016). Whether or not students will continue to exhibit lower automobile use post-graduation is still an open question, and is tied in part to their future residential locations (see Chapters 1 and 14).

The second theme evident in our findings relates to the connection between immigrants and their higher representation in both the "Cars Sometimes" and "No Cars" categories. Specifically, our data indicate that, among respondents, immigrants are more likely to travel by public transit. As previous studies have found, immigrants, especially recent immigrants, are more reliant on public transit than non-immigrants; and, in many areas, immigrants comprise a larger proportion of total transit riders (Blumenberg & Evans, 2010).

Immigrants' lower likelihood to travel by car is shown to be partly a function of their lower income levels, as well as cultural factors (Blumenberg et al., 2012; Simsekoglu et al., 2015; Susilo & Cats, 2014; Polzin et al., 2014). Representation in the "Cars Sometimes" category may also reflect attempts to maximize leverage of motility capital in the absence of economic capital, for example by sharing one vehicle among a large family (Hess, Huang, & Vasic, 2015). Immigrants are somewhat underrepresented in our sample, yet still we find a statistically significant effect. Immigration has been an important force of socio-economic change in both the USA and Canada as it has been the largest component of population growth in major Canadian metropolitan areas. While sometimes attributed to changing preferences among young adults, clearly there is an immigrant dimension to non-automobile-based commuting patterns (believed to be increasing) that requires attention in future research.

The third theme that emerges from our findings is related to lifestyle and household composition. Clearly, one of the starkest contrasts among the groups is the higher share of respondents with children in the household, homeowners, and suburban residential locations in the "Cars Only" category. These are indicators of what we have called "classic suburbanisms" in prior research (Moos et al., 2016). There has been growing interest in the literature on the heterogeneity of suburbs, and the utility of the concept of "suburbanisms" (or suburban ways of living), pointing to a blurring of the suburban–urban dichotomy and increasing shares of non-automobile trips in some suburban areas. Although the strict distinction between the suburban and the urban may be dissolving, it was never as clearly distinct as some assumed.

Nevertheless, the "classic" suburban lifestyle, at least among respondents, remains the most evident determinant of socio-spatial structure. The value-based questions point to the importance the "Cars Only" respondents place on attributes of residential locations that traditionally, and stereotypically, have been associated with the North American suburban ideal such as large yards, quiet and safe neighbourhoods, proximity to good schools and good access to highway/freeways. Socio-spatial structures may be becoming more diverse overall but our findings add evidence to the idea that existing structures are also simultaneously becoming more entrenched (Walks, 2001; Moos et al., 2016).

Finally, the findings do support our hypothesis that mode flexibility is likely an outcome of uneven socio-spatial structures and uneven distribution of motility capital. According to the data, the "Cars Sometimes" group has a higher likelihood of living in urban residential locations, having a higher level of education (one or more university degrees), and holding occupations in education, law, and social science, which have been previously associated with gentrification (Ley, 1996). In terms of income levels, the data also indicate that a higher proportion of respondents who are multi-modal are among the highest income earners in our sample, which indicate that mode choice is more possible among people with higher incomes.

When asked the value questions, the "Cars Sometimes" respondents are also more likely to place high importance on living in vibrant neighbourhoods that are in close proximity to restaurants/cafes/bars, and cultural amenities such as music/arts scene, as well as walkable and cycling-friendly neighbourhoods when deciding on a place to live. The "Car Sometimes" group is also more likely to rank transit, walking and cycling as convenient ways to get around on a daily basis. Additionally, a larger proportion of the "Car Sometimes" group reports that they are homeowners in urban areas, and would like to live in safe and quiet neighbourhoods.

## The motility capital of "Cars Sometimes"

The characteristics of "Cars Sometimes" respondents point to those who are able to afford attributes of a suburban lifestyle in urban locations. These individuals, which we can at least tentatively label as gentrifiers, thus are able to be flexible in their mode choices on a daily basis. They own cars, and often their homes as well, yet value living in vibrant/urbanized neighbourhoods that provide public transport, walkability and cycling infrastructure.

Motility capital in the form of mode flexibility is acquired, in this case, through the workings of the real estate market and planning frameworks that over the past decades have contributed to the gentrification of central areas, particularly in walkable and transit-accessible locations (Quastel et al., 2012). Those who have the means are able to buy into neighbourhoods that offer alternative transport modes. However, there is also an evident demographic and lifestyle dimension that intersects with class. In our sample, it is the non-family households without children that are also more likely to be flexible in terms of daily mode choice.

The ways the acquisition of motility capital is generally conceptualized is through the workings of capitalist class structures; however, the demographic component of housing markets, and residential sorting, has in many ways not received sufficient attention in this regard. Demography, housing tenure, residential location decisions and values are intimately interlinked to produce unique, geographically specific, lifestyles (Ley, 1996; Moos, 2015). Yet still, the degree of motility capital, which we operationalize in our analysis as the ability to commute using multiple modes, is most importantly a function of income, education and occupation as well as student and immigrant status. In other words, the determinants of mode flexibility evidently bear the markers of class.

The analysis of multi-modal behaviour also has utility from the standpoint of policy that aims to reduce automobile use – arguably, those who already commute by modes other than the car might be more likely to do a complete switch (Molin et al., 2016). The findings here suggest that multi-modal behaviour is partly an outcome of higher income earners moving into areas where other modes are also easily accessible. To reduce car use within this group, geographically specific economic disincentives would likely be required – for instance road tolls or congestion charges in central areas. The findings also indicate that those who drive exclusively would take transit if driving becomes costly and/or inconvenient, making a case for policy interventions such as high parking fees and/or gas taxes.

The findings also point to the overall importance of residential location decisions in shaping transport, and in our case mode decisions in the long run. Respondents to our survey that rely exclusively on the car made residential location decisions with preferences for yard space, quiet and safe neighbourhoods and proximity to schools. In a North American context, these attributes are generally associated with low-density suburbs where the car is a necessity (a sentiment also expressed by our respondents). It is unclear whether households are making these decisions with full knowledge about the daily transportation patterns that follow from their location decision.

The findings that households with children present are more likely to be amongst the "Cars Only" group, who are living what we characterize as suburban lifestyles, also raises important policy questions. These same respondents "strongly agree" that having access to a car is important when raising children, and that access to a car is a "necessity" in the neighbourhood where they live. Also, a greater proportion of them report to living in single-family houses. Prior studies support this finding as households with children are more likely to own multiple cars (American Public Transportation Association, 2013). This evidence suggests that the car remains an integral part of their lives.

We can interpret these findings as a continuing preference for suburban neighbourhoods and residential attributes when raising children. Contrary, we can argue that in many cities, urban intensification has not taken the needs and preferences of households with children into account very well. Perhaps both of these interpretations hold merit in different contexts, and for different kinds of households.

Regardless, unless policy and planning directly address the lifestyle and demographic dimensions of residential location decisions, it is unlikely that traditional suburban ideals will fade. Currently, motility capital in the form of access to non-automobile based transport and mode flexibility is a form of class privilege, yet also has clear household and demographic determinants. Mode flexibility is not merely a choice anyone can make, and this needs to be recognized explicitly in planning decisions so as to ensure more equitable distribution of transportation modes that is not solely restricted to privileged individuals (living in small households).

## References

Aarts, H., Verplanken, B., & Van Knippenberg, A. (1998). Predicting behavior from actions in the past: Repeated decision making or a matter of habit? *Journal of Applied Social Psychology*, 28(15), 1355–1374.

American Public Transportation Association (2013). *Millennials & mobility: Understanding the Millennial mindset*. Retrieved from www.apta.com/resources/reportsandpublica tions/Documents/APTA-Millennials-and-Mobility.pdf.

Baum-Snow, N., & Kahn, M. E. (2005). *Effects of urban rail transit expansions: Evidence from sixteen cities, 1970–2000*. Brookings-Wharton Papers on Urban Affairs 2005, pp. 147–206. Retrieved from www.econ.brown.edu/fac/nathaniel_baum-snow/ brook_ final.pdf.

Billings, S. B. (2011). Estimating the value of a new transit option. *Regional Science and Urban Economics*, 41(6), 525–536.

Blumenberg, E., & Evans, A. (2010). Planning for demographic diversity: The case of immigrants and public transit. *Journal of Public Transportation*, 13(2), 23–45.

Blumenberg, E., & Pierce, G., (2014). Multimodal travel and the poor: Evidence from the 2009 National Household Travel Survey. *Transportation Letters*, 6(1), 36–45.

Blumenberg, E., Taylor, B., Smart, M., Ralph, K., Wander, M., & Brumbaugh, S. (2012). *University of California transportation center, What's youth got to do with it? Exploring the travel behavior of teens and young adults*. Retrieved from www.uctc.net/research/ papers/UCTC-FR-2012-14.pdf.

Boschmann, E., & Brady, S. (2013). Travel behaviors, sustainable mobility, and transport-oriented developments: A travel counts analysis of older adults in the Denver, Colorado metropolitan area. *Journal of Transport Geography*, 33, 1–11.

Bourdieu, P. (1984). *Distinction: A social critique of the judgement of taste*. Cambridge, MA: Harvard University Press.

Chatman, D. G., Tulach, N. K., & Kim, K. (2012). Evaluating the economic impacts of light rail by measuring home appreciation: A first look at New Jersey's river line. *Urban Studies, 49*(3), 467–487.

Clifton, K., & Muhs, C. D. (2012). Capturing and representing multimodal trips in travel surveys: Review of the practice. *Transportation Research Record, 2285,* 74–83.

Cortright, J. (2009). *Walking the walk: How walkability raises home values in US cities. CEOs for cities.* Retrived from www.ceosforcities.org/pagefiles/WalkingTheWalk_CEOsforCities.pdf.

Danyluk, M., & Ley, D. (2007). Modalities of the new middle class: Ideology and behaviour in the journey-to-work from gentrified neighbourhoods. *Urban Studies, 44(10),* 2195–2210.

Debrezion, G., Pels, E., & Rietveld, P. (2007). The impact of railway stations on residential and commercial property value: A meta-analysis. *Journal of Real Estate Finance and Economics, 35*(2), 161–180.

Diao, M., & Ferreira, J. (2010). Residential property values and the built environment. *Transportation Research Record: Journal of the Transportation Research Board, 2174,* 138–147.

Gilderbloom, J., Riggs, W., & Meares, W. (2015). Does walkability matter? An examination of walkability's impact on housing values, foreclosures, and crime. *Cities, 42*(PartA), 13–24.

Grube-Cavers, A., & Patterson, Z. (2015). Urban rapid rail transit and gentrification in Canadian urban centres: A survival analysis approach. *Urban Studies, 52*(1), 178–194.

Hess, P., Huang, H. H. W., & Vasic, M. (2015). Automobility, adaptation, and exclusion: Immigration, gender, and travel in the auto-city. In A. Walks (Ed.), *The urban political economy and ecology of automobility: Driving cities, driving inequality, driving politics* (pp. 169–183). New York: Routledge.

Hildebrand, E. (2003). Dimensions in elderly travel behaviour: A simplified activity-based model using lifestyle clusters. *Transportation, 30*(3), 285–306.

Jacques, C., Chakour, V., Mathez, A., Manaugh, K., Barreau, G., Hatzopolou, M., Eluru, & El-Geneidy, A. (2011). *An examination of commuting patterns to McGill University: Results of the 2011 McGill transportation survey.* Office of Sustainability, McGill University, Montreal. Retrieved from http://tram.mcgill.ca/Research/Publications/McGill%20Travel%20Report.pdf.

Jones, C., & Ley, D. (2016). Transit-oriented development and gentrification along Metro Vancouver's low-income SkyTrain corridor. *The Canadian Geographer, 16*(1), 9–22.

Kahn, M. E. (2007). Gentrification trends in new transit-oriented communities: Evidence from 14 cities that expanded and built rail transit systems. *Real Estate Economics, 35*(2), 155–182.

Kaufmann, V. (2002). *Re-thinking mobility.* Aldershot: Ashgate.

Kaufmann, V., Bergman, M. M., & Joye, D. (2004). Motility: Mobility as capital. *International Journal of Urban and Regional Research, 28*(4), 745–756.

Kuhnimhof, T., Buehler, R., Wirtz, M., & Kalinowska, D., (2012). Travel trends among young adults in Germany: Increasing multimodality and declining car use for men. *Journal of Transportation Geography, 24,* 443–450.

Kuhnimhof, T., Chlond, B., & Von der Ruhren, S. (2006). Users of transport modes and multimodal travel behavior steps toward understanding travelers' options and choices. *Journal of Transportation Research Board, 1985,* 40–48.

Ley, D. (1996). *The new middle class and the remaking of the central city.* Oxford: Oxford University Press.

Li, Y. (2003). Evaluating the urban commute experience: A time perception approach. *Journal of Public Transportation, 6*, 41–67.

Meares, W. L. (2014). *The walkable dividend: The impacts of walkability on housing and socio-economic composition in Louisville, KY*. Dissertation, University of Louisville Libraries

Molin, E., Mokhtarian, P., & Kroesen, M. (2016). Multimodal travel groups and attitudes: A latent class cluster analysis of Dutch travelers. *Transportation Research Part A, 83*, 14–29.

Moos, M. (2016). From gentrification to youthification? The increasing importance of young age in delineating high-density living. *Urban Studies, 53*(14), 2903–2920. doi:10.1177/0042098015603292.

Moos, M., & Mendez, P. (2015). Suburban ways of living and the geography of income: How homeownership, single-family dwellings and automobile use define the metropolitan social space. *Urban Studies, 52*(10), 1864–1882.

Moos, M., Mendez, P., McGuire, L., Wyly, E., Kramer, A., Walter-Joseph, R., & Williamson, M. (2016). More continuity than change? Re-evaluating the contemporary socio-economic and housing characteristics of suburbs. *Canadian Journal of Urban Research, 24*(2), 64–84.

Moos, M., Walter-Joseph, R., Williamson, M., Wilkin, T., Chen, W., & Stockmal, D. (2015). *Youthification: The new kid on the block*. Working Paper Series 015-01, University of Waterloo: School of Planning.

Nobis, C. (2007). Multimodality: Facets and causes of sustainable mobility behavior. *Transportation Research Record, 2010*, 35–44.

Patacchini, E., & Zenou, Y. (2005). Spatial mismatch, transport mode and search decisions in England. *Journal of Urban Economics, 58*(1), 62–90.

Pinjari, A. R., Pendyala, R. M., Bhat, C. R., & Waddell, P. A. (2007). Modeling residential sorting effects to understand the impact of the built environment on commute mode choice. *Transportation, 34*, 557–573.

Pivo, G., & Fisher, J. D. (2011). The walkability premium in commercial real estate investments. *Real Estate Economics, 39*(2), 185–219.

Polzin, S., Chu, X., & Godfrey, J. (2014). The impact of millennials' travel behavior on future personal vehicle travel. *Energy Strategy Reviews, 5*, 59–65.

Preston, V., Mclafferty, S., & Liu, X. (1998). Geographical barriers to employment for American-born and immigrant workers. *Urban Studies, 35*(3), 529–545.

Quastel, N., Moos, M., & Lynch, N. (2012). Sustainability as density and the return of the social: The case of Vancouver, British Columbia. *Urban Geography, 33*(7), 1055–1084.

Rayle, L. (2015). Investigating the connection between transit-oriented development and displacement: Four hypotheses. *Housing Policy Debate, 25*(3), 531–548.

Revington, N. (2015). Gentrification, transit and land use: Moving beyond neoclassical theory. *Geography Compass, 9*(3), 152–163.

Roche-Cerasi, I., Rundmo, T., Sigurdson, J., & Moe, D. (2013). Transport mode preferences, risk perception and worry in a Norwegian urban population. *Accident Analysis and Prevention, 50*, 698–704.

Rodríguez, D. A., & Joo, J. (2004). The relationship between non-motorized mode choice and the local physical environment. *Transportation Research Part D – Transport and Environment, 9*, 151–173.

Rundmo, T., Nordfjærn, T., Iversen, H. H., Oltedal, S., & Jørgensen, S. H. (2011). The role of risk perception and other risk-related judgments in transportation mode use. *Safety Science, 49*, 226–235.

Schwanen, T., & Mokhtarian, P. L. (2005). What affects commute mode choice: Neigh-bourhood physical structure or preferences toward neighbourhoods? *Journal of Transport Geography*, *13*, 83–99.

Simons, D., Clarys, P., De Bourdeaudhuij, I., de Geus, B., Vandelanotte, C., & Deforche, B. (2014). Factors influencing mode of transport in older adolescents: A qualitative study. *Transport Policy*, *36*, 151–159.

Simsekoglu, O., Nordfjaern, T., & Rundmo, T. (2015). The role of attitudes, transport priorities, and car use habit for travel mode use and intentions to use public transportation in an urban Norwegian public. *Transport Policy*, *42*, 113–120.

Smith, D. (2005). "Studentification": The gentrification factory? In R. Atkinson & G. Bridge (Eds.), *Gentrification in a global context: The new urban colonialism* (pp. 72–89). New York: Routledge.

Susilo, Y., & Cats, O. (2014). Exploring key determinants of travel satisfaction for multi-modal trips by different traveler groups. *Transport Research Part A*, *67*, 366–380.

Transport for London (2012). *Understanding the travel needs of London's diverse communities: A summary of existing research*. London: Transport for London. Retrieved from http://content.tfl.gov.uk/travel-in-london-understanding-our-diverse-communities.pdf.

Urry, J. (2000). *Sociology beyond societies: Mobilities for the twenty-first century*. London: Routledge.

Urry, J. (2007). *Mobilities*. Cambridge: Polity Press.

US Census Bureau. (2005–2009). *American community survey*. Washington, DC.

Van Nes, R. (2002). *Design of multimodal transport networks: A hierarchical approach*. Netherlands: Delft University Press.

Walks, A. (2001). The social ecology of the post-Fordist/Global city? economic restructuring and socio-spatial polarisation in the Toronto urban region. *Urban Studies*, *38*(3), 407–447.

Whalen, K., Paez, A., & Carrasco, J. (2013). Mode choice of university students commuting to school and the role of active travel. *Journal of Transportation Geography*, *31*, 132–142.

Zax, J., & Kain, J. (1996). Moving to the suburbs: Do relocating companies leave their black employees behind? *Journal of Labor Economics*, *14*, 472–493.

# Appendix

*Table 17.3* Respondents' answer to "how important was each of the following in selecting your current location?" by commuting type

| | Cars only | Cars sometimes | No cars | p-value |
|---|---|---|---|---|
| **Proximity to restaurants/cafes/bars/** | | | | |
| Not important at all | 7.41 | 5.15 | 4.53 | |
| Not very important | 11.73 | 9.44 | 4.53 | |
| Neutral | 14.2 | 7.73 | 7.67 | |
| Somewhat important | 40.74 | 42.06 | 53.66 | |
| Very important | 25.93 | 35.62 | 29.62 | ** |
| **Proximity to public transit** | | | | |
| Not important at all | 28.05 | 3.88 | 1.05 | |
| Not very important | 17.07 | 6.03 | 1.05 | |
| Neutral | 12.8 | 6.47 | 1.05 | |

*Table 17.3* (Continued)

|  | Cars only | Cars sometimes | No cars | p-value |
|---|---|---|---|---|
| Somewhat important | 28.05 | 34.91 | 19.58 | |
| Very important | 14.02 | 48.71 | 77.27 | *** |
| **Walkability of neighbourhood and availability of cycling** | | | | |
| Not important at all | 11.59 | 2.16 | 1.05 | |
| Not very important | 14.63 | 4.74 | 1.75 | |
| Neutral | 15.24 | 7.33 | 3.16 | |
| Somewhat important | 37.8 | 34.05 | 25.96 | |
| Very important | 20.73 | 51.72 | 68.07 | *** |
| **Location with good highway/ freeway access** | | | | |
| Not important at all | 4.91 | 19.31 | 58.54 | |
| Not very important | 13.5 | 22.32 | 21.95 | |
| Neutral | 17.18 | 21.89 | 10.1 | |
| Somewhat important | 44.17 | 29.18 | 8.36 | |
| Very important | 20.25 | 7.3 | 1.05 | *** |
| **Proximity to where my friends live** | | | | |
| Not important at all | 14.02 | 6.01 | 8.01 | |
| Not very important | 12.8 | 12.45 | 11.5 | |
| Neutral | 29.27 | 16.74 | 23.69 | |
| Somewhat important | 28.05 | 48.07 | 43.55 | |
| Very important | 15.85 | 16.74 | 13.24 | ** |
| **Proximity to work/college/ university** | | | | |
| Not important at all | 7.93 | 4.29 | 3.48 | |
| Not very important | 7.32 | 7.3 | 3.48 | |
| Neutral | 15.85 | 11.59 | 12.89 | |
| Somewhat important | 36.59 | 32.62 | 35.89 | |
| Very important | 32.32 | 44.21 | 44.25 | |
| **Proximity to good schools** | | | | |
| Not important at all | 32.72 | 31.3 | 58.1 | |
| Not very important | 19.14 | 20.87 | 12.68 | |
| Neutral | 20.37 | 22.17 | 18.66 | |
| Somewhat important | 17.28 | 16.96 | 6.34 | |
| Very important | 10.49 | 8.7 | 4.23 | *** |
| **Lowest cost option you could find** | | | | |
| Not important at all | 11.04 | 11.11 | 8.36 | |
| Not very important | 14.11 | 21.79 | 15.68 | |
| Neutral | 21.47 | 23.08 | 20.91 | |
| Somewhat important | 31.29 | 28.21 | 37.63 | |
| Very important | 22.09 | 15.81 | 17.42 | |
| **Quiet neighbourhood** | | | | |
| Not important at all | 7.32 | 9.44 | 13.33 | |
| Not very important | 17.07 | 21.03 | 22.81 | |
| Neutral | 18.29 | 24.46 | 22.46 | |
| Somewhat important | 40.85 | 34.33 | 35.44 | |
| Very important | 16.46 | 10.73 | 5.96 | * |

(*Continued*)

*Table 17.3* (Continued)

| | Cars only | Cars sometimes | No cars | p-value |
|---|---|---|---|---|
| **Safety of neighbourhood** | | | | |
| Not important at all | 3.66 | 1.29 | 2.8 | |
| Not very important | 3.66 | 9.44 | 6.29 | |
| Neutral | 8.54 | 16.31 | 15.03 | |
| Somewhat important | 48.17 | 47.21 | 54.9 | |
| Very important | 35.98 | 25.75 | 20.98 | ** |
| **Proximity to a waterfront or other natural amenity** | | | | |
| Not important at all | 22.5 | 16.74 | 16.72 | |
| Not very important | 22.5 | 19.74 | 17.42 | |
| Neutral | 28.75 | 27.47 | 27.87 | |
| Somewhat important | 18.75 | 27.04 | 28.57 | |
| Very important | 7.5 | 9.01 | 9.41 | |
| **Large yard** | | | | |
| Not important at all | 33.54 | 39.48 | 55.24 | |
| Not very important | 16.46 | 24.03 | 22.73 | |
| Neutral | 19.51 | 20.6 | 15.73 | |
| Somewhat important | 23.17 | 12.45 | 4.9 | |
| Very important | 7.32 | 3.43 | 1.4 | *** |
| **Vibrant neighbourhood** | | | | |
| Not important at all | 9.76 | 5.15 | 4.91 | |
| Not very important | 9.76 | 6.87 | 5.96 | |
| Neutral | 29.88 | 12.88 | 12.28 | |
| Somewhat important | 31.71 | 45.49 | 46.67 | |
| Very important | 18.9 | 29.61 | 30.18 | *** |
| **Nightlife** | | | | |
| Not important at all | 27.33 | 18.88 | 17.13 | |
| Not very important | 15.53 | 19.31 | 20.28 | |
| Neutral | 23.6 | 26.61 | 22.38 | |
| Somewhat important | 26.09 | 27.9 | 32.87 | |
| Very important | 7.45 | 7.3 | 7.34 | |
| **Trendy or cool neighbourhood** | | | | |
| Not important at all | 26.22 | 18.03 | 15.68 | |
| Not very important | 16.46 | 20.17 | 18.47 | |
| Neutral | 25 | 24.46 | 23.34 | |
| Somewhat important | 26.22 | 29.61 | 34.49 | |
| Very important | 6.1 | 7.73 | 8.01 | |
| **Cultural amenities nearby and/or music/arts scene** | | | | |
| Not important at all | 20.25 | 13.3 | 7.69 | |
| Not very important | 19.02 | 13.73 | 10.84 | |
| Neutral | 23.93 | 20.17 | 22.73 | |
| Somewhat important | 28.83 | 40.77 | 45.1 | |
| Very important | 7.98 | 12.02 | 13.64 | *** |
| **Proximity to family members** | | | | |
| Not important at all | 28.22 | 31.03 | 42.86 | |
| Not very important | 17.79 | 21.55 | 17.07 | |
| Neutral | 27.61 | 19.4 | 23 | |

*Table 17.3* (Continued)

|  | Cars only | Cars sometimes | No cars | p-value |
|---|---|---|---|---|
| Somewhat important | 18.4 | 20.69 | 14.29 | |
| Very important | 7.98 | 7.33 | 2.79 | ** |
| **Ethnic/racial composition of area** | | | | |
| Not important at all | 35.37 | 34.05 | 39.51 | |
| Not very important | 16.46 | 20.26 | 15.03 | |
| Neutral | 32.93 | 30.6 | 29.37 | |
| Somewhat important | 13.41 | 12.93 | 13.99 | |
| Very important | 1.83 | 2.16 | 2.1 | |

# Part V
# Millennial city futures

# 18 Fun for all ages

## Weaving greenspace, transportation, and housing together in the intergenerational city

*Kyle Shelton and William Fulton*

There is no Millennial city. There are no Millennial communities. There are no Millennial streets. There are millions of Millennials living in our cities, and their preferences are influencing many elements of American cities, but urban spaces are far from theirs alone. While much of this volume grapples with how to understand the changing of the urban guard toward the tastes, experiences, and expectations of the Millennial generation, this essay lays out the argument that in order to plan a city that works for this most-discussed generation, we must start by making a city that works for all generations – while, at the same time, taking advantage of the opportunities created by Millennials' preferences to create a successful multigenerational city.

Cities have always been spaces where people from different racial, national, and economic backgrounds rub shoulders. But in the future, our cities will contain residents from hugely different age brackets. Designing and running a city for only one age group is an approach that too easily ignores the needs of vulnerable populations – the elderly and aging, the young, and those with physical or mental impairments. Building a city for urbanist Millennials too easily ignores the needs of low-wage workers – who staff trendy restaurants but cannot afford to live nearby – and fails to serve the families struggling to reach or remain in the middle class.

The stereotypical image of the Millennial is problematic. A mostly white, tech-saavy, college-educated cohort exists, of course, but so too do millions of other Millennials who do not fit that image. Most non-white Millennials face completely different worlds than their white peers. In many American cities large percentages of Black and Hispanic Millennials must overcome the dual challenges of economic and educational disparities that have limited their opportunities since early in life. Many come from low-income families or communities where college is too rarely a possible path.

As cities and older suburbs become more urban – often in response to Millennials' market preferences – the focus of urban planners is often on three interrelated elements: greenspace, transportation, and housing. All three are vital to the healthy functioning of any urban unit, from neighbourhood to city to metro. All three must be present and in the proper form for the intergenerational, equitable city to emerge. And all three should be woven together.

As our cities urbanize, greenspace and natural areas are becoming harder to conserve, improve, and activate. At the same time, the incredible productivity of well-designed trail systems and public greenspaces attest to their importance for economic and social vitality. A major challenge for planners to grapple with in the future is how to build greenspaces that serve a wide range of ages, activity levels and needs. In cities across America planners, advocates, and citizens are working with city officials to reclaim brownfield sites, reimagine abandoned rights of way, and to redesign our existing greenspaces in ways that work for all residents. Houston and Atlanta have both recentred much of their greenspace plan around emerging trail systems. Both cities are grappling with how to create a system that permits all citizens to move through and utilize these growing amenities.

Most American cities remain dominated by the car, but public transportation, biking, and walking are growing in many locations. In the 2016 election season numerous cities passed major transportation funding referendums, a sign that cities recognize that they must take the lead in creating multi-modal environments. And while transportation priorities may change in the Trump administration, leaders across every level of government – from former Secretary of Transportation Anthony Foxx to Mayor Sylvester Turner of Houston – are beginning to question our continued focus on highway and automobile infrastructure. Beyond these transformations, new technologies and the sharing economy are revolutionizing what car ownership can mean. Uber, Lyft, and other transportation network companies (TNCs) are transforming urban mobility and autonomous vehicles will soon accelerate that transformation.

Just as cities require a wide variety of greenspaces and transportation options, so too does an intergenerational region require housing options for all. This may look like single-family neighbourhoods opening up to multi-family units and mixed-use developments. Or it might be new suburban communities of townhomes, condos, and apartments built around a major walkable amenity. These approaches are being pursued across the USA from a school district in Santa Clara working with developers to create long-term affordable homes for teachers, to Denver and Austin taking different tacks to increase their number of affordable units.

The market preferences of Millennials are a powerful force in bringing about all three transformations. But these market preferences should not be viewed simply as an end in itself. Rather, these preferences should be leveraged to create positive multigenerational change – improving our cities and regions so that everyone, including the vulnerable, the striving, and the invisible members of our metropolitan region can access them. Creating systems that work for each of those groups will create systems that work for all.

The need for comprehensive approaches to housing, transportation, and greenspace is pressing. How we approach these and other topics can redress existing inequality or worsen its impacts. It all depends upon how decisions about these systems are made and for whom we frame the future of our cities.

## Greenspace

Given that fewer and fewer large, central plots remain for the creation of significant park space, cities have begun to think more strategically about how to approach greenspace planning and development. This careful planning is critical because the need for parkspace is huge. According to the Urban Land Institute's *America in 2015* report, nearly 38% of Americans say they are not able to find adequate greenspace for recreation near their homes (Urban Land Institute, 2015). To bring parks into currently underserved areas, many cities are undertaking projects that tie the urban and natural together. Designers are increasingly creating projects that celebrate the urban environment and link the function, shape, and use of greenspaces into the urban features surrounding them (Fulton, 2016).[1]

In most cities, the urban–natural approach has come to fruition in the form of linear parks and trail systems that weave along natural features such as river systems (Houston's bayou system, the Los Angeles River, Louisville's Ohio River waterfront) or along defunct transportation rights of way (Atlanta's Beltline, Chicago's 606, and New York City's High Line).

By their very nature as pathways, linear parks encourage active transportation and can have major health benefits for users – which is part of the reason they are so popular among the Millennials who provide much of the impetus and political support for these facilities (Boarnet, 2006).

In this process it remains essential that designers and planners account for all potential users of a space via its programming, access, and location. With many of these sites becoming major attractions, they must remain accessible and usable for citizens with a variety of abilities.

But Millennial support for linear parks can help create more multigenerational and inclusive cities. Linear parks often cut across neighbourhoods with very different socio-demographic profiles, potentially connecting more users to the space. And the amenities surrounding trails that are likely to attract Millennials – gathering spaces for events, active infrastructure, natural settings, nearby retail and nightlife – are all elements that would be attractive to other subsets of users. A central gathering space or stage can be used as easily for a music festival as for a family-friendly event. Even those elements that do not strike one as intergenerational – say a playground – can be incredibly productive for a wide range of users if a mix of activities and equipment are included. A partnership between Humana and the non-profit group KaBoom! has resulted in the construction of more than 50 such playspaces across the country (Brenoff, 2014). Health and aging researchers have found that intergenerational park and playground spaces can have positive impacts on younger and older users alike (Thang, 2015; Jarrott, Kaplan, & Steinig, 2011; Brown & Henkin, 2015).

Perhaps most important to creating an intergenerational park is ensuring that there are numerous points of access and that those entries are easily traversed by users of all ages and abilities. Ramps, traffic signals at road crossings, adequate lighting, and separate use paths – where walkers/runners and bikers have their own trails – are all steps that can raise the comfort level for all users while not limiting the experience of any.

Houston's Buffalo Bayou Park is an example of a carefully cultivated intergenerational park. Sitting alongside Houston's main bayou (Houston's version of a slow-moving river) – the park was reconstructed in phases over 2014 and 2015 in an effort that cost $58 million. Since its (re)opening the park has been a major success. One of the reasons for that success are its intergenerational design elements. The park includes a number of pieces that are intended to aid the experience of all users – different paths for bicycles and pedestrians, well-lit trails, and numerous access points and levels of accessibility (See Figure 18.1). Beyond those details, the ten-mile park space houses activities that encourage multiple uses and users. The area around the Sabine Bridge just outside of downtown includes a refurbished public cistern that is now in use as an art exhibit, a large skate park, a children's playground, bike rental facility, and a major event space. Dotted by similar collections of activity and accessible to pedestrians, bicyclists, and drivers, the park is open to Houstonians of all abilities and offers a way for many in the city to recreate and interact. With Houston attempting to emulate the success of Buffalo Bayou Park along many of its other bayous as a part of its much publicized "Bayous 2020" effort, the lessons from this showcase park can be applied to many other greenspaces.[2]

Atlanta's Belt Line is another model of this new generation of linear park. The path, designed to take advantage of a defunct rail line encircling central Atlanta, has been a huge success since the first sections opened in 2008. The two completed trail sections have led to increasing real estate values and offered nearby commuters in the traditionally car-based city a new way to move (Fausset, 2016). Planning around the trail system has also contributed to the development of traditional park spaces, with five new parks on or around the trail opening or connecting

*Figure 18.1* Houston's Buffalo Bayou Park features a variety of trail types that encourage intergenerational use and cater to users of varying physical abilities

Source: Photo by Jonnu Singleton, SWA Group. Courtesy of Buffalo Bayou Partnership

since the trail itself has been built. Transportation connections have also been a key consideration of the beltline since its inception. The system is slated to tie into the existing MARTA train and bus routes. The trail is also celebrated for the fact that once it is complete it will connect 45 neighbourhoods across the city, each with their own socio-economic profiles. This idea that the trail can serve as a cross-cutting site of interaction for a wide variety of Atlantans has come under criticism in recent months, however, as initial promises to bring affordable housing to the neighbourhoods adjacent to the Beltline have been slow to come to fruition and concerns about displacements of existing long-term residents grow (Binkovitz, 2016; Saporta, 2016).

## Transportation

The growth of linear parks and pathways as venues for mobility and recreation highlights a much larger transportation trend in cities – a desire to move around in ways that do not only involve the personal automobile. As Millennials have begun to revitalize urban-style neighbourhoods in both cities and older suburbs, they have also begun to change the way people get around as well. Many Millennials prefer walking neighbourhoods; they are less likely to use their own cars all the time and are more likely to use bicycles, public transit, or ridesharing services such as Uber and Lyft. Even less affluent Millennials – many of whom cannot afford cars – use the city differently than previous generations.

Cities across the country are improving bicycling and pedestrian infrastructure to promote active transportation. Many suburban communities are working with Uber and other transportation network companies to help solve first–last mile connection issues between public transit and users' front doors. And, perhaps most importantly, a number of cities are heavily investing in public transportation by passing major tax referendums with the goal of creating the next generation of mobility for our cities.

Many dismiss the idea of a car-free lifestyle as quintessentially Millennial. The stereotype of the Millennial bicyclist/pedestrian/transit rider scooting around a dense city centre without their own car strikes many as unrealistic. And, to be fair, the stereotypical image rings true, Millennials are getting fewer drivers' licences and show a willingness to mode hop (Sivak & Schoettle, 2016; American Public Transportation Association, 2013). But such trends are not the exclusive property of Millennial Americans.

The Urban Land Institute found that many Americans, not just Millennials, desire to live in more walkable spaces where cars are not essential. In its *State of America* report the organization found that fully 50% of American adults want better walkability in their neighbourhoods. And the desire is not limited to wealthier, mostly white communities either. Support for more walkable communities was at 58% among African Americans. Cut another way, the same is true for those Americans who wish to live in places where they do not need cars very often. ULI found that 52% of Americans would like to live in a place where they could rely on cars less. That number is 63% among Millennials, and right at 60% for African Americans, Latinos, and those making less than $50,000 a year. These numbers do not mean that all Americans are ready to give up their cars, but rather that many are open to the creation of cities and streets that give them options to use cars less frequently.

The Millennial push for improved transit and other mobility options, then, represents an immense opportunity for cities to address a need that would benefit residents outside of this generation. Many Millennials are "choice" riders of public transit, meaning they have a choice between using a private car or an alternative form of transportation.

Choice riders are often the target of public transit authorities advertising efforts, but it is important to understand that they are not the bulk of ridership. Census data on transit commuters shows that low-income, non-white riders make up the bulk of public transit ridership (Maciag, 2014). The push for better transit overall – driven partly by Millennials – can serve to impel cities to improve their systems for the benefit of all riders. A key element of this effort is a willingness of choice riders and even non-riders to support the expenditures needed to create a transit system that works well for choice and transit dependent riders equally well.

There has been a groundswell of support for transit in numerous big cities since the early 2000s. To give just two unlikely southwestern examples, Phoenix and Los Angeles – traditionally sprawling car-bound metropolises – have made major investments in transit in recent years.

In Phoenix, voters passed a 35-year transit tax plan intends to raise nearly $32 billion. It will allow the city to expand its light rail system and to focus on connecting activity hubs in downtown Phoenix and nearby Mesa. The region's light rail has been particularly important in helping to create a more vibrant central core as nearly $7 billion in investment has already occurred along the existing 22 miles (Toll, 2015). (See Figure 18.2) Arizona State University's Tempe and Downtown Phoenix campuses are located directly on the line and the university has increasingly placed facilities Downtown because of the convenient rail connection to Tempe. The 2015 vote was not only about transit. The tax is slated to pay for street repaving, sidewalks, bike lanes, and to increase the bus system's frequency.

Los Angeles has undertaken the most ambitious rail transit construction effort in the country, funded by a series of tax measures – including one in 2008 and another in 2016, both of which won the requisite two-thirds vote. The 2008 referendum will provide $40 billion in funding to LA Metro through 2039 and the 2016 vote will bring a projected $860 million per year. The agency has spent these funds on rail construction and rail and bus operations and is doubling the size of the rail transit system from 100 miles to 200 miles. Light rail now connects the University of Southern California to downtown and has reached the beach in Santa Monica; one line currently under construction will improve connections to LAX and the historically African American neighbourhood of Crenshaw. The most recent referendum not only extends spending on light rail and buses, it also helps to establish and expand the city's fledgling bikeshare system, improve the pedestrian realm, and pay for greenspace improvements.

Although the most recent measures were the culmination of years of careful political negotiations, it is clear that market pressure from Millennial

*Figure 18.2* Phoenix's light rail system, run by Valley Metro, already crosses much of the city and is slated to expand into other communities with recent bond approvals

Source: Image courtesy of Valley Metro

Angelenos – especially those in super-hip downtown LA – has helped drive the transit effort. And the resultant system will not only provide a wealth of options to those Angelenos who have a choice about how they move, it will also provide a better system for those whose options are limited.

Thinking about cities' future mobility systems is not complete without thinking about the way that tech-heavy modes, such as car-sharing and ride-sharing, will fit into the city's network. While companies from Uber to Zipcar have revolutionized urban mobility for their many users, their effect on non-choice riders of transit and on the transit system as a whole is less certain. The two major accessibility questions for transportation network companies (TNCs) are how they will provide service to those with special accommodation needs, particularly those requiring on-demand para-transit, and how they will operate in coordination with existing system elements. Both of these questions are at play in agreements between TNCs and local transit authorities. Several authorities have contracted with TNCs such as Uber and Lyft to provide on-demand service calls or to provide first–last mile connections to transit lines. Already a number of critics have accused the TNCs of not providing the same level of service or appropriate service to all riders (Shelton, 2016a). Agencies and TNCs must work together with city officials and residents to make their place within the expanding system of modes work effectively.

## Housing

When it comes to building an interconnected region, transportation is not the only piece that must be carefully calibrated. Improved transit options alone will not make a city accessible to all ages or incomes groups. The way we live in cities must also be re-thought, especially our approach to housing.

The stereotypical urban hipster Millennials are creating market pressure to build higher-density urban dwelling units in walkable, mixed-use neighbourhoods. But Millennials of more modest means – and, indeed, people of all ages – are feeling pressure from upscale urban housing markets. Most cities rely on federal subsidies of some kind to provide affordable housing – either traditional public housing, housing vouchers, or low-income housing tax credits – but such housing is limited to begin with and much of it is evaporating. Nearly 2 million units with some form of public subsidy are at risk of being lost in the next decade (Williams, 2015). Compounding the loss of existing units is the fact that the replacement of naturally occurring affordable housing is paltry because market rate rentals are going quickly. Finally, the construction of new subsidized housing cannot keep up. HUD's low-income housing tax credits, for example, provide at most 100,000 new units per year across the nation (Cadlk, 2016). Cities need to find ways to break from this problematic status quo.

So, in order to house all urban residents, cities have to look a lot different than they have in the past. A housing system that provides options for residents with a variety of incomes and needs requires that cities, developers, non-profits, and neighbourhoods work together. Furnishing homes for the most vulnerable populations, particularly the elderly and low-income residents, is a major challenge. One feasible approach to tackling this issue is to introduce more intergenerational housing options – from mixed-age communities to multigenerational homes – within the existing housing mix. The need to locate that housing in places where residents can benefit from amenities such as greenspaces and useful transportation systems further complicates the process.

An intergenerational approach to housing is one that could address both affordability and availability needs. Multi-generational living is already the norm for 50 million Americans and overall the number of these homes has been increasing since its nadir in 1980 (Ross, 2014). The grouping of several generations within a household can bring a number of benefits from economic savings to health improvements for older members of the household (Pew Research Centre, 2010). Homeshare programmes, where younger individuals live in a home with an older individual are being piloted in numerous countries (Reed, 2015). The younger resident is often given a subsidy either in the form of lower rent or aid with paying for school in exchange for helping their housemate and socializing with them (Generations United and the Generations Initiative, 2013). Another approach to housing older populations would be to move toward intergenerational or co-housing communities. Here older residents maintain their own homes, but are tied closely to support structures of family or private service providers. These arrangements are often made through accessory dwellings unit or "granny flats" on a family member's property or through co-housing/

cooperative housing arrangements where common amenities and facilities are shared (Durett, 2009).

One reason why intergenerational housing is more viable than ever is that the living preferences of older Americans are shifting just as much as those of Millennials. Older Americans are on the search for more amenities and greater accessibility to services (Wasik, 2016). Empty nesters and older Americans living in suburbs, groups that represent 35% of the suburban population, are looking to avoid becoming stranded in car-dependent areas or "graying suburbs" (Urban Land Institute, 2015). And such situations can be a financial strain on both the residents and older suburbs with few financial resources. Encouraging older residents to relocate to more central areas could have benefits for both individual and community alike.

As cities work to provide housing options for graying populations they must also confront the issue of affordable housing for lower-income Americans, including low-income Millennials. In order to accommodate the needs of all residents a bevy of strategies must be undertaken. These should include ways of bringing lower-income residents into high opportunity areas and finding ways to keep long-term residents in transitioning communities. Like cities, suburbs must also tackle the issue of affordability, especially as they grow more diverse and more urban. As suburban town centres and job centres grow the low-wage workers who do much to operate those areas will need affordable housing that is close to jobs. In most suburbs those options are sparse.

In both the urban and suburban setting, leaders must find politically acceptable ways to mix housing types, affordability, and populations. In many suburbs, density is not seen as a plus. But in Sugar Land, Texas, for example, a new master planned community, Imperial Sugar Land, includes nearly 40 live–work units that will permit small business owners to live above their businesses (Shelton, 2016b). The pilot approach should allow officials to show the benefit of creating mixed-use spaces and achieve density without completely overturning the expectations of existing residents.

In hot real estate markets employers with a high proportion of middle-income workers – school districts, police departments, and cities – have sought ways to help employees secure housing. One of the best examples of this work comes from Santa Clara where the school district worked with a developer to build the Casa del Maestro project, which provided more than 30 affordable housing units specifically aimed at new teachers in the school district. The units allowed the young teachers to live within the district, to save money for a future home purchase, and to help the district retain staff (Rosan et al., 2012).

In Austin, Texas, the city has worked with non-profit Guadalupe Neighborhood Development Corporation to find ways to create affordable housing in Austin's booming Eastside. In addition to capitalizing on Austin's new regulations that allow the building of accessory dwelling units, the organization has instituted a community land trust. This tool, which aims to reduce the housing costs for lower-income residents by placing land into a trust and reducing tax burdens for residents, has been used successfully in communities around the USA. But only rarely has it been used in hot real estate markets as a way to help existing landowners remain in their homes. The efforts of the group are helping to keep dozens of lower-income families near downtown Austin and will allow them to benefit from the amenities and jobs available to them as the community continues to change.

As we mentioned earlier, part of the key to making successful multigenerational communities is not just creating specific pieces – greenspace, transportation, housing – but undertaking efforts to integrate them. This is happening more and more with urban housing along transit corridors. The City of Denver has included housing alongside its transportation development for the past several years. After regional voters approved the FasTracks transportation programme in 2004, officials recognized that to take advantage of a growing light rail network, housing and development would need to be built around the developing lines. To that end, in 2010, a number of partners created the city transit-oriented development (TOD) fund with the mission of building 1,000 affordable housing units as a part of a larger TOD initiative. The fund was created by Enterprise Community Partners, which sought investment from the City of Denver, the Urban Land Conservancy, and several other groups. The fund has been used to leverage additional financing for projects that work to accomplish the affordable housing goals of the region and have thus far helped to locate needed housing in highly desirable parts of the city (Urban Land Conservancy, n.d.; US Department of Housing and Urban Development, n.d.). (See Figure 18.3)

*Figure 18.3* Denver's Evans Station Lofts were one of the first projects built with the city's TOD fund

Source: Photo Urban Land Conservancy and Medici Development Group

The efforts of cities like Austin and Denver just scratch the surface of what will be required as more and more affordable housing stock is lost or redeveloped across the nation. Cities should approach housing development efforts and incentives with an eye toward first effectively providing safe and accessible housing to the elderly and the low-income, rather than to building units intended for only Millenials or for higher-income residents. If done strategically, as done in Denver, units built for the most vulnerable can also meet the needs of market-rate renters and buyers.

## Conclusion

The market preferences of Millennials provide a powerful force in shaping the future of American cities and metropolitan areas. But these preferences must be leveraged to create not just playgrounds for affluent young folks, but a high quality of life for people of all ages, incomes, and backgrounds. Planners must continue to think about how best to integrate greenspace, transportation, and housing into solutions that function well for all urban residents. Approaching all three subjects as interconnected can lead to far more holistic and effective outcomes in each venue.

Transit-oriented development projects are a great example of how this approach can be effectively deployed. TOD concretely ties housing and transportation together and, if approached strategically, greenspace connections can be brought into these projects from the beginning. But the building of effective, long-term projects such as these requires sustained commitments from officials, developers, and residents alike. The same approaches should be taken with mobility and greenspace development. Mobility systems should be as diverse as possible and geared toward providing all residents with as many options as possible to connect to high activity centres, homes, and other amenities. Greenspaces, whether connective linear greenspaces or major parks, must be programmed for all users and developed in ways that augment other efforts to improve quality of life.

Simply bringing these three elements into concert will not create an equitable or intergenerational city. In order to build systems that work for all residents, plans must be carefully oriented to the diverse needs of the population (see City of Austin, 2016).[3] Rather than catering investments and major projects toward meeting the needs and expectations of Millennials, we should approach investments with an eye toward making them work for all residents. This broader frame will take more care and investment, but will also likely result in a city that is sustainably intergenerational. Such a city will pay dividends for Millennials, Baby Boomers, and generations that come after them.

## Notes

1 William Fulton discussed the idea that cities are entering into an era of planning and design where the urban and the natural are merging at the Cultural Landscape Foundation's national conference in March 2016.
2 For more on this initiative visit the Houston Parks Board website, http://houstonparks board.org/bayou-greenways-2020/.
3 One option would be for Cities to pursue the creation of intergenerational plans such as the City of Austin's, produced in 2016.

## References

American Public Transportation Association (2013). *Millenials and mobility: Understanding the Millenial mindset*. Retrieved from www.apta.com/resources/reportsandpublications/Documents/APTA-Millennials-and-Mobility.pdf.

Binkovitz, L. (2016, September 28). Atlanta BeltLine creator resigns citing affordability, equity concerns. *The Urban Edge*. Retrieved from https://urbanedge.blogs.rice.edu/2016/09/28/atlanta-beltline-creator-resigns-citing-affordability-equity-concerns/#.WMBeRW-Qzug.

Boarnet, M. (2006). Planning's role in building healthy cities. *Journal of the American Planning Association, 72*(1), 5–9.

Brenoff, A. (2014, June 4). Playgrounds for seniors improve fitness, reduce isolation. *Huffington Post*. Retrieved from www.huffingtonpost.com/2015/06/04/playgrounds-for-seniors_n_7452270.html.

Brown, C., & Henkin, N. (2015). *Intergenerational approaches to building healthy communities*. Philadelphia, PA: The Intergenerational Center, Temple University.

Cadlk, E. (2016). *The low-income housing tax credit, enterprise community partners*. Retrieved from www.enterprisecommunity.org/policy-and-advocacy/policy-priorities/low-income-housing-tax-credits.

City of Austin (2016). *Age-friendly Austin action plan*. Retrieved from www.austintexas.gov/edims/document.cfm?id=260993.

Durett, C. (2009). *The senior cohousing handbook: A community approach to independent living* (2nd ed.). Gabriola Island, BC. New Society Publishers.

Fausset, R. (2016, September 11). A glorified sidewalk, and the path to transform Atlanta. *The New York Times*. Retrieved from www.nytimes.com/2016/09/12/us/atlanta-beltline.html.

Fulton, W. (2016, March 31). What era of design is Houston in? *OffCite*. Retrieved from http://offcite.org/what-era-of-design-is-houston-in/.

Generations United and the Generations Initiative (2013). *Out of many, one*: *Uniting the changing faces of America*, p. 8. Retrieved from www.gu.org/RESOURCES/PublicationLibrary/OutofManyOne.aspx.

Jarrott, S., Kaplan, M., & Steinig, S. (2011). Shared sites: Avenues for sharing space, Place and life experience across generations. *Journal of Intergenerational Relationships, 9*(3), 343–348.

Maciag, M. (2014, February 25) Public transportation's demographic divide. *Governing*. Retrieved from www.governing.com/gov-data/Public-Transportation-Commuting-in-US-Cities-Map.html.

Pew Research Center (2010, March 10). *The return of the multi-generational family household*. Retrieved from www.pewsocialtrends.org/2010/03/18/the-return-of-the-multi-generational-family-household/.

Reed, C. (2015, April 5). Dutch nursing home offers rent-free housing to students. *PBSNewsHour.*

Rosan, R., Rosan, C., Ross, L., Robles, S., Schmitz, A., & Thoerig, T. (2012). *Housing America's workforce: Case studies and lessons from the experts*. Washington, DC: Urban Land Institute.

Ross, L. (2014). *Residential futures II: Thought-provoking ideas on what's next for multi-generational housing and intergenerational communities*. Washington, DC: Urban Land Institute.

Saporta, M. (2016, September 26). *Ryan Gravel and Nathaniel Smith resign from BeltLine partnership board over equity concerns*. Saporta report. Retrieved from http://saportareport.com/ryan-gravel-nathaniel-smith-resign-beltline-partnership-board-equity-concerns/.

Shelton, K. (2016a, September 22). Can public transit and ride-share companies get along? *The Conversation.* Retrieved from http://theconversation.com/can-public-transit-and-ride-share-companies-get-along-64269.

Shelton, K. (2016b). *Building stronger suburbs: Adaptability and resilience best practices from suburban Houston.* Kinder Institute for Urban Research at Rice University and Urban Land Institute Houston.

Sivak, M., & Schoettle, B. (2016). *Recent decreases in the proportion of persons with a driver's license across all age groups: A report of the University of Michigan Transportation Research Institute.*

Thang, L. L. (2015). Creating an intergenerational contact zone: Encounters in public spaces within Singapore's public housing neighborhoods. In R. Vanderbeck & N. Worth (Eds.), *Intergenerational spaces* (pp. 17–32) London: Routledge.

Toll, E. (2015, July 28) Valley metro: Development along light rail tops $8 billion. *Phoenix Business Journal.* Retrieved from www.bizjournals.com/phoenix/news/2015/07/28/valley-metro-development-alonglight-rail-tops-8.html.

US Department of Housing and Urban Development (n.d.). *Innovative partnership funds transit-oriented housing in Denver.* Retrieved from https://archives.huduser.gov/scrc/sustainability/newsletter_120612_3.html

Urban Land Conservancy (n.d.). *Denver transit-oriented development fund.* Retrieved from www.urbanlandc.org/denver-transit-oriented-development-fund/.

Urban Land Institute (2015). *America in 2015: A ULI survey of views on housing, transportation, and community.* Washington, DC: The Urban Land Institute, p. 17.

Wasik, J. (2016, October 4). The future of retirement communities: Walkable and urban. *New York Times.* Retrieved from www.nytimes.com/2016/10/15/business/the-future-of-retirement-communities-walkable-and-urban.html?_r=0.

Williams, S. (2015). *Preserving multifamily workforce and affordable housing: New approaches for investing in a vital national asset.* Washington, DC: Urban Land Institute, p. 4. Retrieved from http://uli.org/wp-content/uploads/ULI-Documents/Preserving-Multifamily-Workforce-and-Affordable-Housing.pdf.

# 19 What does the march of the Millennials[1] mean for the future American city?

*Alan Mallach*

The movement of those born since the mid-1970s, known as the Millennial Generation, into North American cities has received widespread attention. Until recently the Millennials have been more a subject for bloggers and journalists rather than as a subject for research.[2] A few papers have looked generally at Millennial migratory patterns (Cortright, 2015, Mallach, 2014b), and a few at their impact on specific cities (Piiparinen, Russell, & Post, 2015 on Cleveland; Eichel, 2014 on Philadelphia; Moos, 2016), but this volume may well be the first to take a systematic look at a phenomenon that has already dramatically changed the course of North American cities, and that holds powerful yet uncertain implications for their future.

For purposes of this chapter, I look at Millennials as those young adults who have come of age, in the sense of beginning their lives as independent adults, since 2000. More specifically, I look at a distinct subset, the roughly one-third of Millennials who have received a university degree. Not only is it this subset that has implicitly or explicitly been the subject of virtually all of the media attention given to the Millennial migration, but – problematic as it may be – the bachelor's degree has become such a distinct social and economic class distinction in American society that it can be seen as a reasonable proxy for middle- or upper-income status; in the context of largely lower-income older cities, they make up a distinctly new element in those cities' economic and demographic mix.[3] When I refer to Millennials here without otherwise modifying the term, I am referring to this subgroup alone, not the entire cohort.

In this chapter, I explore the implications that may arise from the arrival of this new element in American cities in order to identify critical unanswered questions, and to suggest public policy perspectives that, while recognizing that change is inevitable, are sensitive to both the positive features of that change and the problems it may trigger. While the principal thrust of the chapter is interpretive and arguably speculative, I will set the stage for my exploration of these questions with a brief quantitative overview of the magnitude and the spatial features of this migration, looking at a number of cities that have become important Millennial destinations.

My focus is on central cities, rather than metropolitan areas. While I will look at "magnet" cities like San Francisco that were already major loci for highly-educated young adults long before the Millennial generation, I will devote more attention to older industrial cities or "legacy" cities[4] like Pittsburgh or Baltimore, where change is more recent, and more directly attributable to the "March of the Millennials". While there are many other types of cities in the USA, my purpose is not to do a comprehensive analysis of the many possible permutations of Millennial population movement, but to identify and interpret what I see as the most critical patterns. I will not look at Sunbelt cities, since the evidence to date suggests that the Millennial migration, which is such a powerful factor in coastal and Midwestern cities, has had far less effect on Sunbelt cities like Phoenix, Las Vegas or San Antonio (also see Chapter 10).[5] That, of course, could change in the future.

## Tracking the Millennial migration

While the evidence for a wave of well-educated Millennials into urban areas is compelling, there are important variations between cities in the features of Millennial migration. The nation's major older cities fall into three distinct categories, as seen in Table 19.1.

*Magnet cities* like San Francisco or Boston had disproportionately large young adult populations long before the current wave, but have seen continued growth to the point where roughly one out of six residents of these cities today is a college-educated Millennial. In 1990, by contrast, legacy cities as a group had far fewer educated young adults; since then, however, the trajectories of different cities have diverged sharply. *Reviving legacy cities* like St. Louis or Pittsburgh are drawing large Millennial in-migration to where they make up today a much larger share of the city's population than in the nation as a whole, but *struggling legacy cities* like Cleveland, Detroit or Buffalo are still lagging.[6] That said, as Table 19.1 shows, even the last group of cities are seeing *some* Millennial in-migration, and contain some areas, albeit small ones, that have become Millennial concentrations.

One salient difference between the magnet cities and the legacy cities is that in legacy cities, as a general proposition, the in-migration of Millennials represents the entire growth taking place in those cities; in other words, while cities like Baltimore or St. Louis are seeing significant Millennial in-migration, they are continuing to experience a negative migration balance among other age groups. Despite the growth of some cities' Latino populations, but for the Millennial in-migration these cities for the most part would still be seeing the population decline they experienced in the 1970s and 1980s continue. The picture in magnet cities is very different; here, Millennial in-migration is only part, albeit a large part, of an overall pattern of population growth. Although both Boston and Washington DC steadily lost population after 1950, since hitting bottom both cities have regained roughly 40% of their total population loss, and on the basis of current trends, may return to their historical peak population between 2020 and 2030.

Thus, the evidence is compelling that the in-migration of educated Millennials into central cities is taking place on a scale large enough to shift the social and economic dynamics of these cities in ways that are unlikely to be merely transitory. Before examining the implications of those shifts, however, the spatial distribution of Millennial in-migration and in particular, the highly concentrated population patterns typical of the in-migration to legacy cities, as distinct from greater Millennial dispersion in magnet cities, needs to be added to the picture.

## The spatial distribution of Millennial in-migrants

The Millennials who are moving to central cities are not so much moving to those cities as moving to a relatively small number of distinct areas within those cities, which are in many respects atypical of those cities as a whole. Central cities in the USA are not on the whole high-density, mixed use places. On the contrary, with a tiny handful of exceptions, they are largely characterized by relatively homogenous neighbourhoods of single family (or in Boston or Newark, two and three family) housing interspersed with low density commercial corridors (Mallach, 2016). By contrast, surveys have shown that the Millennial population is drawn to high-density, mixed-use and walkable areas provided with more transit services than most parts of American cities can offer (Nielsen, 2014, Lachman & Brett, 2015). These areas, which are only a small part of most cities' land area, are typically located in or near their downtowns, or close to major universities or medical centres.

The reality is consistent with the survey findings, while varying by city type, as shown in Figure 19.1 (also see Moos, 2016). The figure shows those census tracts in which over half of the 25+ population has a BA or higher degree *and* in which 20% or more of the total population is between 25 and 34 in 2015 in three cities. In Seattle, reflecting both their large population share, and the extent to which highly educated Millennials have long since been a major part of the city's demographic mix, those tracts make up over one-third of the city. In Detroit, at the other end of the continuum, only a bit more than 1% of the city's census tracts (four tracts) meet these criteria; one is in the city's downtown, while two are in the adjacent Midtown area, nestled between Wayne State University and the Detroit Medical Center. Baltimore, a more robustly reviving legacy city, falls midway between Seattle and Detroit. Its Millennial population is concentrated in 27 census tracts (13%), most of which are in the city's downtown and around the Inner Harbor, with the rest in a secondary concentration to the north around the Johns Hopkins University campus. Over half of the city's college-educated Millennial population lives in these 27 tracts, compared to only 9% of the city's non-Millennial population. Educated Millennials make up nearly one-third of the population of these tracts, compared to 4% of the population in the rest of the city.

While the spatial distribution of these cities' Millennial populations has expanded as their population has grown, it has expanded only into areas contiguous to previously established concentrations. This is not only consistent with research that has found that lower value areas adjacent to higher value areas are more likely to see upward movement in house prices (Guerrieri, Hartley, & Hurst, 2013), but reflects our understanding of Millennial location preferences, in that areas adjacent to high-density areas are likely to have more of the attributes sought by Millennials than more outlying neighbourhoods.

The pattern can be seen clearly in Philadelphia, where Figure 19.2 compares areas of Millennial concentration (using the same criteria as Figure 19.1) in 2000

Table 19.1 Migration trends for population 25–34 with BA or higher degree in selected cities

| City | Population 25–34 with BA or higher degree as a percentage of total city population | | | | Average annual net migration 2010–2014 | | | |
|---|---|---|---|---|---|---|---|---|
| | 1990 | 2000 | 2010 | 2014 | Total net migration | 25–34 with BA+ | All other migration | 25–34 BA+ share |
| **Magnet cities** | | | | | | | | |
| Boston | 9.9% | 10.9% | 12.7% | 15.8% | +9573 | +6368 | +3205 | 66.5% |
| Washington DC | 8.0% | 8.9% | 13.8% | 15.5% | +17626 | +5368 | +12258 | 30.5% |
| San Francisco | 9.7% | 14.2% | 14.8% | 15.8% | +14895 | +4323 | +10572 | 29.0% |
| Seattle | 10.0% | 12.0% | 13.0% | 14.9% | +13435 | +4947 | +8488 | 36.8% |
| **Reviving legacy cities** | | | | | | | | |
| Baltimore | 3.6% | 4.0% | 6.2% | 8.2% | +445 | +3111 | (−2666) | >100% |
| Philadelphia | 4.0% | 3.9% | 5.7% | 7.1% | +8572 | +8582 | (−10) | >100% |
| Pittsburgh | 5.0% | 6.1% | 8.3% | 11.7% | +137 | +2642 | (−2505) | >100% |
| St. Louis | 4.1% | 4.1% | 7.4% | 9.1% | (−1875) | +1311 | (−3186) | >100% |
| **Struggling legacy cities** | | | | | | | | |
| Buffalo | 4.1% | 3.5% | 5.5% | 5.4% | (−652) | (−231) | (−421) | NA |
| Cleveland | 2.0% | 2.6% | 2.9% | 3.1% | (−1824) | +98 | (−1922) | >100% |
| Detroit | 1.7% | 1.7% | 1.3% | 2.1% | (−8382) | +1284 | (−9666) | >100% |
| Milwaukee | 3.7% | 3.8% | 4.5% | 4.9% | +1202 | +703 | +499 | 76.5% |
| USA | 4.0% | 3.9% | 4.1% | 4.5% | | | | |

Souce: US Census of Population, One Year American Community Survey

and 2015.[7] In 2000, Millennials were clustered tightly in and around Center City, with a small pocket in West Philadelphia, just west of the University of Pennsylvania campus. By 2015, the areas of concentration had grown larger, but were not markedly different. The West Philadelphia pocket had grown from two to five census tracts, while Millennial populations had filled in gaps in Center City and moved into areas contiguous to Center City; Fairmount and Spring Garden to the north; Northern Liberties and Fishtown in the northeast; and, to a more limited extent, South Philadelphia.

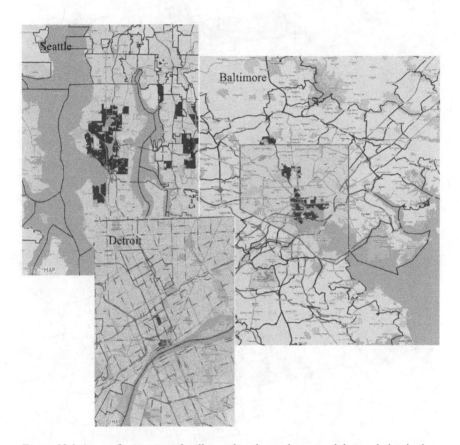

*Figure 19.1* Areas of concentrated college-education and young adult population in three cities in 2014

Source: Five-year 2014 American Community Survey; maps by PolicyMap

Note: The Seattle and Baltimore maps are at approximately the same scale, showing (nearly) the entire city, but the Detroit map is at a much larger scale, and shows only a small part of the city.

*Figure 19.2* Expansion of areas of millennial concentration in Philadelphia 2000–2014

Source: 2000 Census and 2014 American Community Survey; map by PolicyMap.

The patterns of Millennial concentration, and gradual expansion into contiguous areas from small central – usually downtown – cores, seen in the context of other social and economic dynamics currently affecting central cities, has powerful implications for the future of those cities. As I will discuss in the next section, it is not only contributing to significant shifts in these cities' social and political climate, but also to increasing polarization, both economic and racial, in these cities.

## Prosperity, poverty and polarization

The growing polarization of American society between affluent and poor, and the "hollowing of the middle class", has been widely studied and reported (c.f. Autor, Katz, & Kearney, 2006; Bluestone, 1995; Peck, 2011). Closely linked to this trend has been the phenomenon dubbed economic sorting, the increasing tendency of people of similar economic status to congregate in economically homogenous neighbourhoods (Reardon & Bischoff, 2011; Bozza, Cutsinger, & Galster, 2006). These patterns not only exist, but can be seen in exaggerated form in central cities; while 31% of all Milwaukee households in 1970 were middle-income household (with incomes between 80% and 120% of the citywide median), by 2013 the percentage of the city's households that were middle-income had dropped to 15%. The increase in households, moreover, was concentrated at the edges of the income range, among households whose incomes were either under 50% or over 150% of the citywide median; that is, the poorest or, at least relatively speaking, most affluent households.

Not only did the size of the city's middle class diminish sharply, but, in keeping with the findings of Bischoff and Reardon (2013), the population tended to concentrate more and more in either low or high income areas. Between 1999 and 2013, the number of census tracts in Milwaukee where the tract median was within 80% to 120% of the citywide median declined from 30% to 19% (Mallach, 2015a).

Over and above the effect of the Millennial movement into increasingly affluent, homogenous areas, the magnitude of this shift in central cities is linked to two salient trends, both of which I have addressed in previous research. The first is the accelerated out-migration of the African-American middle class from urban neighbourhoods (Mallach, 2015a) leading to increased poverty concentration in urban areas (Jargowsky, 2015). The second is the uncoupling of urban economies from their cities' resident populations. While many central cities are seeing job growth, fewer and fewer of those jobs are going to residents of the same cities, while the overall number of job-holders living in these cities is declining precipitously (Mallach, 2014a).

In this context, the introduction of a large new and different demographic in the form of highly-educated and generally affluent Millennials has tended not only to increase the already-strong tendency toward polarization, but, because of the concentrated nature of their migration, to give it a pronounced, highly visible spatial focus. For example, Washington Avenue in St. Louis, the city's onetime garment district, is an east–west avenue of late 19th and early 20th century industrial buildings that have been restored in recent years and converted to residential use with ground floor stores, restaurants and entertainment venues. The glittering focal point of St. Louis' Millennial population, it runs through the city's Central Corridor only three short blocks south of Delmar Boulevard, across from which lies the city's deeply poor and disinvested – and predominately African-American – Northside. In Philadelphia, Girard Avenue divides the Millennials of Fairmount from the poor African-American community of North Philadelphia. Similar, although not always as neatly configured, contrasts can be found in all cities experiencing significant Millennial migration.

The revived precincts where Millennial occupancy is clustered offer a distinct series of visible cultural signifiers, which vary little if at all from city to city. These signifiers include restored 19th and early 20th century office and industrial buildings in central city downtowns, such as those on Washington Avenue, in Cleveland's Warehouse District, or downtown Detroit; brick or brownstone row houses in nearby neighbourhoods, as in Brooklyn, Baltimore, St. Louis or Philadelphia, distinguished by the pristine character of their power-washed facades; bars, restaurants and coffeehouses, sharing eclectic but recognizable menus and aesthetic; and increasingly, stores and services catering to Millennial tastes and pocketbooks, such as vintage clothing stores, fitness centres, gourmet emporia, outdoor outfitters or bicycle stores.

While there is a tendency to categorize this simplistically as gentrification, it is a more complex phenomenon that can perhaps be better explained in part as "youthification" (Moos, 2016). If gentrification is to be understood as the movement of affluent populations into lower-income neighbourhoods, gradually resulting in the displacement of the lower-income population, the Millennial in-migration into many (if not most) of the areas where they have located should not appropriately be characterized as gentrification. Indeed, what is notable about so many of those areas is that they had little or no previous residential population, such as Midtown Detroit, the Warehouse District, or Washington Avenue. In-migrating Millennials to these areas have sometimes been accommodated by new construction, but more often by conversion of formerly vacant non-residential buildings for residential use. While gentrification is widespread in magnet cities like Washington DC, in most legacy cities it tends to be a limited phenomenon, dwarfed by the extent of the simultaneous decline in the city's single-family neighbourhoods, particularly those of predominantly African-American character (Mallach, 2015b).

The contrast of affluence and poverty is also racial in nature. African-Americans make up 10% of the total pool of college graduates, a far smaller percentage than the African-American share of most central city populations in the Northeast and Midwest. One product of the dominant Millennial role in urban in-migration is that white in-migration, measured as a percentage of the same-race population base, has come to substantially exceed African-American in-migration in almost all large central cities, as shown in Table 19.2, which compares average annual in-migration from outside the state[8] to the reviving and the struggling legacy cities for the period of 2011–2015. Notably, the disparity between white and African-American in-migration is markedly greater among the reviving than among the struggling cities.

*Table 19.2* Average annual in-migration from outside the state for reviving and struggling legacy cities by race 2011–2015

| Category | White | | | African-American | | |
|---|---|---|---|---|---|---|
| | Base population | Average annual in-migration | Percent of base | Base population | Average annual in-migration | Percent of base |
| Reviving legacy cities | 1,057,351 | 46,769 | 4.42% | 1,268,122 | 18,368 | 1.44% |
| Struggling legacy cities | 528,291 | 13,637 | 2.58% | 1,064,104 | 12,103 | 1.14% |

Source: 2015 Five-Year American Community Survey

While white in-migrants to central cities tend to be more highly educated than white out-migrants, the opposite is true among African-Americans, reflecting the middle-class exodus. This has led to growing racial disparities in educational attainment and income, reversing the trends of the 1990s when racial gaps in both education and income narrowed in many central cities.

The regrowth of urban white populations, reversing a seemingly inexorable long-term trend toward larger African-American central city majorities, raises potentially important social and symbolic implications. While white non-Latino populations are still dropping overall in most legacy cities, as the in-migration of Millennials has yet to offset the continued erosion of older white residents, the change in Washington DC may be a potential harbinger of future trends; between 2000 and 2014, the white non-Latino population in the District increased by nearly 80,000, while the African-American population dropped by over 20,000. Washington lost its absolute Black majority in 2012, and at the current rate of change, non-Latino white residents will exceed African-Americans by 2026. This evolution, characterized by one writer as the transformation of "Chocolate City"[9] to "Latte City" (Dvorak, 2015), has been widely noted in the media and the blogosphere. Similar concerns have also emerged in New Orleans, another historically majority African-American city undergoing Millennial-driven population shifts.

These changes raise serious questions for African-Americans, including important albeit intangible ones about their relationship to their city and its identity, as Natalie Hopkinson writes, "In Washington, we were not 'minorities,' with the whiff of inferiority that label carries; we were 'normal'" (2012). They also raise more concrete ones about the distribution of political power and the role of government. Since economic power in cities has remained concentrated largely in white hands, many African-Americans have seen their political power as a critical counterbalancing force; now, that too has been called into question.

While some of this may seem unduly abstract, it has tangible implications. As with cities in the Global South, where the abrupt juxtaposition of wealth and poverty can lead to conflict, and in turn to stigmatization and oppressive policing, increases in such outcomes are a very real possibility in the remade American city, rendered more likely by those cities' rigid spatial dichotomies of revival and decline. Much attention was given, in the wake of Freddie Gray's death in police custody in Baltimore, to the disparity between the neglect of his impoverished, deteriorated neighbourhood and the investment taking place around the Inner Harbor not far away. It is not, I believe, too far a leap to suggest that a relationship may exist between the Millennial-driven revival of cities like Baltimore and Chicago, the increasing polarization of these cities, and the apparent increase in police violence directed against African-Americans in these same cities.

## The future of the American city: Millennial playground or home sweet home?

The March of the Millennials to America's older cities is very much a work in progress. While it has already had a powerful effect on many cities, it is a relatively new phenomenon seen from the perspective of long-term social and economic change, and its future trajectory remains highly uncertain. Just as no one predicted the contours of today's American cities 25 years ago, one cannot know what they will be 25 years into the future. Over and above the effect of changes in the national and global economic and political environments, at least three distinct factors specific to migration and demographic trends will affect the extent to which the future city will resemble that of today, and to what extent it will be different. First, of course, is the future behaviour of the members of the Millennial generation themselves, in particular the extent to which their love affair with urbanism will continue as they enter their child-rearing years (Chapters 9); second, whether future generations, whatever they may be called, will share the Millennial generation's desire to partake of what urban life offers; and, third, the size of future cohorts or generations, whatever their attitudes or preferences.

One can hypothesize three alternative future city scenarios, based on alternative outcomes reflecting the behaviour of the Millennial and subsequent generations (Chapter 14). The *Sustained Growth* scenario assumes that increasing numbers of highly educated Millennials will choose to remain in central cities as they start to form families and raise children, and beyond; and that, as they do so, the number of young people in the cities is replenished by future cohorts, perhaps along with growing numbers of empty nester and retired older households. Under this scenario, growing Millennial demand is likely to spill over into many single family urban neighbourhoods currently unaffected by the Millennial migration. Central cities are likely to gain population, while becoming increasingly white and more affluent; while this scenario may lead to some economically integrated neighbourhoods, it will also result in increased economic pressure on lower-income households.

The *Stability* scenario assumes that the preponderant tendency of Millennials as they form families will be to follow their parents' lead and suburbanize, although they may be more likely to favour higher-density suburbs with walkable downtowns, good schools and high levels of public safety, to exurban large-lot suburbs. As they leave, however, they will be replaced by roughly similar numbers of the members of future cohorts, along with some number of older childless households. Under this scenario, while a certain amount of neighbourhood flux, as some new areas are "discovered" and others decline, will take place, the overall spatial distribution of market demand in cities is likely to remain roughly constant, and less change will take place in the overall social landscape of the city.

The *Reversal* scenario assumes not only that the preponderant tendency of the Millennial generation will be to suburbanize, but that future demand for urban living from subsequent cohorts, whether because of smaller numbers, changes in preferences, or both, will reduce the demand that cities have experienced in recent years. Under this scenario, demand for the expensive housing, goods and services that have emerged to serve the Millennial market over the past decade or so may decline, and cities may see some retrenchment and potential market decline in some neighbourhoods that have seen more recent Millennial in-migration.

In the absence of a crystal ball, any of these scenarios are at least plausible. Moreover, one might well find that different cities follow different future scenarios. The Millennial and subsequent similar generations are likely to continue to grow in Washington DC or Seattle, for example, given their highly developed amenity base and the durability of demand from those cities' economies for highly-skilled, highly-educated, workers, while shrinking in others. The extent to which one or another scenario may unfold is likely to be affected as well by the nature and extent of job growth in different cities; although Millennial migration is far from being entirely dependent on job growth, it is certainly affected by it. The uncertainty in this respect is increased by the extent to which many cities' economic base has become heavily dependent on higher education and health-care, both of which are susceptible to unpredictable changes associated with technology, public policy and other factors.

Although one cannot predict the behaviour of Millennials from that of previous generations, the evidence suggests that the majority of highly educated young adults who lived in cities ten or 15 years ago did indeed leave as they got older; and, moreover, that there is little evidence for the proposition that a significant return of either Empty Nesters or Baby Boomers to central cities is yet taking place. Figure 19.3 shows the increase in college-educated adults by age group from 2000 to 2015 as a percentage of the city's 2000 population in three major Millennial destinations, Boston, Pittsburgh and Baltimore. The increase in the 25–34 age group is far greater than that of the older cohorts; indeed, to the extent that cohorts aged 45 and older saw growth, it is attributable far more to national trends associated with increased diffusion of higher education from the 1950s through the 1980s than to local dynamics.[10] It is worth noting, however, that Boston, which has had a disproportionate share of young college-educated adults for far longer than either Baltimore or Pittsburgh has seen much greater growth in older college-educated cohorts than either of those cities.

As was true of previous generations, some Millennial couples will undoubtedly remain in the cities and raise their children there, under any scenario. The critical question is whether, and under what circumstances, a significantly larger share would do so than in earlier years. It is likely that Millennials are more open to remaining in cities than earlier cohorts (Chapter 14). Whether they will actually do so is likely to depend heavily on the extent to which they believe the city offers the features they consider necessary for successful family life; attractive

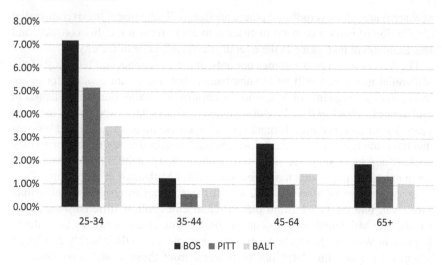

*Figure 19.3*  Population with BA or higher degree by age cohort in 2000 and 2014 in Boston and Pittsburgh

Source: Census of Population 2000, 1-year American Community Survey 2014

houses in walkable neighbourhoods offering good schools, safety, and public services.

Prediction, especially about the future,[11] is difficult, but I would suggest that the most likely future scenario, barring unpredictable large-scale economic shocks, will be closest to the stability scenario described above, but tending more toward growth than reversal. Although the growth in the number of highly-educated Millennials who remain in the city to raise families is likely to be incremental rather than dramatic, some increase is likely to take place, leading in turn to further expansion of the areas of Millennial concentration into predominately single-family areas currently occupied by less affluent households outside the cities' central cores. This is supported by anecdotal accounts suggesting that the proliferation of school options (magnet schools, charter schools, inter-district transfers, etc.) in many cities is reducing the role of education as a barrier to urban retention of more affluent child-rearing households.[12]

Demographic trends, however, may place a damper on that scenario. Demographer Dowell Myers (2016) has pointed out that the number of people born in 1990 (1990 birth cohort), and who turned 25 in 2015, was the largest single year birth cohort since the peak of the baby boom, nearly 30 years earlier. With smaller birth cohorts in subsequent years, the cohorts that will be graduating from college in the 2020s will be substantially smaller than those swelling the ranks of Millennials today. For this and other reasons,[13] Myers has suggested that American cities may be approaching what he has called "peak millennial" (2016). If his thesis is sound, and assuming no change in preferences, the post-Millennial migration may slow, offsetting much of the likely increase in the number of Millennials remaining in the cities. However, several critiques raising serious questions about the validity of Myers' thesis (and the assumptions behind his empirical analysis) have recently appeared (Cortright, 2017; Chapter 12, this volume).

Some cities are likely to be exceptions. A handful of cities, most particularly Washington, New York, Seattle and San Francisco, serve unique economic functions that are, at least at present, not fungible with other cities or regions. In the absence of fundamental change in the global or national economy, growth in those functions should drive continued in-migration. These cities, particularly the last two, already retain more young in-migrants from pre-Millennial generations (and/or draw more older adults) than cities with more recently emerging migration patterns like Pittsburgh. To the extent that the ability to meet demand is not arbitrarily constrained by exclusionary land use regulations, demand is likely to prompt continued redevelopment and population growth, much of it likely to take place in areas previously occupied by lower income households, or in previously non-residential areas.

In either event, the pressures currently driving increased polarization are unlikely significantly to abate, and may grow. Increased numbers of affluent parents may lead to additional spheres of potential conflict, such as the emergence of de facto two-tier public school systems catering to their children (Cucchiara, 2013). These cities' political and civic leadership will then have to determine whether they will continue to pursue what might be considered the path of least resistance; that is, to cater to their growing educated, affluent populations to the real or perceived disadvantage of the rest; or whether they will confront and attempt to address polarizing forces through intentional strategies. Even if they do so, the question remains whether their efforts will result in any meaningful change to those cities' trajectories.

## Urban policy, Millennials and the future American city

The combination of economic and demographic change that led to the Millennial migration, which has in turn so dramatically changed the dynamics of so many American cities, was not the product of any explicit, or even coherent, governmental policy, least of all at the level of local government. In retrospect, most of the public sector initiatives of the 1980s and 1990s, including the construction of stadiums, arenas and convention centres, or the subsidization of downtown office buildings with tax incentives, appear to have had little effect on their cities' trajectories. Similarly, while governments have encouraged the growth of medical centres and university complexes, which arguably have turned out to be important Millennial magnets, these activities were largely reactive and seen in narrow economic development terms, rather than in the framework of any transformative demographic strategy.

More recently, local governments have widely tried to capitalize on the Millennial migration by facilitating redevelopment through tax incentives, infrastructure investments, and other forms of subsidy areas that have become Millennial destinations, often angering advocates for lower-income communities, to whom the use of public funds for upscale housing, cultural amenities and commercial development in the face of the needs of lower-income communities is inappropriate or worse. Most municipal efforts to facilitate redevelopment have been reactive, however, responding to increased demand and developer interest, rather than strategic; moreover, they have often been driven by developer pressure (Chapter 10), rather than by a rational calculus of the development's need for the incentives or the relationship between the cost and the benefit to the community. An arguable case in point is Baltimore's recent commitment of $660 million in public financing for Under Armour founder Kevin Plank's "plan to overhaul an industrial area in the city's southern end into a mix of housing, offices and retail around a new corporate campus for the athletic giant", representing only part of the $1.1 billion in public funds the developer is seeking (O'Connell, 2016).

Given the extent that public actions have been directed toward reinforcing market trends that emerged independently, it is unsurprising that many have been seemingly successful in achieving that goal; it is equally unsurprising that their overall effect has been to reinforce the underlying trends toward greater inequality and polarization in the spatial and economic fabric of the city. Similarly, should city government policies continue in much the same direction, it is likely that inequality will continue to grow, and that while the areas of Millennial concentration may continue gradually to expand, much of the rest of the city may become increasingly impoverished, blighted and desperate. I hesitate to pronounce whether such a state of affairs is or is not sustainable, in a strict sense; one can point to many cities in the Global South such as Lagos (Kim & Short, 2008) or Jakarta (Smith, 1996) that have long survived with far greater levels of inequality and polarization, although arguably at the price of practices or conditions that many in the USA and other Western nations might find unacceptable. For many reasons, however, which do not need to be belaboured here, I consider it profoundly undesirable, whether or not sustainable.

This is not to suggest that there are not legitimate reasons for current urban policies. Central cities, particularly legacy cities, are resource-poor, and despite the Millennial migration, most still house disproportionate concentrations of the poor. In 2014, the poverty rate in Boston was 40% higher than the national level, and in Baltimore and Pittsburgh 50% higher. It is in a city's rational interest to foster greater investment and a more economically diverse population base; assuming, moreover, that some investment may not come about through purely market means and that the ultimate return to the city exceeds the cost, it is not unreasonable for a city to invest public funds, either through direct investment or tax incentives, for that purpose.

Looked at through a wider lens, however, these policies raise difficult equity questions. In her book *The Just City*, Fainstein defines equity as "a distribution of both material and nonmaterial benefits derived from public policy that does not favor those who are already better off at the beginning" (2010, p. 36). From this social justice perspective, public policy in many if not most American cities fails the smell test.

Going forward, the most fundamental challenge facing urban policymakers is a twofold one. First, how to allocate public resources more equitably; and, second, closely related to the first, how to create meaningful pathways of opportunity within the city that enable the larger and less affluent part of the city's population to benefit from the revitalization and investment that is taking place, without dis-couraging investment or migration. To try to halt or even slow in-migration to and investment in struggling cities in the ostensible interest of equity would be counter-productive. Without the resources and opportunities created by invest-ment and in-migration, there will be fewer resources available to allocate, and fewer opportunities open to city residents.

Policymakers need to be both more rigorous and more transparent about the allocation of public resources. I am not arguing that public resources should not be spent to encourage projects that further Millennial or other upscale in-migration; I would argue, however, that use of public resources for that purpose should be subject to both a determination that the expenditures are needed to achieve a legit-imate public benefit, that the magnitude of the benefits is commensurate with the cost, and that the distribution of the benefits is equitable. Those determinations, moreover, need to be transparent, and their calculus visible in such a way that they strengthen rather than undermine trust in local government, a commodity that is currently in short supply. Moreover, local governments have an obligation to ensure that, to the extent realistically feasible, the interests of lower income resi-dents are recognized and addressed where they may be affected not only by public expenditures, but by the larger process of transformation taking place.

This could take many forms. Where public resources such as tax abatements or infrastructure investments are being used to support new residential development, those developments could be required to include some percentage of housing affordable to lower-income families and individuals, as has been initiated, albeit on a modest scale, in Washington DC and San Francisco. Similarly, Philadelphia and Cook County, Illinois (which contains Chicago) have adopted rules to pro-tect lower-income homeowners from the effect of sudden property tax increases in neighbourhoods facing sharply rising property values. Schools designed to attract or retain child-rearing Millennial families could be intentionally organized to serve an economically and racially diverse population, as is the case with the highly-successful City Garden charter school in St. Louis (Huck, 2016).

There is reason to believe that many Millennials could be recruited to be part of efforts to foster greater inclusion and equity in the cities to which they have moved. Millennials, as a group, share high levels of concern about social justice and equity, as reflected in their engagement in the Occupy and Black Lives Matter movements; as a recent study concluded:

> The youngest members of this generation are just 16 years old. Where their predecessors had barely begun to care about causes at their age, these young people have spent every minute of their lives amid a generation of individu-als eager to connect, get involved with and give to causes they're passionate about (Achieve, 2016).

As other research has found, however, the pathways through which Millenni-als are engaged in social causes, and the manner in which they engage in those causes, are very different from those of older generations (Achieve, 2013). A major challenge for local governments and non-profit organizations seeking to engage Millennials in creating more equitable outcomes as their cities change will be to engage them on their own terms, rather than expect them to follow more traditional models of engagement.

The creation of pathways of opportunity, however, for the city's population may be the single most important task, over and above the basic delivery of public services, facing local government. There is no shortage of jobs, in the numerical sense, in American cities. Even Detroit, perhaps the nation's most economically distressed major city, contains 60,000 more jobs than there are workers who live in the city. Of the 234,000 jobs in the city, however, fewer than 60,000 are held by Detroit residents; the vast majority are held by suburban commuters. Even a modest increase in the number of city residents holding those jobs could have a major impact on social and economic conditions in the city. While the elite jobs in the education and medical sectors that dominate urban economies require advanced degrees, both sectors offer thousands of jobs at all skill and educational levels, as well as – particularly in healthcare – potential job ladders where incremental skill acquisition can lead to significant upward mobility.

The scope of the challenge has been extensively delineated (Giloth & Meier, 2012 among others). It demands a very different approach to building human capital than what is prevalent in most communities, which can be characterized as fragmented and short-term oriented, with large numbers of public sector, non-profit and corporate players operating largely on their own and often driven by short-term funding to focus on short-term goals. These players include public school systems, and increasingly, charter schools; community colleges and vocational schools; public, non-profit and for-profit entities engaged in job training and placement; Workforce Development Boards; and, of course, the major employers of the city and region. While many individual players are dedicated and capable, and their organizations effective within their compass, their systems, cultures and incentives often tend to work against either effective collaboration or sustained long-term investment in building human capital. They do not add to up a process capable of leading to the level of fundamental change that is needed.

Creating a process to bring multiple agencies and organizations together to create a long-term transformational approach to realizing the potential of the human capital already in our cities, particularly in light of the already large amount of resources being devoted to education, training and skill development in these cities, should certainly be within our capabilities as a society. A number of promising initiatives have emerged in recent years, including an innovative partnership between the Wisconsin Department of Corrections and Employ Milwaukee (the Milwaukee Area Workforce Development Board) to both reduce recidivism and increase the employment potential of people returning to the Milwaukee area from prison (Justice Center, 2016).

Finally, whether the future follows the trajectory of sustained growth, stability or of reversal, I believe that the Millennial migration has already had a significant impact on American cities, one that is likely to have long-term consequences. Notwithstanding the concerns that have been expressed in the preceding pages, the fact that hundreds of thousands of talented, energetic and highly skilled young people are choosing to make America's cities – including many all but given up for dead a few short decades ago – their home represents a remarkable opportunity for building a better and more just city. Rather than either demonize them or treat them as a cash cow to be milked, local governments should reach out to this generation, engage them constructively in the challenges of the city in which they live, and harness their energy to make it a better place.

# Notes

1  With acknowledgements to the 2013 *Washington Post* series (Chang, Tucker, Goldstein, Yates, & Davis, 2013), which, to the best of my knowledge, coined the phrase "march of the millennials".

2  While much of the blogging and reportage is superficial, and often misleading, in terms of such elementary errors as failing to distinguish between data for metropolitan areas and for central cities, it has yielded some valuable work, including the series by Chang and her colleagues (2013) mentioned above.

3  Failure to distinguish between this subset and the larger group of Millennials is another weakness of some writings, particularly those that claim to debunk the assertion that there is a significant shift in Millennial migration patterns compared to earlier generations; as in Casselman (2015).

4  The term legacy cities is increasingly being used to describe those older industrial cities, such as Pittsburgh or Detroit, that have both largely lost their industrial base and a significant percentage (typically 20% or more) of their peak population (American Assembly, 2012).

5  As of 2014, none of these three cities had a Millennial population share greater than the national average. Although the Millennial share had grown modestly in San Antonio between 2000 and 2014, it had not increased at all over this period in the other two cities, in dramatic contrast to the picture in either magnet or legacy cities.

6  This may be changing, at least in Detroit; the visible increase in downtown residential units, restaurants, coffee places and other Millennial signifiers in Detroit in the last two to four years, too recent to be reflected in available demographic data, is stunning.

7  The city has some smaller outlying areas of Millennial concentration not shown in Figure 19.2, the neighbourhoods of Manayunk and Mt. Airy in the northwestern part of the city.

8  Migration from outside the state is more significant as an indicator of change, in that migration from other counties within the same state, which is largely made up of migration from within the same metropolitan area, is more likely to contain a significant element of population "churning", in the sense of back-and-forth population movement.

9  The origins of the term are obscure, but it was made famous by George Clinton's 1975 funk classic, "Chocolate City". It is most often used to refer to Washington DC, but has been used elsewhere.

10  Nationally, more than twice as many people over 65 in 2014 held a BA or higher degree than the number of those over 65 in 2000, reflecting the extent to which the former were more likely to have been educated in the 1960s and 1970s, benefiting from the massive expansion of higher education after World War II.

11  The saying, "prediction is difficult, especially when it's about the future" has been variously attributed to Nils Bohr, Yogi Berra, and others.

12  This is based on recent interviews by the author in Detroit and St. Louis.

13  In his paper, Myers also makes important points about the potential effect of other factors as job growth, household formation and housing production on Millennial urban concentration, all of which are worth serious discussion but cannot be addressed in the limited confines of this chapter.

# References

Achieve (2013). *The 2013 Millennial impact report: Connect, involve, give*. Washington, DC: The Case Foundation.

Achieve (2016). *Cause, influence & the workplace: The Millennial impact report retrospective: Five years of trends*. Washington, DC: The Case Foundation.

American Assembly (2012). *Reinventing America's legacy cities: Strategies for cities losing population*. New York: The American Assembly.

Autor, D. H., Katz, L. F., & Kearney, M. S. (2006). The polarization of the US Labor Market. *American Economic Review, 96*(2), 189–194.

Bischoff, K., & Reardon, S. F. (2013). *Residential segregation by income, 1970–2009*. Providence RI: US 2010 Project, Brown University.

Bluestone, B. (1995). *The polarization of American society: Victims, suspects, and mysteries to unravel*. New York: Twentieth Century Fund Press.

Booza, J. C., Cutsinger, J., & Galster, G. (2006). *Where did they go? The decline of middle-income neighborhoods in metropolitan America*. Washington, DC: Metropolitan Policy Program, The Brookings Institution.

Casselman, B. (2015, March 20). Think Millennials prefer the city? Think Again. *Fivethirtyeight.com*. Retrieved from http://fivethirtyeight.com/datalab/think-millennials-prefer-the-city-think-again/.

Chang, E., Tucker, N., Goldstein, J., Yates, C., & Davis, M. (2013, October 18). March of the Millennials. *Washington Post*, p. 3.

Cortright, J. (2015, March 23). *Twenty-somethings are choosing cities*. Really Portland, OR: City Observatory. Retrieved from http://cityobservatory.org/twenty-somethings-are-choosing-cities-really/.

Cortright, J. (2017, January 24). Here's what's wrong with that "Peak Millennials" story. *Citylab*. Retrieved from www.citylab.com/housing/2017/01/flood-tide-not-ebb-tide-for-young-adults-in-cities/514283/?utm_source=nl__link1_012517.

Cucchiara, M.B. (2013). *Marketing Schools, Marketing Cities: Who Wins and Who Loses When Schools Become Urban Amenities*. Chicago, IL: University of Chicago Press.

Dvorak, P. (2015, October 15). From chocolate city to latte city: Being black in the New D.C. *Washington Post*. Retrieved from www.washingtonpost.com/local/from-chocolate-city-to-latte-city-being-black-in-the-new-dc/2015/10/15/c9839ce2-7360-11e5-9cbb-790369643cf9_story.html.

Eichel, L. (2014). *Millennials in Philadelphia: A promising but fragile boom*. Philadelphia, PA: Pew Charitable Trusts.

Fainstein. S. (2010). *The just city*. Ithaca, NY: Cornell University Press.

Giloth, B., & Meier, J. (2012). Human capital and legacy cities. In A. Mallach (Ed.), *Rebuilding America's legacy cities: New directions for the industrial heartland*. New York: The American Assembly.

Guerrieri, V., Hartley, D., & Hurst, E. (2013). Endogenous gentrification and housing price dynamics. *Journal of Public Economics, 100*, 45–60.

Hopkinson, N. (2012, June 23). Farewell to chocolate city. *The New York Times*.

Huck, C. (2016). *Testimony: Improving school integration for equity, not diversity*. New York: The Century Foundation.

Jargowsky, P. (2015). *The architecture of segregation*. New York: The Century Foundation.

Justice Center, Council of State Governments (2016). *Executive summary: Increasing job readiness and reducing recidivism among people returning to Milwaukee County, WI, from Prison*. Washington DC: Council of State Governments.

Kim, Y-H., & Short, J. R. (2008). *Cities and economies*. New York: Routledge.

Lachman, M. L., & Brett, D. L. (2015). *Gen Y and housing: What they want and where they want it*. Washington, DC: Urban Land Institute.

Mallach, A. (2014a). The uncoupling of the economic city: Increasing spatial and economic polarization in American older industrial cities. *Urban Affairs Review, 51*(4), 443–473.

Mallach, A. (2014b). *Who's moving to the cities, who isn't: Comparing American cities*. Washington, DC: Center for Community Progress.

Mallach, A. (2015a). *What drives neighborhood trajectories in legacy cities? Understanding the dynamics of change*. Working Paper, Lincoln Institute of Land Policy, Cambridge, MA.

Mallach, A. (2015b). *Gentrification and decline in a legacy city: Looking at Milwaukee 2000–2012*. Washington, DC: Center for Community Progress.

Mallach, A. (2016). Is the urban middle neighborhood an endangered species? Multiple challenges and difficult answers. In P. Brophy (Ed.), *On the edge: America's middle neighborhoods*. New York: The American Assembly.

Moos, M. (2016). From gentrification to youthification? The increasing importance of young age in delineating high-density living. *Urban Studies, 53*(14), 2903–2920.

Myers, D. (2016). Peak Millennial: Three reinforcing cycles that amplify the rise and fall of urban concentration by Millennials. *Housing Policy Debate, 26*(6), 928–947.

The Nielsen Company (2014). *Millennials: Breaking the myths*. New York: The Nielsen Company.

O'Connell, J. (2016, September 20). Baltimore approves $660 million for under armour development. *Washington Post*.

Peck, D. (2011, September). Can the middle class be saved? *The Atlantic*.

Piiparinen, R., Russell, J., & Post, C. (2015). *The fifth migration: A study of Cleveland Millennials*. Cleveland, OH: Cleveland State University.

Reardon, S. F., & Bischoff, K. (2011). *Growth in the residential segregation of families by income*. Providence, RI: US2010 Project, Brown University.

Smith, D. A. (1996). *Third world cities in global perspective*. Boulder, CO: Westview Press.

# 20 (Millennial) cities of tomorrow

*Tara Vinodrai, Markus Moos*
*and Deirdre Pfeiffer*

We began this book noting the growing interest in the Millennial generation. There are plenty of claims in the popular press about the attitudes and preferences of this group, as well as many assumptions about where Millennials want to live, work and play. Yet, there remain few sustained and systematic investigations that unpack the locational choices, experiences and preferences of this group and consider what this means for how we understand and plan our cities.

We argued that a generational lens is critical for understanding how our cities have evolved and to best understand the experiences of Millennials in contemporary urbanism. To do so, we brought together a diverse group of North American scholars from planning, geography and the broader social sciences with interests related to demographic and generational change, housing, transportation, and labour markets to provide theoretically and empirically grounded perspectives on how Millennials shape and are shaped by the cities in which they live. In this chapter, we provide a synthesis of their scholarship, identifying common themes and points of divergence. We offer lessons for planners and policymakers, and trace an agenda for future research.

## Unpacking Millennial mythologies

As we noted at the outset, the Millennial generation is confronted by several contradictions and mythologies. Front and centre amongst these contradictions is the relationship between education and the anticipated advantages that it affords. The Millennial generation is the most educated generation, yet – in aggregate – Millennials fare worse in economic terms, experiencing lower incomes, higher debt loads, and greater labour market and housing affordability challenges. Our contributors interrogated the contradictions and mythologies associated with the Millennial generation by analyzing a range of quantitative and qualitative data using a diverse set of analytical tools and techniques. Four cross-cutting key themes and observations arise by considering this body of scholarship in its entirety.

First, a number of chapters consider *intergenerational differences*. In Chapter 4, Fry finds that US Millennial households fare relatively well in terms of wealth and household income levels compared to previous generations. However, he does find greater diversity in the outcomes amongst Millennials. In particular, educated Millennials in the USA are faring relatively well – even better than their predecessors at a similar stage. By contrast, Walks et al. (Chapter 5) observe greater wealth inequality amongst younger generations in Canada compared to earlier generations. On the housing front, Mawhorter (Chapter 11) finds that US Millennials have generally fared worse that the Boomer generation and have reduced levels of headship, homeownership, and housing affordability.

Second, issues related to *intra-generational inequality* arise in several chapters. Both Fry and Mawhorter find that Millennials with lower levels of educational attainment fare worse than others in their generation; in other words, there are widening inequalities *within* the Millennial generation. Moos (Chapter 9) and Worth (Chapter 7) focus on wealth, specifically examining different forms of intergenerational wealth transfers. Worth highlights the privilege afforded to Millennials who receive financial and other forms of support from their families allowing them better access to education, jobs and the housing market. Moos focuses on intergenerational wealth transfers in support of access to home ownership. Both chapters suggest deepening inequalities within the Millennial generation based on a birth lottery. Ralph's research (Chapter 15) on transportation behaviour amongst young people also highlights high levels of disparity within this cohort, articulated through car-free living by necessity in transit poor (yet affordable) suburbs rather than car-free living by choice in transit-rich (and expensive) urban neighbourhoods.

Third, *variations amongst metropolitan regions* were highlighted by several authors. Building on Myers' (2016) claim that we have reached peak millennial, Henry and Moos (Chapter 12) argue that while this may be true at the national level in the US and Canada, there are substantial variations across US and Canadian metropolitan areas. They emphasize the importance of local context for understanding the generationed dynamics of cities and the need to pay attention to both domestic migration and immigration, noting national differences in the latter. Mallach (Chapter 19) reinforces the divergent pathways of different US cities in terms of levels of Millennial migration. In Chapter 13, Jiao and Wegmann demonstrate that urban vacation rentals are spiky in terms of the cities and neighbourhoods where there are higher levels of uptake and where vacation rental markets may be displacing housing.

Returning to Walks et al.'s (Chapter 5) analysis of wealth accumulation, they find that there are even bigger differences between cities than there are between generations, leading the authors to conclude that "one's chances at wealth accumulation, and therein one's financial well-being, to a large degree depend on where one lives". This suggests a need for careful, comparative work examining different urban contexts to best understand the dynamics of wealth accumulation. Pfeiffer and Pearthree's chapter does just this with respect to housing. They show that – at least in the cases of Phoenix and Houston – Millennials are choosing urban living, but that real estate developers are active agents in promulgating the re-population of the urban core and central city.

Finally, several of our authors tackle the question of *Millennial preferences*, adding depth to an ongoing debate about whether or not Millennials prefer urban living over suburban living, and its associated transportation patterns. Chapters by Ralph (Chapter 15), Agarwal (Chapter 16) and Moos et al. (Chapter 17) look at transportation decisions. Agarwal's findings in Canada's largest city, Toronto, align with a commonly held view of Millennials. Young people are getting licences later and rely less on vehicles. However, as Ralph finds in the USA, most young people drive and – despite a popular narrative of individuals using multiple modes of transportation in a given day – this is quite rare. Moreover, those young people who are car-less are not so by choice, but rather due to economic constraints, thus debunking Millennial myths. Moos et al. (Chapter 17) find similar patterns: Millennials who do not use cars are often students or immigrants, groups that may face socio-economic challenges. Moreover, Millennial multi-modal users (using different modes on different days of a week) are those who have the economic means to live in high-cost urban neighbourhoods, where they own cars but also have access to transit systems and other amenities. Finally, those using cars exclusively articulate lifestyle preferences that align with traditional, stereotypical notions of suburban living.

Pfeiffer and Pearthree's detailed study shows that new housing in central city and urban core locations was being chosen by Millennials more than other groups. Mallach (Chapter 19) points to the role of highly educated and affluent Millennials in shaping central cities, but notes that this in-migration of educated Millennials is fraught with challenges related to class and race. Indeed, Mallach suggests that the educated Millennials' preference for urban city living may be leading to greater polarization within the city and rising levels of spatial inequality and racial segregation. Similarly, in Chapter 14, Moos and Revington confirm Millennial preferences for urban living, although with some important caveats related to gender, ethnicity, and desires to raise children that highlight the importance of accounting for differences within the Millennial generation. They are also careful to note that preferences do not always translate into outcomes, and that – moreover – the distinction between urban and suburban living in terms of transportation choices, walkability and amenities is becoming blurred.

## Generating cities, generationed cities: towards a city for the ages?

Many of our planning and policy approaches have either purposefully or indirectly contributed to a generationed city, one where neighbourhoods – and whole cities – are become more similar in terms of their age structures. Throughout this volume, our contributors have marshalled empirical evidence from a variety of sources (public and private datasets, surveys, interviews) to develop lessons for planners and policymakers interested in city building and making places that are durable for current and future generations. Here we build on the arguments made by Shelton and Fulton (Chapter 18) by focusing explicitly on local level planning interventions that address the unique challenges faced by Millennials and avoid generating cities that are too highly generationed. Overall, planning a Millennial city for all ages and for the ages requires a holistic and integrative approach to planning for transportation, housing and other infrastructure.

Many of our contributors identify housing availability and affordability challenges. Mawhorter (Chapter 11) advocates for building more housing to address these shortfalls in the existing housing stock. In doing so, we should avoid the pitfalls of further youthification, by planning for a mix of housing appealing to different generational needs. Pfeiffer and Pearthtree also emphasize that planners will need to be attuned to whether or not the appropriate zoning, transportation infrastructure and services are in place to support new housing developments. Moreover, they suggest coordinated approaches between urban and suburban locations are needed to ensure diverse housing choices and affordability, rather than encouraging the concentration of wealth in particular pockets of the city. Moos et al. (Chapter 17) also emphasize that planners and policymakers need to consider demographic and lifestyle factors when considering residential loca- tion decisions. For instance, Millennials who pursue having a family will require access to social infrastructure such as schools and daycares.

When it comes to transportation, Ralph identifies the dual challenge of address- ing: 1) latent demand for multi-modal or car-free travel through programmes and policies that support the development of appropriate transportation infrastructure (e.g. transit, bike lanes); and 2) providing access to transportation networks for those who are unable to afford a car. Moos et al. (Chapter 17) extend this argu- ment, suggesting that to encourage drivers to switch to transit would require pol- icy interventions such as high parking costs, tolls or gas taxes, which may provide the appropriate economic incentive.

Finally, as noted by several of our contributors (Geobey, Shearmur, Vinodrai, and Jiao and Wegmann), new forms of work and emerging technological plat- forms are reconfiguring the urban economy through the dislocation of work (the "gig economy") and the reimagining of services and infrastructure through the use of technology platforms (the sharing or platform economy), which raise chal- lenges for planners. Here our authors suggest that planners must be attuned to and call for research to inform decision making to ensure that our cities can adapt and respond to these changes. Moreover, as Geobey emphasizes, planners and other urban policymakers would be wise to view these changes through the lens of insti- tutional experimentation, rather than locking into a particular form of regulation and governance too early in a fast-evolving field. And, importantly, as Shearmur notes, we must question our notions of an economy with fixed work locations.

## Moving the agenda forward

This volume has covered immense territory with respect to developing an empiri- cally grounded and theoretically informed set of perspectives on Millennials and the generationed city. In this final section, we identify potentially fruitful avenues for future research. While our contributors call for and provide a roadmap to fur- ther research in their specific domain, our intent here is to call attention to several issues that that push a broader research agenda forward and further enhance our knowledge and understanding of the Millennial city.

### Global Millennial cities

At the outset of this book, we were careful in advocating for the utility in study- ing the Millennial generation (or any generation for that matter) to note that context – especially local and national geographic context – mattered. Thus, we restricted our analysis of generational change to the USA and Canada, with the understanding that lessons might – with care – be applied beyond North America to other Western contexts (e.g. UK, Australia), and that important differences exist even between the Canadian and US contexts. However, a recent article in *The Economist* (2016a) notes that the global young adult population is currently at about 1.8 billion people, all of whom are growing up in a similar context in terms of technological development and global media, despite of course important national differences. *The Economist* adopts a somewhat different definition of the Millennial generation (the group between ages 15 and 30; those born between 1987 and 2002) compared to that of our contributors (generally, those born between 1980 and 1997). Nonetheless, they provide an important insight.

According to *The Economist*'s (2016a, 2016b) estimate, more than 85% of the world's young people live in developing countries. From this, two important observations can be made. First, the Millennial research agenda must go global. Comparative urban research that seeks to understand the commonalities and differences in the relationships between Millennials and their cities could do much to help us contextualize the North American experience and add depth and richness to our understanding of the generationed city. Such an agenda would have to take into account national institutional differences that shape economic circumstances, housing and labour markets and mobility, alongside specific local urban governance mechanisms.

In this volume, we observed some discernible differences even between Canada and the USA, especially shaped by sharply divergent national policy approaches in areas such as immigration, which influences the urban social geography of cities. This brings us to our second observation. In the face of declining birth rates in North America, population growth is likely to come from immigration. And some of the most mobile individuals are likely to be those in younger generations. Flows of Millennials or subsequent generations from other cities around the world will undoubtedly introduce different ideas and perspectives on urban living with the potential to shape the urban fabric of US and Canadian cities. Understanding the Millennial city at both source and destination will be critical. This will not come without its challenges, especially at a time when immigration has been highly politicized. We return to this issue below.

## Politics and the Millennial city

We cannot disregard the role of politics at the local, national and global scale. One key assumption regarding the Millennial generation is that they are not politically engaged. And moreover, that Millennials are being disenfranchised by the voting patterns of older generations. With respect to the former, there may be real challenges to participation in local decision making and planning processes. Noting that many development decisions are made at the local level, Mawhorter (Chapter 11) identifies the following challenge:

> Young people . . . have relatively little influence in local political processes. [They] are less likely to vote, tend to make frequent moves, are often at a

demanding and time-intensive phase of their lives when it is difficult to participate in local meetings, and may have fewer financial resources to support
local politicians or advocacy groups.

She notes that Millennials and Baby Boomers will have to build local alliances to
address issues related to ensuring that there is appropriate and affordable housing stock if Millennials are to enjoy the same housing opportunities as earlier
generations.

Beyond simply considering Millennial's engagement in urban planning processes, there are broader issues at stake. As emphasized by Filion and Grant
(Chapter 2), shared events and experiences can lead to a generational habitus
or identity that allow for collective action and societal transformations. For the
Millennials, they point to the Great Recession and the outcome of the most recent
presidential election in the USA as being defining generational experiences. The
2016 US presidential election revealed clear generational and geographic differences in voting patterns, with the majority of voters in large cities voting for
Hillary Clinton (Democrat) rather than Donald Trump (Republican). Similarly,
the youth vote stacked up in favour of Clinton, although more youth voted for
third-party candidates than ever before, a signal that youth are not beholden to the
political status quo (Richmond, Zinshteyn, & Gross, 2016). And while there are
clear generational differences in aggregate, it is still important to remember that
there are differences within this group in terms of race, class and gender (Hendrickson & Galston, 2016).

Generationed political differences are not isolated to the USA. In the United
Kingdom, the highly divisive "Brexit" vote, in which then Prime Minister David
Cameron put the question of whether or not the UK should remain in or leave
the European Union (EU) to a referendum, brought geographic and generationed
political differences to the fore. Voters in the UK's large urban centres over-
whelming voted in favour of remaining in the EU; so too, did young voters: 73%
of 18- to 24-year-olds voted to remain, as did 62% of 25- to 34-year-olds (Kelly,
2016). But, the overall vote went narrowly in favour of leaving. And while initial
reporting on the Brexit vote suggested youth apathy as being responsible for a
"leave" outcome, more recent analysis suggests that youth turnout reached an
all-time high (Helm, 2016). The contradictions of the Millennial generation once
again rear their head.

In both the case of the US presidential election and Brexit, the statistics and
popular discourse underscore strong generational and geographic divides, in
which the majority of urbanites and Millennials favoured outcomes that ultimately
were on the losing side of history. And in each case, the rallying cry was that older
generations voted against the interests of the Millennial and future generations
deepening an already wide generational divide. The fallout from these political
shifts underway will undoubtedly (re)frame the urban experiences of Millennials,
as well as the generation close on their heels that are now teenagers or just entering adulthood. This could very well be the defining moment of the Millennial city.

The outcomes of the recent federal election in the USA and, for that matter, in
Canada too, are leading to shifts in policy directions in both countries. While it is

too early to know the full extent of the impact these political changes will have on the Millennial city and beyond, it is clear that – at least in the US – the institutions, norms and rules of the game that have shaped Millennial cities and lives are being undone and re-written at a rapid pace. Several points are worth considering. In the instance of immigration – an issue where there are longstanding policy differences between the USA and Canada – the early signals indicate that there will be a substantial widening of these differences, as the Trump administration rolls out a highly punitive and restrictive anti-immigrant agenda. What is also apparent is that this is only the beginning of a reconfiguration of the roles that cities might play. Particularly in the USA, questions are already arising about how cities and their university and college campuses (themselves important spaces for youth) navigate their roles as sanctuaries and as arrival points for newcomers in a rapidly (and radically) changing policy landscape.

Broader questions arise about how US federal government spending will be reduced and/or (re)directed in the areas of health, environment, defence and infrastructure and how this will affect the prosperity and possibilities of the Millennial city and different social groups within it. Moreover, sweeping political changes may hamper or endanger the public data that planners and urban scholars use to understand the generationed city; a challenge that is not new in the Canadian context (Vinodrai & Moos, 2015). And while the use of other datasets (such as large-scale surveys or web-scraped data as used in several chapters in this volume) can enrich our knowledge, they are not a substitute for high quality, public data. The lack of such data undermines the ability of planners and policymakers to assess their own cities and impedes evidence-based decision making. New leadership at agencies such as the US Department of Housing and Urban Development will undoubtedly lead to programmatic changes that affect housing availability and affordability, reshaping the city for generations to come. Business decisions too may lead to disinvestment, relocation and a further reconfiguring of labour markets, even in the youth-oriented, innovation-driven tech sector centred on Silicon Valley (Hager, 2017; Loizos, 2017); these decisions will likely be most acutely felt by Millennials.

In short, national political changes will have a long-lasting impact on urban economies, housing and labour markets, and mobilities – meant both as the flows of people within and between cities, as well as the social and residential mobility experiences of individuals and groups. Will these policy shifts reinforce the existing geographies of the Millennial city or lead to new locational patterns? How will different groups within the Millennial generation fare? How will local planners and policymakers respond? For planners and urban scholars more broadly, there is much to pay attention to as these changes unfold across the restless urban landscape.

## Millennials (and planning) leaving their mark

Generations leave their mark on the urban spatial structure. The Greatest Generation, which came of age during World War II, and their Baby Boomer children have left in their wake a sea of sprawling suburbs, linked to declining or booming centre cities by vast highways, and growing anxieties about climate change and persisting socio-economic inequalities. What mark will Millennials leave on US and Canadian cities? This volume has shown that Millennials are having an undeniable impact, but how these actions will transform urban spatial structures remains unclear. Millennials have already brought greater flexibility and diversity to housing and labour markets, transportation systems, and public spaces. They have tremendous capacity as the most educated and racially and ethnically diverse generation yet to occupy and shape urban regions. Planners and policymakers have the opportunity to be co-producers of the next urban transformation, alongside Millennials; and – in fact – many policymakers themselves are and will be Millennials moving forward! The time to harness the Millennial generation's creativity in planning for more equitable and prosperous cities is now.

## References

The Economist (2016a, January 23). The millennial generation: Young, gifted and held back. *The Economist*. Retrieved from www.economist.com.

The Economist (2016b, January 23). The young: Generation uphill. *The Economist*. Retrieved from www.economist.com.

Hager, M. (2017, January 30). B.C. tech sector to get surge of talent fleeing Silicon Valley: Insiders. *The Globe and Mail*, Retrieved from www.theglobcandmail.com/.

Helm, T. (2016, July 10). EU Referendum: Youth turnout almost twice as high as first thought. *The Guardian*. Retrieved from www.theguardian.com/.

Hendrickson, C., & Galston, W. A. (2016, November 21). *How Millennials voted this election* [Web log post]. Retrieved from www.brookings.edu/blog/fixgov/2016/11/21/how-millennials-voted/.

Kelly, J. (2016, June 24). Brexit: How much of a generation gap is there. *BBC News Magazine*. Retrieved from www.bbc.com.

Loizos, C. (2017, January 31). *Silicon Valley is making plans to move foreign-born workers to Canada* [Web log post]. Retrieved from https://techcrunch.com/2017/01/31/in-silicon-valley-plotting-to-get-foreign-born-workers-to-vancouver/.

Myers, D. (2016). Peak Millennials: Three reinforcing cycles that amplify the rise and fall of urban concentration by millennials. *Housing Policy Debate, 26*(6), 928–947.

Richmond, E., Zinshteyn, M., & Gross, N. (2016, November 11). Dissecting the youth vote: Americans under 30 years old leaned left in this election, but not to the extent that they have in the past. *The Atlantic*. Retrieved from www.theatlantic.com/.

Vinodrai, T., & Moos, M. (2015). Do we still have quality data to study Canadian cities. In P. Filion, M. Moos, T. Vinodrai, & R. Walker (Eds.), *Canadian cities in transition: Perspectives for an urban age* (5th ed., pp. 436–439). Toronto: Oxford University Press.

# Index